Introduction to Geospatial Technologies

Fourth Edition

Bradley A. Shellito

Youngstown State University

w.h.freeman
Macmillan Learning

Vice President, STEM: Ben Roberts
Editorial Program Director: Andrew Dunaway
Program Manager: Jennifer Edwards
Editorial Assistant: Alexandra Hudson
Marketing Manager: Shannon Howard
Marketing Assistant: Savannah DiMarco
Senior Media Project Manager: Chris Efstratiou
Director, Content Management Enhancement: Tracey Kuehn
Managing Editor: Lisa Kinne
Senior Content Project Manager: Harold Chester
Senior Workflow Project Supervisor: Susan Wein
Senior Workflow Project Manager: Paul Rohloff
Production Supervisor: José Olivera
Project Management: Lumina Datamatics, Inc.
Permissions Manager: Jennifer MacMillan
Photo Editor: Sheena Goldstein
Director of Design, Content Management: Diana Blume
Cover and Interior Design: Lumina Datamatics, Inc.
Art Manager: Matthew McAdams
Illustrations: Dragonfly Media Group; Maps.com
Composition: Lumina Datamatics, Inc.
Printing and Binding: LSC Communications
Cover Photos: Pete Saloutos/Image Source/Getty Images; Aerial Archives/Alamy; Ink Drop/Shutterstock
Title Page Photo: olenadesign/Shutterstock
Icon Photo Credit: VitaminCo/Shutterstock

Library of Congress Control Number: 2017950414
ISBN-13: 978-1-319-06045-9
ISBN-10: 1-319-06045-5

Printed in the United States of America

First printing

W. H. Freeman and Company
One New York Plaza
Suite 4500
New York, NY 10004-1562
www.macmillanlearning.com

Dedication

This book is dedicated to my parents, David and Elizabeth, who taught me that with a lot of hard work—and a little luck—all things are possible.

Neal P. McNally

Bradley A. Shellito is a geographer whose work focuses on the application of geospatial technologies. Dr. Shellito has been a professor at Youngstown State University (YSU) since 2004, and was previously a faculty member at Old Dominion University. He teaches classes on GIS, remote sensing, GPS, and 3D visualization, and his research interests involve using these concepts within a variety of real-world issues. His second book, *Discovering GIS and ArcGIS*, was also published by Macmillan Learning. He also serves as YSU's Principal Investigator in OhioView, a statewide geospatial consortium. A native of the Youngstown area, Dr. Shellito received his bachelor's degree from YSU, his master's degree from the Ohio State University, and his doctorate from Michigan State University.

Contents in Brief

Contents

CHAPTER 11
Images from Space

Satellite Remote Sensing, Satellite Orbits, Sensor
Resolutions, the Landsat Program, High-Resolution
Satellite Sensors, Small Satellites, and Using
Satellites for Monitoring   393

CHAPTER 12
Studying Earth's Climate and Environment from Space

NASA and NOAA Satellites, MODIS, VIIRS, LST,
SST, CERES, AIRS, OMI, OMPS, and Jason-3   433

Preface

Why I Wrote *Introduction to Geospatial Technologies*

When people ask me what I teach, I say "geospatial technology." The usual response to this statement is a blank stare, a baffled "What?" or a variation on "I've never heard of that." However, if I say I teach "technologies like GPS, taking images from satellites, and using online tools like Google Earth or MapQuest," the response generally improves to "GPS is great," or "Google Earth is so cool," or even "Why do I get the wrong directions from that thing in my car?" Although geospatial technologies are everywhere these days—from software to Websites to smartphones—it seems that the phrase "geospatial technology" hasn't really permeated into common parlance.

I hope that this book will help remedy this situation. As its title implies, the goal of this book is to introduce several aspects of geospatial technologies—not only what they are and how they operate, but also how they are used in hands-on applications. In other words, the book covers a little bit of everything, from theory to application.

In a sense, the book's goal is to offer students an overview of several different fields and techniques and to provide a solid foundation on which further knowledge in more specialized classes can be built, such as those delving further into geographic information systems (GIS) or remote sensing. Whether the book is used for a basic introductory course, a class for non-majors, or as an introduction to widely used geospatial software packages, this book is aimed at beginners who are just starting out. Youngstown State University (YSU) offers an introductory class titled "Geospatial Foundations," but similar classes at other universities may have names such as "The Digital Earth," "Introduction to Geospatial Analysis," "Survey of Geospatial Technologies," "Introduction to GIS," or "Computer Applications in Geography." All of these courses seem aimed at the audience for which *Introduction to Geospatial Technologies* was written.

Organization of the Book

This book is divided into four main parts.

Part 1: Geospatial Data and GPS focuses on geospatial technology as it relates to spatial measurements and data.

➤ Chapter 1, "It's a Geospatial World Out There," introduces some basic concepts and provides an overview of jobs, careers, and some key technologies and applications (such as Google Earth).

➤ Chapter 2, "Where in the Geospatial World Are You?," explains how coordinates for location-based data and measurements from a three-dimensional (3D) world are translated into a two-dimensional (2D) map on a computer screen.

➤ Chapter 3, "Getting Your Data to Match the Map," discusses reprojection and georeferencing, important information when you're using any sort of geospatial data.

➤ Chapter 4, "Finding Your Location with the Global Positioning System," introduces GPS concepts. Taking a hand-held receiver outside, pressing a button, and then having the device specify your precise location and plot it on a map sounds almost like magic. This chapter demystifies GPS by explaining how the system works, why it's not always perfectly accurate, and how to get better location accuracy.

Part 2: Geographic Information Systems (GIS) contains the following chapters.

➤ Chapter 5, "Working with Digital Geospatial Data and GIS," serves as an introduction to GIS, examining how real-world data can be modeled and how GIS data can be created and used.

➤ Chapter 6, "Using GIS for Spatial Analysis," covers additional uses of GIS, including querying a database, creating buffers, and geoprocessing.

➤ Chapter 7, "Using GIS to Make a Map," offers instruction on how to classify your data and how to transform GIS data into a professional-looking map.

➤ Chapter 8, "Getting There Quicker with Geospatial Technology," discusses concepts related to road networks, such as how streets, highways, and interstates are set up and used in geospatial technology; how computers translate a set of letters and numbers into a map of an actual street address; how programs determine the shortest route from point *a* to point *b*.

Part 3: Remote Sensing examines issues related to remote sensing.

➤ Chapter 9, "Remotely Sensed Images from Above," focuses on aerial photography. It explains how the field started over 150 years ago with a man, a balloon, and a camera, and how it continues today with unmanned aircraft systems flying almost everywhere. This chapter also describes how to interpret features in aerial imagery visually and how to make accurate measurements from items present in an image.

➤ Chapter 10, "How Remote Sensing Works," delves into just what remote sensing is and how it works, and exactly how that image of a house is acquired by a sensor 500 miles away. This chapter also discusses all of the things that a remote sensing device can see that are invisible to the human eye.

➤ Chapter 11, "Images from Space," focuses on the field of satellite remote sensing and how satellites in orbit around Earth acquire images of the ground below.

➤ Chapter 12, "Studying Earth's Climate and Environment from Space," discusses how we can use the technologies onboard a series of environmental satellite observatories that orbit the planet and continuously transmit data back to Earth about the land, seas, and atmosphere.

Part 4: Geospatial Applications focuses on individual topics in geospatial technology that combine GIS and remote sensing themes and applications.

➤ Chapter 13, "Digital Landscaping," describes how geospatial technologies model and handle terrain and topographic features. Being able to set up realistic terrain, landscape features, and surfaces is essential in mapping and planning.

➤ Chapter 14, "See the World in 3D," delves into the realm of 3D modeling, shows how geospatial technologies create 3D structures and objects, and then explains how to view or interact with them in programs such as Google Earth.

➤ Chapter 15, "Life in the Geospatial Cloud and Other Current Developments," wraps things up with a look at the influence and advantages of the cloud, information regarding organizations and educational opportunities within geospatial technologies, and a look ahead to the future of the field.

Geospatial Lab Applications

Each chapter of *Introduction to Geospatial Technologies* covers one aspect of geospatial technology with an accompanying Geospatial Lab Application. The goal of these lab applications is not to teach software, but to help students work directly with the chapter's concepts. Each lab application uses freely available software that can be downloaded from the Internet or accessed through a Web browser. These software packages include:

➤ ArcGIS Online
➤ Google Earth Pro
➤ MapTiler
➤ MultiSpec
➤ QGIS
➤ SketchUp
➤ Trimble GNSS Planning Online

Three of the chapters in Part 2 (Geographic Information Systems) offer three versions of the lab application. Instructors can choose to use either the free QGIS, ArcGIS Pro, or ArcMap. The QGIS and ArcGIS Pro lab applications are printed in the book while the ArcMap version is available on the book's Website. The labs provide hands-on application of the concepts and theories covered in each chapter—it's one thing to read about how 3D structures can be created and placed into Google Earth Pro, but it's another thing entirely to use SketchUp and Google Earth Pro to do exactly that. Each lab application has integrated questions that students must answer while working through the lab. These questions are designed both to explore the various topics presented in the lab and also to keep students moving through the lab application. Note that words or phrases highlighted in purple in the labs indicate menu items or icons that are clicked on or specific items that are typed in during the lab.

Some labs use sample data that comes with the software when it's installed; others require students to download sample data for use in the lab. Each lab provides links to a Website from which you can download the software. (The Website will also provide information regarding the necessary hardware or system requirements. Not all computers or lab facilities work the same way, therefore be sure to check the software's Internet resources for help on installing the software.) The Instructors' section of this book's catalog page also offers a "tech tips" section with some additional information related to installing or utilizing some of the software or information about compatibility, browsers, and the like.

The lab applications for each chapter are set up as follows:

➤ Chapter 1: This lab introduces the free Google Earth Pro as a tool for examining many facets of geospatial technology.

➤ Chapter 2: Students continue using Google Earth Pro, investigating some other functions of the software as they relate to coordinates and measurements.

➤ Chapter 3: Students use the free version of the MapTiler program to match a graphic of a campus map with remotely sensed imagery and real-world coordinates.

➤ Chapter 4: This lab uses Trimble GNSS Planning Online and some other Web resources to examine GPS planning and locations. It also provides suggestions for expanding the lab activities if you have access to a GPS receiver and want to get outside with it.

➤ Chapter 5: This lab introduces basic GIS concepts using QGIS. Alternate versions of the lab use either ArcGIS Pro or ArcMap.

➤ Chapter 6: This lab continues investigating the functions of QGIS (or ArcGIS Pro or ArcMap) by using GIS to answer some spatial analysis questions.

➤ Chapter 7: This lab uses QGIS (or ArcGIS Pro or ArcMap) to design and print a map.

➤ Chapter 8: This lab uses Google Maps and Google Earth Pro to match a set of addresses and investigate shortest paths between stops on a network.

➤ Chapter 9: This lab tests students' visual image interpretation skills by putting them in the role of high-tech detectives who are trying to figure out just what a set of aerial images are actually showing.

➤ Chapter 10: This lab is an introduction to MultiSpec, which allows users to examine various aspects of remotely sensed imagery obtained by a satellite.

➤ Chapter 11: This lab continues using MultiSpec by asking students to work with imagery from the Landsat 8 satellite and investigate its sensors' capabilities.

➤ Chapter 12: This lab uses NASA data and resources, Google Earth, and Google Earth Pro to examine phenomena such as hurricanes, fires, and pollution on a global scale.

➤ Chapter 13: This lab uses Google Earth Pro to examine how terrain is used in geospatial technology (and film a video of flying over 3D-style terrain). It also uses the various terrain functions of Google Earth Pro for work with several digital terrain modeling features.

➤ Chapter 14: This lab introduces 3D modeling. Starting from an aerial image of a building, students design a 3D version of it using SketchUp, and then look at it in Google Earth Pro.

➤ Chapter 15: This lab utilizes Esri's free ArcGIS Online to stream data from the cloud, create Web maps, wrap things up, and look at many of the book's concepts combined in a single package.

Additional Features

In addition to the lab applications, each chapter contains several **Hands-On Applications**, which utilize free Internet resources to help students explore the world of geospatial technologies further and get directly involved with some of the chapter concepts. There's a lot of material out there on the Internet, ranging from interactive mapmaking to real-time satellite tracking, and these Hands-On Applications introduce students to it. In the fourth edition, each Hands-On Application has a set of Expansion Questions for students to answer while working with that Application's Web resources.

Each chapter also has one or more boxes titled **Thinking Critically with Geospatial Technology**. These boxes present questions to consider regarding potential societal, privacy, design, or ethical issues posed by geospatial technologies and their applications. The questions presented in these boxes are open-ended and are intended to stimulate discussion about geospatial technologies and how they affect (or could affect) human beings. For instance, how much privacy do you really have if anyone, anywhere, can obtain a clear image of your house or neighborhood and directions to drive there with just a few clicks of a mouse?

Lastly, each chapter ends with two boxes. The first of these, chapter **Apps**, presents some representative apps for a mobile device related to the chapter's content that you may wish to investigate further. For instance, Chapter 9's *UAS Apps* box showcases apps related to flying drones, mapping from drones, and accessing real-time information for drone no-fly zones for your smartphone or your tablet. Note that at the time of writing, all of these representative apps were free to obtain and install.

The second section, **Social Media**, highlights some representative Facebook, Twitter, and Instagram accounts, as well as YouTube videos and blogs, that are relevant to the chapter's topics. For instance, Chapter 11's *Satellite Imagery in Social Media* box features Facebook and Twitter accounts from satellite imagery sources such as DigitalGlobe or the USGS updates on Landsat, as well as videos of satellite imagery applications. (Note that all of these apps and social media accounts are examples, not recommended products.)

New to This Edition

The fourth edition contains multiple key updates. Each chapter has something new within it, whether it's a newly added or revised text and social media section, Hands-On Application, or Lab Application. The Lab Applications have been updated to use current software and techniques as things have changed since the third edition. This also includes all-new Lab Applications that utilize ArcGIS Pro (Chapters 5, 6, and 7), MapTiler (Chapter 3), and the new browser version of Google Earth (Chapter 12). Also, Chapter 9 features all-new images for interpretation in its Lab Application.

There are many other updates and revisions throughout each chapter. For instance, Chapter 9 showcases much more about drones, how they're being used for research and commercial purposes, and what restrictions are placed on them, while Chapter 14 discusses how we can take drone imagery and build 3D models from it. The four remote sensing focused chapters (9, 10, 11, and 12) have been expanded to include many topics about the state of remote sensing today, including Landsat 9, WorldView-4, and new small satellite developments coming from Terra Bella and Planet Labs.

Chapter 12 has been almost entirely rewritten to present an overview of selected satellite missions that study climate and the environment from NASA, NOAA, ESA, and JAXA. Then it covers examples of environmental monitoring with remote sensing, focusing on applications of land cover, wildfires, soil moisture monitoring, land surface temperature measurements, sea surface temperature measurements, studying Earth's heat/energy budget, air quality, water quality, and monitoring Earth's sea level.

Several chapters also feature new and expanded Thinking Critically with Geospatial Technology sections. Also, throughout the book there are all-new Hands-On Applications that use new Web resources, including Google Earth Engine, AirMap, Cesium, NASA WorldView, the USGS Historical Topographic Map Viewer, and various state and county GIS Web resources.

Ancillary Materials and Student and Instructor Resources

The catalog page **www.macmillanlearning.com/catalog/shellitoigt4e** offers a set of valuable resources for both students and instructors.

For **students**, the catalog page offers a multiple-choice self-test for each chapter, as well as an extensive set of references, categorized by topic, to provide further information on a particular topic. There is also a set of links to the free software packages needed to complete the lab activities, as well as the datasets required for specific lab applications. A set of world links is also provided.

For **instructors**, the catalog page offers an instructor's manual, which provides teaching tips for each chapter on presenting the book's material, a set of "tech tips" related to software installation and usage, a set of key references for the chapter materials, helpful tips for ArcGIS Online, and an answer key for all the lab activities. A test bank of questions is also provided, as well as material that could be used for student projects.

Acknowledgments and Thanks

Books like this don't just spring out of thin air—I owe a great deal to the many people who have provided inspiration, help, and support for what would eventually become this book.

Alexandra Hudson, my editor, has offered invaluable help, advice, patience, and guidance throughout this entire project. I would also very much like to thank Jennifer Edwards, Sheena Goldstein, Harold Chester,

Paul Rohloff, and Diana Blume at W. H. Freeman and Misbah Ansari and Martha Solonche at Lumina Datamatics for all of their extensive "behind the curtain" work that shaped this book into a finished product.

I want to thank Sean Young for his very helpful comments and feedback on the previous editions and for his close reading and expert review of the labs and text in this fourth edition. Sherry Rogelberg gave great representation and advice. The students who contributed to the development of the YSU 3D Campus Model (Rob Carter, Ginger Cartright, Paul Crabtree, Jason Delisio, Sherif El Seuofi, Nicole Eve, Paul Gromen, Wook Rak Jung, Colin LaForme, Sam Mancino, Jeremy Mickler, Eric Ondrasik, Craig Strahler, Jaime Webber, Sean Welton, and Nate Wood) deserve my sincere thanks. Many of the 3D examples presented in Chapter 14 wouldn't have existed without them. Others who supported this book in various ways include: Jack Daugherty, for tech support, assistance with the labs, and help with the GeoWall applications; Lisa Curll, for assistance with the design and formatting of the lab applications; Grant Wilson, for his insightful technical reviews of the ArcGIS lab applications; Margaret Pearce, for using an earlier draft of the manuscript with her students at the University of Kansas and for her extremely useful comments and feedback; and Hal Withrow, for invaluable computer tech assistance.

I offer very special thanks to all of my professors, instructors, colleagues, and mentors, past and present (who are too numerous to list), from Youngstown State University, the Ohio State University, Michigan State University, Old Dominion University, OhioView, and everywhere else, for the help, knowledge, notes, information, skills, and tools they've given me over the years. I am also deeply indebted to the work of Tom Allen, John Jensen, Mandy Munro-Stasiuk, and the members of SATELLITES for some methods used in some of the chapters and labs.

Finally, I owe a debt of gratitude to the colleagues who reviewed the original proposal and various stages of the manuscript for the first, second, third, and fourth editions. Thank you for your insightful and constructive comments, which have helped to shape the final product:

Robbyn Abbitt, *Miami University*
Amy Ballard, *Central New Mexico Community College*
Chris Baynard, *University of North Florida*
Robert Benson, *Adams State College*
Margo Berendsen, *University of Wyoming*
Edward Bevilacqua, *SUNY College of Environmental Science and Forestry*
Krystopher Chutko, *University of Saskatchewan*
Julie Cidell, *University of Illinois*
W. B. Clapham Jr., *Cleveland State University*
Craig Coburn, *University of Lethbridge*
Russell G. Congalton, *University of New Hampshire*
Jamison Conley, *West Virginia University*

Kevin Czajkowski, *University of Toledo*
Nathaniel Dede-Bamfo, *Texas State University*
Adrienne Domas, *Michigan State University*
Christine Drennon, *Trinity University*
Charles Emerson, *Western Michigan University*
Jennifer Fu, *Florida International University*
Juan Gonzalez, *San Bernardino Valley College*
Nandhini Gulasingam, *DePaul University*
Victor Gutzler, *Tarrant County College Southeast*
Paddington Hodza, *University of Wyoming*
Jung Eun Hong, *University of West Georgia*
Ulrike Ingram, *Kennesaw State University*
Melanie Johnson, *Paul Smith's College*
Marilyne Jollineau, *Brock University*
Jessica K. Kelly, *Millersville University*
Sara Beth Keough, *Saginaw Valley State University*
James Kernan, *SUNY Geneseo*
Kimberley Britt Klinker, *University of Richmond*
Michael Konvicka, *Lone Star College*
Eunmok Lee, *Miami University*
James Lein, *Ohio University*
James Leonard, *Marshall University*
Russane Low, *University of Nebraska–Lincoln*
Chris Lukinbeal, *Arizona State University*
Marcos Luna, *Salem State University*
John McGee, *Virginia Tech*
George M. McLeod, *Old Dominion University*
Bradley Miller, *Michigan State University*
Trent Morrell, *Laramie County Community College*
Nancy Obermeyer, *Indiana State University*
Tonny Oyana, *The University of Tennessee*
Margaret Pearce, *University of Kansas*
Kimberly Perez, *Tidewater Community College*
Hugh Semple, *Eastern Michigan University*
Thomas Sigler, *University of Queensland*
Anita Simic, *Bowling Green State University*
Brian Tomaszewski, *The Rochester Institute of Technology*
Cuizhen (Susan) Wang, *University of South Carolina*
Clayton J. Whitesides, *Coastal Carolina University*
Shuang-Ye Wu, *University of Dayton*
Sean Young, *University of Arkansas for Medical Sciences*
Donald Zeigler, *Old Dominion University*
Sally Zellers, *University of Central Missouri*
Yuyu Zhou, *Iowa State University*
Arthur Zygielbaum, *University of Nebraska–Lincoln*

A Rapidly Changing Field

As Chapter 15 points out, geospatial technology has become so widespread and prevalent that no book can cover every concept, program, or online mapping or visualization tool (as much as I'd like this one to). I hope that the students who use this book will view the concepts and applications presented herein as an introduction to the subject—and that this will motivate them to take more advanced courses on the various aspects of geospatial technology.

One thing to keep in mind: In such a rapidly advancing field as geospatial technology, things can change pretty quickly. New satellites are being launched and old ones are ending their mission lives. Websites get updated and new updates for software and tools are released on a regular basis. As of the writing of this book, all of the Web data, software, and satellites were current, but if something's name has changed, a Website works differently, or if a satellite isn't producing any more data, there's probably something newer and shinier to take its place.

I'd very much like to hear from you regarding any thoughts or suggestions you might have for the book. You can reach me via email at **bashellito@ysu.edu** or follow me on Twitter **@GeoBradShellito**.

Bradley Shellito
Youngstown State University

Accessing Data Sets for Geospatial Lab Applications

Some of the Geospatial Lab Applications in this book use data that comes with the software, sample data that gets installed on your computer when you install the software itself, or data that you'll create during the course of the lab. However, the lab applications for Chapters 3, 5, 6, 7, 8, 9, 10, 11, 12, and 15 require you to download a set of data that you'll use with those labs.

The lab applications will direct you to copy the dataset before beginning the lab. Each dataset is stored in its own folder online. To download these folders, please visit **www.macmillanlearning.com/catalog/shellitoigt4e**. Under "Student Options," you'll find access to student resources, including "Lab Data Sets."

This book was not prepared, approved, or endorsed by the owners or creators of any of the software products discussed herein. The graphical user interfaces, emblems, trademarks, and associated materials discussed in this book remain the intellectual property of their respective owners:

ArcGIS for Desktop 10.5 (Esri), ArcGIS Online (Esri), ArcGIS Pro 2.0 (Esri), Google Earth Pro 7.3 (Google), Google Maps (Google), SketchUp Make (Trimble), MapTiler (Klokan Technologies), MultiSpec, QGIS 2.18, Trimble GNSS Planning Online.

It's a Geospatial World Out There

1

An Introduction to Geospatial Technologies, Geospatial Jobs, Geospatial Data, Volunteered Geographic Information, Geolocation, and Google Earth

Have you ever done any of the following?

➤ Used a smartphone, tablet, or other mobile device to find your location coordinates, get directions, or to look for the nearest restaurant or gas station?

➤ Used an online mapping service like MapQuest, Google Maps, or Bing Maps to find directions (and the best route) to a destination or to print a map of an area?

➤ Used an app to request an Uber vehicle for transportation and viewed the location of the cars on the app?

➤ Used an in-car navigation system (say, one from Garmin, Magellan, or TomTom) to navigate to or from a specific location?

➤ Used social media (such as Facebook or Twitter) to add your location information to a post or tweet?

➤ Used a Global Positioning System (GPS) receiver while hiking, jogging, hunting, fishing, golfing, or geocaching?

➤ Used a Web resource to find a map of your neighborhood so that you can compare nearby housing values or see exactly where your property ends and your neighbor's begins?

➤ Used a drone to fly over an area and take a photo or video of something on the ground below it?

➤ Used a virtual globe program (like Google Earth) or an online map to look at photos or images of your home, street, school, or workplace?

If so, then congratulations—you've used geospatial technologies. Whenever you're using any sort of technology-assisted information (on a computer, smartphone, or tablet) concerning maps, locations, directions, imagery, or analysis, you're putting geospatial technology applications to use.

Geospatial technology has become extremely widespread in society, with a multitude of uses in both the private and public sectors. However, more often than not, if you tell someone you're using geospatial technology, you'll be asked, "What's that?"

 ## What Is Geospatial Technology?

geospatial technology a number of different high-tech systems that acquire, analyze, manage, store, or visualize various types of location-based data

geographic information system (GIS) computer-based mapping, analysis, and retrieval of location-based data

remote sensing acquisition of data and imagery from the use of satellites or aircraft

satellite imagery digital images of Earth acquired by sensors onboard orbiting spaceborne platforms

aerial photography acquisition of imagery of the ground taken from an airborne platform

Global Positioning System (GPS) acquisition of real-time location information from a series of satellites in Earth's orbit

Although geospatial technology is being used in numerous fields today, the term "geospatial technology" doesn't appear to have seeped into everyday usage. Terms such as "satellite images" and "Google Earth" and acronyms such as GIS and GPS are growing increasingly commonplace, yet the phrase "geospatial technology" seems relatively unknown, even though it incorporates all of these things and more. **Geospatial technology** describes the use of a number of different high-tech systems and tools that acquire, analyze, manage, store, or visualize various types of location-based data. The field of geospatial technology encompasses several fields and techniques, including:

➤ **Geographic Information System (GIS)**: Computer-based mapping, analysis, and retrieval of location-based data

➤ **Remote sensing**: Acquisition of data and imagery from the use of satellites (**satellite imagery**) or aircraft (**aerial photography**)

➤ **Global Positioning System (GPS)**: Acquisition of real-time location information from a series of satellites in Earth's orbit

There are numerous related fields that utilize one or more of these types of technologies. For instance, an in-car navigation system already contains extensive road-network data, mapped out and ready to use, which includes information about address ranges, speed limits, road connections, and special features of roads (such as one-way streets). It also requires the mapping of points of interest (such as gas stations or restaurants), and should be capable of referencing new user-defined destinations. It also has to be able to plot the car's real-time position in relation to these maps and may even have a feature that shows a representation of the surrounding landscape as taken from an overhead viewpoint. Many of these types of systems combine different geospatial technologies to work together in one application.

Who Uses Geospatial Technology?

Geospatial technology is used in a wide variety of fields (**Figure 1.1**), including federal, state, and local government, forestry, law enforcement, public health, biology, and environmental studies (see *Hands-on Application 1.1: Industries Using Geospatial Technology* for a look at industries employing people in these fields). As long as the job field involves the utilization of some sort of information or data that has a location associated with it, chances are that some sort of geospatial technology is being used. Geospatial technology has been heralded by the U.S. Department of Labor as one of the main emerging and evolving job fields in the United States, with enormous growth potential. For instance, the Department of Labor's O*NET (Occupational Information Network) OnLine

FIGURE 1.1

Examples of geospatial technology in action on the job. *[(top left) AP Photo/ Wilfredo Lee (top right) Ron Nickel/Perspectives/Getty Images (center left) Joe Raedle/ Getty Images (center right) Bob Nichols/USDA NRCS (bottom left) AP Photo/U.S. Geological Survey, Dr. Dan Dzurisin (bottom right) Justin Sullivan/Getty Images]*

Hands-On Application 1.1

Industries Using Geospatial Technology

Geospatial technology is being used in a variety of applications in numerous different fields today. For a deeper look at some of these applications, visit **www.esri.com/industries**, which is run by Esri (we'll discuss more about Esri in Chapter 5, but the short version is that it is the market leader in GIS). This site lists dozens of different fields that are using geospatial technology, and describes how GIS (and other Esri products) are being utilized in them. Examine a few that are connected to your own fields of interest. For instance, if your interest is in law enforcement, examine some of the "Public Safety" applications. If you're involved in environmental science, examine some of the "Natural Resources" applications, then describe how GIS is being utilized in some real-world, on-the-job applications.

Expansion Questions

➤ How is GIS being utilized in some real-world, on-the-job applications in fields of interest to you?

➤ Who in these fields is using GIS with what kinds of applications and why are they using GIS?

Hands-On Application 1.2

Jobs in the Geospatial Field

Businesses are hiring in the geospatial field. For examples of current job openings, visit some of the following Websites:

1. Geosearch: **www.jobs.geosearch.com/JobSeeker/Jobs.aspx**

2. GIS Careers: **www.giscareers.com**

3. GIS Jobs.com: **www.gisjobs.com**

4. The GIS Jobs Clearinghouse: **www.gjc.org**

5. My GIS Jobs: **www.mygisjobs.com**

These are just a sampling of Websites on which employers post job openings worldwide. Examine several jobs from areas near where you are (or where you'd like to go to).

Expansion Questions

➤ What kinds of jobs are being advertised?

➤ For what kinds of job qualifications, training, and skill sets are employers looking?

➤ What kinds of starting salary ranges are employers offering?

utility contains job descriptions in fields such as "Geospatial Information Scientists and Technologists," "Remote Sensing Scientists and Technologists," and "Geographic Information Systems Technicians" (see *Hands-on Application 1.2: Jobs in the Geospatial Field* for more information about types of jobs).

The following are just a handful of examples of fields that utilize geospatial technology.

Archeology

The ability to pinpoint the location of artifacts uncovered on a dig, construct a map of the area, and then search for patterns on the site are all archeological functions that can be rendered quickly and efficiently with geospatial technology (**Figure 1.2**). Archeologists can utilize historical maps, current aerial photography or satellite imagery, and location information obtained on the site throughout the course of their work.

City Planning

Utilities, wastewater, green space, traffic, roads, zoning, and housing are all matters of concern to urban planners. Geospatial technology provides a means of working with all of these entities together for planning purposes. Strategies for smart urban growth and the management and updating of city resources can be examined through a variety of different applications.

Environmental Monitoring

Processes that affect Earth's environment in many different ways can be tracked and assessed using geospatial technology. Information about land-use change, pollution, air quality, water quality, and global temperature levels is vital to environmental research, ranging from the monitoring of harmful algae blooms to studies in climate change (see **Figure 1.3**).

Forestry

All manner of forest monitoring, management, and protection can be aided through the use of geospatial technology. Modeling animal habitats and the pressures placed upon them, examining the spatial dimensions of forest

FIGURE 1.2

Using GIS for archeological mapping of Clay Tobacco Pipestems in Virginia. *[Colonial National Historical Park, National Park Service]*

FIGURE 1.3

Global carbon monoxide concentrations as monitored by NASA remote sensing satellites. *[NASA GSFC Scientific Visualization Studio, based on data from MOPITT (Canadian Space Agency and University of Toronto)]*

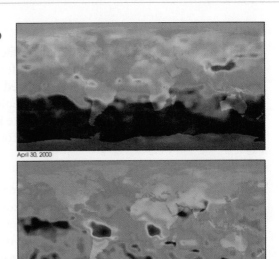

April 30, 2000

October 30, 2000

Carbon Monoxide Concentration (parts per billion)
50   220   390

FIGURE 1.4

Geospatial technology used for assessing the risk of wildfires in Virginia. *[Courtesy Virginia Department of Forestry]*

fragmentation, and managing fires are among the many different ways that geospatial technology is utilized within the field of forestry (**Figure 1.4**).

Homeland Security

Geospatial technology is a key component in examining vulnerable areas with regard to homeland security. Risk assessment of everything from evacuation plans to smoke-plume modeling can be examined using geospatial technology.

Disaster mitigation and recovery efforts can be greatly enhanced through the use of current satellite imagery and location-based capabilities.

Law Enforcement

The locations of various crimes can be plotted using GIS. Law enforcement officials can use this information to analyze patterns (**Figure 1.5**) and to determine potential new crime areas. Geospatial technology can be used in several other ways beyond mapping and analysis; for instance, police departments can gather high-resolution aerial photography of locations of patrols to provide further information about potentially dangerous areas.

Health and Human Services

Geospatial technology is used in a variety of health-related services. For example, monitoring of diseases, tracking sources of diseases, and mapping health-related issues (such as the spread of H1N1 or other influenza; see **Figure 1.6**) are all tasks that can be completed using geospatial technology applications.

FIGURE 1.5

A GIS map showing crime locations in Minneapolis. *[Minneapolis Police Crime Analysis Unit]*

FIGURE 1.6

A CDC map examining the spatial distribution of influenza activity. *[Centers for Disease Control and Prevention]*

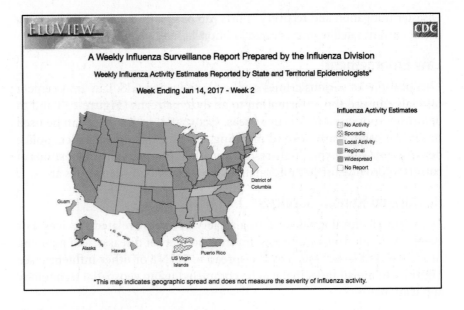

Military and Intelligence

Geospatial technology plays a key role in today's military, defense, and intelligence worlds. Satellite imagery and aerial photography have long been used for intelligence gathering and reconnaissance, practices that continue today with an increased use of Unmanned Aerial Systems (UAS) and high-resolution satellite sensors. The field of Geospatial Intelligence (**GEOINT**) encompasses all types of remotely sensed imagery, GIS, and GPS data for information gathering and analysis related to national security and defense applications.

GEOINT Geospatial Intelligence

Real Estate

Through geospatial technology, realtors and appraisers (as well as home buyers and sellers) can create and examine maps of neighborhoods and quickly compare housing prices and values of nearby or similar properties. Other features of a property can be examined by viewing high-resolution aerial images, showing the topography, the terrain, and even if it's located on a floodplain. You can also examine where a property is located in relation to schools, highways, wastewater treatment plants, and other urban features.

All of these areas and many, many more are reliant on the capabilities of geospatial technology. No matter the field, what distinguishes geospatial technology from other types of computer systems or technologies is that it explicitly handles **geospatial data**.

geospatial data items that are tied to specific real-world locations

What Is Geospatial Data?

Geospatial data (also often referred to as "spatial data") refers to location-based data, which is at the heart of geospatial technology applications. This ability to assign a location to data is what makes geospatial technology different from other systems. You're using geospatial concepts anytime you want to know "where" something is located. For instance, emergency dispatch systems can determine the location of a call to 911. They also have access to information that tells them where the local fire stations, ambulance facilities, and hospitals are. This type of information allows the closest emergency services to be routed to its location.

When the data you're using has a location that it can be tied to, you're working with geospatial data. This isn't just limited to point locations—the length and dimensions of a hiking trail (and the locations of comfort stations along the trail) are examples of real-world data with a spatial dimension. Other kinds of data, like the boundaries of housing parcels in a subdivision or a satellite image of the extent of an area impacted by a natural disaster, would fall into this category. Geospatial technology explicitly handles these types of location-based concepts.

However, not all of the data in the world is geospatial data. For instance, data regarding the location of a residential parcel of land containing a house and the parcel's dimensions on the ground would be geospatial information. However, other data, such as the names of the occupants, the assessed value of the house, and the value of the land is not geospatial information. A benefit of using geospatial technology is that this type of **non-spatial data** can be linked to a location. For instance, say you sell a lot of used books on eBay. You could make a map showing the locations of each of the purchasers of your online sales—this would be geospatial information. You could then link related non-spatial data (such as the title of the book that was sold or the price it was purchased for) to those locations, creating a basic database of sales.

Geospatial information can also be gathered about other characteristics, such as the landscape, terrain, or land use. Remote sensing provides images of the ground that serve as "snapshots" of particular locations at specific times, which can generate other forms of spatial information. For instance, a satellite image of Las Vegas, Nevada, can help ascertain the locations of new housing developments. Examining imagery from several dates can help track locations where new houses are appearing (**Figure 1.7**). Similarly, geospatial technology can be used to determine where, spatially, new urban developments are most likely to occur, by analyzing different pressures of development on these areas.

non-spatial data data that is not directly linked to a geospatial location (such as tabular data)

1972

1986

1992

2000

FIGURE 1.7

Satellite imagery (from 1972, 1986, 1992, and 2000) showing the growth of Las Vegas, Nevada, over time.

[USGS]

The National Map an online basemap and data viewer with downloadable geospatial data maintained and operated by the USGS and part of the National Geospatial Program

At the national level, the USGS maintains a large geospatial data distribution program called **The National Map**. Users of The National Map can access multiple types of data, including Digital Elevation Models (see Chapter 13), hydrography (see Chapter 5), transportation (see Chapter 8), land-cover data (see Chapter 5), orthoimagery (see Chapter 9), boundary files, and many more, all in formats that GIS software can easily read. The National Map is just one part of the larger coordinated vision that comprises the National Geospatial Program, and it serves as a source of geospatial data for the United States (see **Figure 1.8** and *Hands-on Application 1.3: The National Map* for more information about The National Map). You can think of The National Map as a "one-stop-shopping" venue for free geospatial data—numerous types of datasets are available for the entire United States.

Many types of decision making are reliant on useful geospatial information. For instance, when a new library is proposed to be built in a particular area, the choice of location is going to be important—the new

Hands-On Application 1.3

The National Map

You can access The National Map online at **https://viewer.nationalmap.gov/launch**. From the available choices, select the option for Download GIS Data. A new Web page will open with the Basic Viewer, the tool for accessing and downloading GIS data via The National Map. To begin, first use the Search box and type in an area of interest (go to Reno, Nevada) and click Go. The Viewer will shift to a map of your chosen location, and from the Data options on the right-hand side you can select which geospatial data layers you'd like to download for that place. For instance, choose Boundaries—National Boundary Dataset—a clickable graphic of the boundaries layer will appear which is what you can download. Also click on Hydrography (NHD) and Watersheds (WBM) for similar options. Under the Hydrography options you'll be prompted to choose which layer you want—for instance, click

on the HU8 Subbasin—and you'll see those water bodies and rivers that you can download appear on the map. Once you've chosen the layers you want, click on Find Products. A set of downloadable options will appear—click on Results for each of them and you'll find download links for each dataset. The National Map allows you to download on demand—by choosing to download you'll receive a zipped file with your data directly from the Website.

Expansion Questions

➤ Search for your own local area—what products are available for you to download?

➤ What types of data can be accessed via The National Map besides boundaries and hydrography? Why does this make The National Map so versatile as a one-stop data destination?

site should maximize the number of people who can use it but minimize its impact on other nearby sites. The same type of geospatial thinking applies to the placement of new schools, fire stations, and fast-food chain restaurants. Geospatial information fuels other decisions that are location dependent. For instance, historical preservation efforts seek to defend historic Civil War battlefield sites from urban development. Knowing the location of the sections of land that are under the greatest development pressure helps in channeling preservation efforts to have their greatest impact. With more and more geospatial technology applications becoming available, these types of geospatial problems are being taken into consideration (see *Hands-on Application 1.4: Mapping Data Layers Online*; *Hands-on Application 1.5: Examining Real Estate Values Online*; and *Hands-on Application 1.6: Geospatial Technology and Emergency Preparedness* for examples).

Hands-On Application 1.4

Mapping Data Layers Online

Online mapping applications that combine a number of different types of spatial data into a series of map layers are becoming increasingly common. A good, easy-to-use resource for viewing spatial information online is the Vermont Interactive Map Viewer, a utility for mapping various layers of data about the state of Vermont. Go to **http://maps .vermont.gov/vcgi/html5viewer**. Under the Layers list on the left-hand side of the Map Viewer, you'll see multiple options for various data layers to show on the map—with a checkmark in the layer's box it will be displayed on the map.

By expanding the options for Base Maps you'll see that you have several options as your basemap (the initial layer of the map on top of which other GIS layers will be placed). Also keep in mind that layers at the top of the list will be drawn on top of layers at the bottom, so if you place a checkmark in a layer and it doesn't appear make sure a basemap option above it on the list isn't checked.

Next, zoom in closer to a major city (such as Burlington) by typing it into the search box and then choosing from the options that appear in the table in the lower half of the screen. At a close-in level, you'll be able to examine other clickable layers from the map like contours, boundaries, facilities, and water resources such as private wells. In addition, the pull-down menus and buttons at the top of the screen will allow you to draw on the map to highlight things, make measurements, or switch to street-level photography on a road.

Expansion Questions

➤ With the Cached Basemap layer also turned on, how is the city of Burlington represented differently using the Bing Aerial, Lidar Hillshade, and Topo Basemap (ESRI) layers?

➤ Why are certain layers (such as buildings, facilities, and landmarks) only visible as you zoom in closely to areas? Why would they not be visible when you are zoomed further out to view a larger area of Vermont?

Hands-On Application 1.5

Examining Real Estate Values Online

Zillow is an online tool used to examine housing prices in cities across the United States. You can examine the prices of houses for sale in any neighborhood and see—spatially—where housing prices change. In other words, Zillow provides online mapping related to real estate prices and appraisal. To check it out, visit **www.zillow.com**. Starting with the Buy option on the main page, go to the Search box and type: Virginia Beach, Virginia. On the next screen a map will appear—zoom in on some of the residential neighborhoods, and you'll see markers in blue (auction sites) and red (homes for sale). Also, under the Listing Type pull-down menu, you'll have options to add purple markers (homes for rent) and yellow markers (recently sold).

Selecting a marker will give you the sale (or sold) price, along with Zillow's estimated market value. This tool allows you to select any location and examine information (such as housing prices and other details) that refers to this location.

Expansion Questions
➤ Use Zillow to check out your own local area. What does the housing market look like for your own area (in terms of how many houses are for sale and, more importantly, where are there any patterns or clusters of houses for sale)?

➤ How do the asking prices of the houses for sale compare with their estimated values?

Hands-On Application 1.6

Geospatial Technology and Emergency Preparedness

Geospatial technologies are used in many applications related to disaster management and emergency preparedness. A fun (but still useful) example of this is available in the Map of the Dead online at **www.mapofthedead.com/map**, which includes a free mapping tool intended to be used to help one survive the "zombie apocalypse." When the map opens, search for or zoom in on your own local area—you'll see the delineation of urban areas, as well as locations useful in a disaster or emergency, such as gas stations, police stations, hospitals, and grocery stores. By clicking on the Key button, you can access a list showing what each symbol represents and by clicking on the Place Names button you can obtain more information about the locations. All of this location information was taken from users of the Foursquare app (see page 19) and translated to places on the map.

While surviving fictional outbreaks of the walking dead makes for good entertainment, this kind of Web mapping tool highlights how geospatial tech-

nology can be used to enable individuals to find the locations of necessary resources quickly in the event of an actual emergency, such as a flood, tornado, or hurricane. Along these lines, Google has set up an online emergency preparedness tool for the state of Florida at **www.google.org/crisismap/florida_emergency_preparedness**, which shows current weather radar and traffic conditions, shelters, and evacuation routes in order to aid in emergency preparation and disaster management for the state.

Expansion Questions
➤ From examining both maps, what types of resources would be best to highlight on an emergency preparedness mapping tool and why?

➤ How can these Web resources be used or adapted for other types of disaster mitigation or emergency preparation (such as for floods, fires, or earthquakes)?

What Does All This Have to Do with Geography?

The first sentence for the entry "geography" in *The Dictionary of Human Geography* describes geography as "the study of the Earth's surface as the space within which the human population lives." To simplify a definition further into words of one syllable, geography is not just "where things are" but the study of "why things are where they are." Geography deals with concepts of the spatial characteristics of our planet and the spatial relationships and interactions of the people and features that occupy it. The notion of "Geographic Information Science" has been around for years as the study of the multiple concepts that surround the handling of spatial data, and today the use of geospatial technologies is tightly interwoven with the discipline of geography. See *Hands-on Application 1.7: The Geospatial Revolution Will Be Televised* for several examples of geospatial technologies being used for solving geographic problems.

At the higher education level, courses in GIS, remote sensing, GPS, and various applications of these technologies are key components of a geography curriculum. Certificate programs in Geographic Information Science (GISci) or Geospatial Technologies are also being offered increasingly by colleges and universities. Geospatial degree programs at the bachelor's, master's, and doctoral levels have been developed at numerous universities (see Chapter 15 for more information about geospatial technology in education). Even in basic activities such as collecting data in the field for a class project or producing a map of results, geospatial technologies have become an integral part of the geography discipline.

 ## Hands-On Application 1.7

The Geospatial Revolution Will Be Televised

Penn State University's Geospatial Revolution Project showcases a series of excellent videos that detail how geospatial technology affects our everyday lives, especially when it comes to concerns such as monitoring climate change and tracking disease. Visit **http://geospatialrevolution.psu.edu** and watch the available videos. There are additional videos aimed at a K–12 audience (but also present relevant information for all education levels) in the For Educators section as well.

Expansion Questions

➤ How can geospatial technologies be used for aiding in monitoring climate change conditions?

➤ How have geospatial technologies helped aid workers in dealing with food shortages or in mitigating the effects of disease outbreaks?

In a 2004 article, Jack Dangermond, CEO of the Environmental Systems Research Institute, Inc. (Esri; see Chapter 5), heralded GIS as a way of "speaking the language of geography," and in a 2009 article noted that GIS was seen as "geography in action." Both of these are great descriptions—for example, when you use an in-vehicle navigation system to navigate your way through unfamiliar territory or use an online tool to examine housing values for your neighborhood, you're using basic geographic concepts of space and place through technology. Geospatial technology gives us the tools to apply location-based principles to real-world situations more easily. For example, how is the new freeway bypass going to affect the local community in regard to traffic flow, new commercial development, and individual properties? Or, what are the potential impacts of the site chosen for a new casino development in relation to land use, economic and social effects, and the draw of gamers from far outside the local area? Throughout this book, you'll be dealing with geospatial technology concepts and applications, using some sort of geospatial data, measuring spatial characteristics, or examining spatial relations.

One of the most welcome recent developments is the opportunity for people without geospatial training to get involved in this field by applying their geographic knowledge. With geospatial tools becoming more widely utilized, people now have the chance to create their own geospatial content. Whether they're creating new maps, updating existing maps, tying digital pictures to locations on a map, or adding to geocaching databases, more and more people are using geospatial technologies to add to or enhance geospatial data and knowledge. Dr. Michael Goodchild coined the term "volunteered geographic information" (**VGI**) to give a name to data generated in this way.

A **wiki** is a Web utility that allows users to create, manage, update, and change the content of a Website freely. Possibly the best known one is *Wikipedia*, an online encyclopedia that anyone can edit. By extending this wiki concept to geospatial data, individuals can contribute their own maps, images, and updates to geospatial content available on the Internet, thereby generating VGI. Online projects such as The National Map Corps or OpenStreetMap (see **Figure 1.9** and *Hands-on Application 1.8: User-Generated Geospatial Content Online* for more information) allow users to generate or contribute geospatial content as volunteered geographic information.

For example, Google Maps allows users to contribute to the content of the Website by adding places, locations, and hiking trails, or by helping to improve on directions given by the system. Similarly, the USGS has established The National Map Corps to allow users to make edits and updates to the data of The National Map. After all, who would know more than you about new businesses that have opened, new roads that have been put in, or local hiking trails in your area? By contributing your geographic knowledge to sources such as OpenStreetMap, you're helping to build an

VGI volunteered geographic information, a term used to describe user-generated geospatial content and data

wiki a database available for everyone to utilize and edit

FIGURE 1.9

Detroit, Michigan, and Windsor, Canada, as shown in OpenStreetMap.

Hands-On Application 1.8

User-Generated Geospatial Content Online

Volunteered geographic information (VGI) represents user-generated geospatial content online. Use your Web browser to investigate some of these sites (and see how your own local area is represented with each one):

➤ OpenStreetMap: **www.openstreetmap.org**

➤ The National Map Corps: **www.edits .nationalmap.gov/tnmcorps**

➤ Wikimapia: **www.wikimapia.org**

In addition, check out the Humanitarian OpenStreetMap Team Website at **www.hotosm.org.** This group uses OpenStreetMap capabilities to be able to edit the maps for a number of different projects, such as mapping populated places

in Africa and South Asia as part of an effort to combat malaria.

Expansion Questions

➤ From examining your local area on OpenStreetMap, Wikimapia, and The National Map Corps Websites, what kinds of things could you potentially update or add using your own local geographic knowledge (and have any of these additions or updates been already made to one site but not another)?

➤ What types of current projects are ongoing with the Humanitarian OpenStreetMap Team Website, and how can you contribute to them with your own geospatial content?

extensive database of geospatial content as well as improving available online resources used by others.

It's becoming easier to apply geographic principles using geospatial technologies. For instance, on May 4, 2012, the *Saugerties Times* published a call for volunteers to take part in the launch of a free smartphone app for mapping the spread of invasive species, and explained that the collected data would be utilized in a New York invasive species database. Activities such as these are referred to as **crowdsourcing**, in which untrained volunteers collect data, content, or observations as part of a larger project (often referred to as **citizen science**). In geospatial terms, crowdsourcing projects allow huge numbers of people to contribute VGI to ongoing projects. Geospatial technologies are enabling everyone to become involved with collecting and analyzing geographic data. In particular, as mobile technologies and smartphones become more and more widespread, geospatial technology applications are ending up, quite literally, in the palm of your hand. There are numerous apps available that use geospatial technologies, and a technique called geolocation lies at the heart of many of them.

crowdsourcing the activities of untrained volunteers to create content and resources that can be utilized by others

citizen science the activities of untrained volunteers conducting science

What Is Geolocation?

Geolocation is the technique of determining where something is in the real world. By using geolocation, your smartphone, tablet, laptop, or desktop computer can determine what its current location is, give you the coordinates (such as latitude and longitude; see Chapter 2), and then plot that information on a map. This position can be found in several different ways—by using the IP address of your computer, your wi-fi connection, the cell towers your phone's signal is pinging off, or by the GPS receiver in your mobile device. Of course, once your phone, tablet, or computer knows where it is, it can share that location information with others via the Internet.

geolocation the technique of determining where something is in the real world

For instance, Websites often ask you if they can use your location. When you use a Website such as fandango.com to look up movie times or chilis.com to find information about a Chili's restaurant, these Websites will ask if they can use your computer's location. If you agree to share your location, you will receive information about movie theatres nearby or address information about the Chili's closest to you. Websites such as these can leverage knowledge about where you are (i.e., geolocation) to provide services to you based on your location.

This is done through the **W3C Geolocation API**, a set of commands and techniques used by programmers to obtain the location information of your phone, tablet, or computer (see Chapter 15 for more about APIs). Basically, with a few lines of computer code, a Web or app developer can use geolocation to determine where your computer or mobile device is and

W3C Geolocation API the commands and techniques used in programming to obtain the geolocation of a computer or mobile device

geotag connecting real-world location information to an item

in turn use that location in its services. For an example of geolocation in action, see *Hands-on Application 1.9: Where Am I Right Now?*

Often an app or device will give you the ability to **geotag** your media (such as a photo) by assigning your location to it through geolocation. For instance, when you take a photo with your phone, you may have the ability to include the location information of where you took the photo. Thus, you can connect the photos to their locations on a map. This use of geolocation is becoming a large component of social media as well. Whenever you use the Facebook app on your smartphone to make a post or send a tweet with the Twitter app, you have the option of adding your current location to that post or tweet. Geotagged tweets can then be mapped to see the locations from which people are sending tweets.

For instance, in the aftermath of the January 7, 2015, terrorist attack on the offices of the *Charlie Hebdo* newspaper in Paris, people worldwide tweeted "Je suis Charlie" (which is French for "I am Charlie") to show support for the murdered journalists. When the people sending the tweets included their location information with the tweet, those locations can be mapped. **Figure 1.10** shows an example of these worldwide geotagged "Je suis Charlie" tweets.

FIGURE 1.10

Geotagged tweets of "Je suis Charlie" sent in the aftermath of the terrorist attacks on the *Charlie Hebdo* newspaper on January 7, 2015.

Hands-On Application 1.9

Where Am I Right Now?

To see the W3C Geolocation API in action, go to **http://whereamirightnow.com**. When prompted by your browser, indicate that you will allow your location to be shared with the Website. Give it a few seconds and the Website will display a map of your current location (as determined using the W3C Geolocation API). This feature will also work on a smartphone or tablet using its Web browser as well.

Expansion Questions

➤ What kind of information did the Website display about your current location?

➤ How accurate was the Website in identifying where you were using the W3C Geolocation API? What methods did geolocation use to find you?

Through geolocation, a mobile device can know where it is. The "Find My iPhone" service uses geolocation to track your iPhone or iPad and displays its location on a map. Geolocation fuels many other common applications used on mobile devices today. For instance, the popular Yelp app allows users to see nearby restaurants' locations and read (or write) reviews for them. The Foursquare City Guide app works in a similar way—it collects crowdsourced information about restaurants, shopping, and nightlife from users and then can provide that information to you for places near your location. The Foursquare Swarm app allows users to share their location with others (referred to as a "check in") at different venues, such as shops or clubs. Even online dating uses geolocation; the popular Tinder app will show you how far away other users of the dating site are from your location. Beyond geolocation, there are many different free geospatial technology apps available today, and a widely used example is Google Earth.

What Is Google Earth?

An example of a widespread geospatial technology application is the popular **Google Earth** software program. Google Earth is a **virtual globe**—a program that provides an interactive three-dimensional (3D) representation of Earth. Google Earth brings together several aspects of geospatial technology into one easy-to-use program. You can examine satellite imagery or aerial photography of the entire globe, zoom from Earth's orbit to a single street address, navigate through 3D cities or landscapes, measure global distances, and examine multiple layers of spatial data together, all with an intuitive interface.

Google Earth a freely available virtual globe program first released in 2005 by Google

virtual globe a software program that provides an interactive three-dimensional map of Earth

Google Earth was first launched in 2005, but in 2017 Google significantly revised its groundbreaking yet aging software. If you want to put the software onto your computer and use multiple layers and various analytical features, you can freely download Google Earth Pro (which we will use in several Geospatial Lab Applications in this book—see **Figure 1.11** for an example of Google Earth Pro's layout). However, Google Earth itself now doesn't need to be downloaded and installed on a computer. Instead, Google Earth can now run in a Web browser or as a app on your smartphone or tablet and allows you to tour the world, viewing satellite imagery and 3D visualizations across the globe (see **Figure 1.12**).

Both the Pro version for your computer and the regular browser version of Google Earth rely on a high-speed Internet connection to send the data from Google servers to your computer. For instance, when you use Google Earth to fly from space to your house, the imagery you see on the screen is being streamed across the Internet to you. The imagery on Google Earth is "canned"—that's to say, it's fixed from one moment in time, rather than being a live view. For instance, if the imagery of your house used by Google was obtained in 2015, then that's the snapshot in time you'll be viewing, rather than a current image of your house. However, Google Earth Pro has a feature that enables you to examine past images. For instance, if imagery of your town was available in 2010, 2012, and today, you can choose which of these images you want to view.

Geospatial Lab Application 1.1: Introduction to Geospatial Concepts and Google Earth Pro at the end of this chapter walks you through several different ways of utilizing Google Earth Pro. It assumes no prior experience with Google Earth Pro, however, even veteran users may find some

FIGURE 1.11

A view of downtown Boston, Massachusetts, from Google Earth Pro.

new tricks or twists in how geospatial data is handled. You'll use a variety of Google Earth Pro tools to investigate several of the avenues this book will explore, from aerial image interpretation and 3D representations of terrain and objects to ways of determining the shortest path between two locations.

FIGURE 1.12

The Google Earth app running on a tablet. *[© Iain Masterton/Alamy]*

 ## Thinking Critically with Geospatial Technology 1.1

What Happens to Privacy in a Geospatial World?

Think about this scenario. If you own property (a house or a business), then that information is part of the public record with the county assessor's office and likely available online. Someone can type your address into an online GIS (or mapping Website such as MapQuest or Google Maps) and get a map showing how to travel to your property. In addition, the dimensions, boundaries, and a high-resolution image of your property can be obtained easily. It is possible that other information can be acquired, such as a close-up photo of your house taken from the street, and accessed directly from the Website. All of this is free, publicly available information that can be had with just a few clicks of a mouse. Anonymity about where you live or work is suddenly becoming a thing of the past. What does the widespread availability (and access to) this type of spatial information mean with regard to privacy? If anyone with an Internet connection can obtain a detailed image of the house you live in, has your personal privacy been invaded?

Chapter Wrap-Up

So what are you waiting for? Move on to this chapter's lab (see *Geospatial Lab Application 1.1: Introduction to Geospatial Concepts and Google Earth Pro*) and start getting to work. This chapter's lab will have you doing all sorts of things with Google Earth Pro while touching on ideas that will be explored in every chapter of this book. There are also questions throughout the lab for you to answer based on the tasks you'll be doing. Also, check out the Geospatial Apps box for a sample of downloadable apps (including a Google Earth app) for your mobile device. For further ways to get connected with geospatial technology applications through blogs, Facebook, Twitter, and YouTube, check out the options in *Geospatial Technologies in Social Media*.

Important note: The references for this chapter are part of the online companion for this book and can be found at **http://www.macmillanlearning.com/catalog/shellitoigt4e**

Geospatial Apps

Here's a sampling of available representative geospatial technology apps for your smartphone or tablet. Note that some apps are for Android, some are for Apple iOS, and some may be available for both.

➤ **3D Geo Globe:** An app that will determine your coordinates and plot them on a map of Earth

➤ **Field Photo:** An app that allows you to take photos with your smartphone's camera, geotag them, then upload them to a crowdsourced database for use in environmental studies

➤ **Foursquare City Guide:** An app that provides information on restaurants, shopping, and nightlife near your location

➤ **Foursquare Swarm:** An app that allows you to "check in" your location with your smartphone and share that information with others

➤ **Google Earth:** An app that allows you to use Google Earth on your mobile device

➤ **ViewRanger:** An outdoor app that allows you to view your location on a variety of different maps

➤ **Where Am I At?:** An app that determines where you are and shows your location on a map

➤ **Yelp:** An app for locating restaurants near your location and showing reviews of them

➤ **You Need a Map:** An app that displays several different types of maps and map layers and uses your own location for detailed geospatial mapping

➤ **Zillow Real Estate:** An app that allows you to view maps and information about homes for sale

Geospatial Technologies in Social Media

Here's a sampling of some representative Facebook and Twitter accounts, along with some YouTube videos and blogs related to this chapter's concepts.

f On Facebook, check out the following:

➤ Geoawesomeness:
    www.facebook.com/GeoAwesomeness

➤ Google Maps:
    www.facebook.com/GoogleMaps

➤ OpenStreetMap:
    www.facebook.com/OpenStreetMap

🐦 Become a follower on Twitter of:

➤ **Directions Magazine** (geospatial news magazine): @directionsmag

➤ **Geoawesomeness** (location-based news): @geoawesomeness

➤ **GIS and Science:** @GISandScience

➤ **GIS GeoTech News:** @gisuser

➤ **Google Earth:** @googleearth

➤ **MondeGeospatial:** @MondeGeospatial

➤ **OpenStreetMap:** @Openstreetmap

📷 Become an Instagram follower of:

➤ Geoawesomeness: @geoawesomeness

You Tube On YouTube, watch the following videos:

➤ **Introduction to The National Map** (a USGS video about The National Map): **www.youtube.com/watch?v=ISzUlINbB4o**

➤ **Jobs in Geography** (a video sponsored by the Association of American Geographers): **www.youtube.com/watch?v=mJz0YTHvCY8**

➤ **Weird Earth: Overview** (a video of some weird and wonderful places around the planet as seen by geospatial technologies): **www.youtube.com/watch?v=DDfSxMCof2w**

For further up-to-date info, read up on these blogs and news sites:

➤ **Directions Magazine** (online geospatial news source): **www
.directionsmag.com**

➤ **The Google Earth Blog** (a blog about Google Earth and its
applications): **www.gearthblog.com**

Key Terms

aerial photography (p. 2)
citizen science (p. 17)
crowdsourcing (p. 17)
GEOINT (p. 8)
geospatial data (p. 8)
geospatial technology (p. 2)
Geographic Information System
 (GIS) (p. 2)
geolocation (p. 17)
geotag (p. 18)
Google Earth (p. 19)

Global Positioning System (GPS)
 (p. 2)
The National Map (p. 10)
non-spatial data (p. 9)
remote sensing (p. 2)
satellite imagery (p. 2)
VGI (p. 15)
virtual globe (p. 19)
wiki (p. 15)
W3C Geolocation API (p. 17)

Introduction to Geospatial Concepts and Google Earth Pro

This chapter's Geospatial Lab Application will introduce you to some basic concepts of geospatial technology through the use of the Google Earth Pro software program. This lab also provides an introduction on how to use Google Earth Pro and will help familiarize you with many of its features. Although the Geospatial Lab Applications in later chapters will use a variety of software packages, you'll also be using Google Earth Pro in many of them. The investigations in this introductory lab may seem pretty basic, but labs in later chapters will be more in-depth and will build on concepts learned here.

Objectives

The goals for you to take away from this lab are:

➤ To familiarize yourself with the Google Earth Pro (GEP) environment, basic functionality, and navigation using the software

➤ To use different GEP layers and features

Using Geospatial Technologies

The concepts you'll be working with in this lab are used in a variety of real-world applications, including the following:

➤ Cable and network news, which frequently incorporate Google Earth imagery or flyovers into news segments to show areas of interest, to demonstrate the spatial relationships or distances between cities visually, or to make visual links between locations of news items

➤ K–12 teachers, who utilize Google Earth imagery (overhead views, 3D models, flyovers, or directions) to help illustrate a variety of concepts in class, including geography, history, and science

Obtaining Software

The current version of Google Earth Pro (7.3) is available for free download at **www.google.com/earth/download/gep/agree.html**. You'll

[ZUMA Press, Inc. / Alamy]

be using Google Earth Pro instead of the regular Google Earth in this lab (and others in the book) to take advantage of its many additional features.

Important note: Software and online resources sometimes change fast. This lab was designed with the most recently available version of the software at the time of writing. However, if the software or Websites have changed significantly between then and now, an updated version of this lab (using the newest software) is available online at **http://www.macmillanlearning .com/catalog/shellitoigt4e**.

Lab Data

There is no data to copy in this lab. All data comes as part of the GEP data that is installed with the software or is streamed across the Internet when using GEP.

Localizing This Lab

Although this lab visits popular locations in a tour around South Dakota, the techniques it uses can be adapted easily to any locale. Rather than using GEP imagery and features of South Dakota, find nearby landmarks or popular spots and use GEP to tour around your local area.

1.1 Starting to Use Google Earth Pro (GEP)

1. Start **Google Earth Pro**. (the default install folder is called Google Earth Pro). GEP will usually open with a startup tip box, which you can close. Also, make sure the GEP window is maximized, or else some features may not show up properly.

A popular use of GEP is to view the ground (whether physical landscapes or structures) from above by using the aerial photography or satellite imagery streamed in from Google's servers. The **Search** box is used for this.

2. In the **Search** box, type in **Mitchell, SD**. Next, click on the **Search** button.

3. GEP will rotate and zoom the view down to eastern South Dakota. One of the popular attractions in Mitchell is the Corn Palace, a historic building of fanciful design that now serves as a civic arena, presenting cultural and sports events and attracting the attention of millions each August when the murals that cover the building's façade (which are completely constructed from corn) are reconfigured to illustrate the theme of that year's Corn Palace Festival. More information about the Corn Palace is available at **www.cornpalace.com**.

4. In the **Layers** box, place a checkmark next to **Photos**. This enables linking of locations on the ground to photos that users have taken. The locations that have linked photos have small blue squares over

the imagery. You'll see a large photo symbol appear with numerous smaller photo locations clustered together in one place—that would be the Corn Palace. Double-click on the larger symbol to see some photos (taken from the ground) of the Corn Palace.

5. In GEP, you can rotate and zoom your view to get a closer look at the imagery on the ground. Move the mouse to the upper right-hand side of the screen and a set of controls will appear. These fade out when not in use and reappear when the mouse rolls over them.

6. There are five controls. The first one (the ring with the N at the top) allows you to change the direction you're facing. Grab the **N** with the mouse (by placing the pointer on **N** and holding down the left mouse button) and drag the **N** (that is, slowly rotate it) around the ring (this is commonly called "grab and drag"). You'll see the view change (north will still be the direction in which the N is pointed). Clicking on the **N** will return the imagery so that the north direction is facing the top of the screen.

7. The second control is the **Look** control (the arrows surrounding the eye). By selecting one of the directional arrows and pressing it with the mouse, you can tilt your view around the terrain as if you were standing still and looking about. You can also grab the control and move it (by holding down the left button on your mouse), to simulate looking around in various directions.

8. The third control is the **Move** control (the arrows surrounding the hand). By grabbing this control, you can pan around the imagery in whichever direction you choose.

9. The fourth control is the **Street View** icon (the yellow figure of a human—referred to as the "pegman"—standing on a green circle). This icon will appear when Google Street View imagery is available to see on the ground. We'll spend more time on Google Street View in Chapter 8, as it's a utility that allows you to see what places look like as if you were standing on the street in front of them. To use Street View, you would grab the icon from the controls with the mouse and drag it to a street that's visible in the view to enter the Street View mode. This control will only be visible if there are streets in the view through which you can enter Street View mode (which is why you won't initially see it when you start GEP and are looking at the entire planet).

10. The last control is the **Zoom Slider**. By pushing forward (moving the bar toward the plus sign), you zoom in, and by pulling backward (moving the bar toward the minus sign), you zoom out. By default, when GEP zooms in, it tilts the terrain to give you a perspective view, rather than looking from the top directly down. To always use the top-down option rather than tilting the ground when zooming, select the

Tools pull-down menu and choose **Options**. In the dialog box that appears, click on the **Navigation** tab, and then select the radio button that says "**Do not automatically tilt when zooming.**" To have GEP tilt while zooming, select one of the other radio buttons.

11. You can also use the mouse for navigating around GEP. Grab the imagery and drag it to sweep across the map. To zoom in and out, roll the mouse wheel back and forth. Pan and zoom over to the location of the Corn Palace. To confirm that you're in the right place, type the address of the Corn Palace into the **Search** box as follows: **604 North Main Street, Mitchell, SD 57301** (GEP will automatically adjust to that location).

12. For now, turn off the **Photos** options by removing the checkmark from the box.

13. Zoom in on the Corn Palace so that it fills most of the view and you can make out some details about it.

14. Grab the **Street View** icon from the controls and drag it to the street right in front of the Corn Palace. You'll see the street turn blue (this indicates that Street View imagery is available for this street). Release the icon on the street and the view will shift from an aerial view to imagery that looks like you're actually standing on Main Street in front of the Corn Palace.

15. In this Street View, use the mouse to pan around the image (you can see 360 degrees around your position) until you can see the entrance of the Corn Palace. You can pan and zoom with the mouse as necessary (and also click on the road—or elsewhere in the image—to advance your view).

16. Examine the Corn Palace and its surrounding area. Then answer Question 1.1. To return to the aerial view, click on the **Exit Street View** text in the upper right corner of the Street View.

> **Question 1.1** From viewing the Corn Palace from above and in front, what details from the aerial view can help identify just what the building is? (You may need to switch back and forth between the aerial view and the street view to answer this question.)

17. When dealing with remotely sensed imagery, it's important to consider that you're viewing the ground from above and that many features you can view from a street level are difficult to perceive from an aerial view. Chapter 9 delves further into interpreting aerial images and offers some strategies and tips for doing so, while Chapter 10 deals with how this type of remotely sensed imagery is acquired. Chapters 11 and 12 explore the acquisition of imagery from various satellites.

18. Here's another way of looking at the Corn Palace—a 3D representation. From the GEP **Layers** options, select **3D Buildings**. A 3D representation of the Corn Palace will appear on top of the imagery of the building. Use the zoom and tilt controls and navigate around the building to get a better look at the 3D model of the Corn Palace, including its façade and turrets. If the 3D model of the Corn Palace doesn't appear, select the **Tools** pull-down menu, select **Options**, then select the **3D View** tab and make sure a checkmark is not in the **Use 3D imagery (disable to use legacy 3D buildings)** box.

19. After viewing the 3D model of the Corn Palace, turn off the option for **3D Buildings** when you're done. There are numerous buildings and objects that have 3D representations designed for use in GEP. In Chapter 14, you'll investigate more about working with 3D versions of geospatial data, as well as constructing your own 3D buildings and viewing them in GEP.

1.2 Finding Your Way Around with Google Earth Pro

1. From the **Layers** box, turn on the **Roads** layer. You'll see major roads (interstates, state highways) identifiable as their labels appear, and local roads will also be identifiable when you zoom in close.

2. We're going to leave the Corn Palace behind and continue west through South Dakota to Wall Drug, a famous tourist location. More information on Wall Drug can be found at **www.walldrug.com**.

3. In the **Search** box, click on the **Get Directions** option.

4. In the **A** option (this is where you're traveling from), type in the Corn Palace's address as follows: **604 N Main St, Mitchell, SD 57301**.

5. In the **B** option (this is where you're traveling to), type in Wall Drug's address as follows: **510 Main Street, Wall, SD 57790**.

6. Click the **Get Directions** button. GEP will zoom out to show you the path (in purple) it has calculated for driving distance between the Corn Palace and Wall Drug, while the **Search** box will fill with directions featuring stops and turns along the way.

> **Question 1.2** By viewing the expanse of South Dakota between Mitchell and Wall, you'll realize there are many possible routes between the two. Why do you suppose GEP chose this particular route to go from the Corn Palace to Wall Drug?

> **Question 1.3** Based on the route that GEP calculated, what is the driving distance (and approximate time equivalent) to get from the Corn Palace to Wall Drug?

7. This capability to take a bunch of letters and numbers and turn them into a mapped location and to calculate the shortest driving distance between points are further developed in Chapter 8. Matching addresses and calculating shortest routes are key reasons for utilizing geospatial technologies, and the Chapter 8 lab will go into more depth on how these types of operations are performed.

8. In the **Search** box, scroll down through the list of directions and click on the Wall Drug address next to **B**. Google Earth will zoom to Wall Drug's location. Press the **X** at the bottom of the **Search** box to clear the directions and remove the route.

1.3 Using Google Earth Pro to Examine Landscapes

1. There's a lot more to the South Dakota landscape than tourist attractions. Wall Drug is located in the town of Wall, which is named for the line of magnificent land formations nearby and lies directly adjacent to Badlands National Park. The Badlands are located to the south and southwest of Wall. For more information about the Badlands visit **www.nps.gov/badl/index.htm**.

2. To see the boundaries of Badlands National Park in Google Earth Pro, go to the **Layers** box and expand the option for **More** (click on the triangle to the left of the name). In the options that appear under the expanded heading, put a checkmark in the **Parks/Recreation Areas** option. Zoom out from the boundaries of Wall, and to the south you'll see the northernmost boundary of Badlands National Park highlighted in green.

3. Zoom out so that you can see the entire park in the view. New icons for the locations of Visitors Centers and other features will also appear.

4. Pan over to the eastern edge of the park and you'll see a large Question Mark icon indicating a park entrance as well as a green arrow indicating an overlook.

5. Make sure that the option for **Terrain** in the **Layers** box is turned on. Zoom in very closely to the point representing the overlook. At the bottom of the view you'll see numbers representing the latitude and longitude of the point, as well as the real-world elevation of that spot.

> **Question 1.4** What is the elevation of this particular overlook in Badlands National Park?

6. The imagery in GEP is placed on top of a model of Earth's vast array of terrains and landscapes (hills, valleys, ridges, and so on). It's very difficult to make out the pseudo-three-dimensional (3D) nature of the terrain model from above, so use the **Zoom Slider** to tilt the view down (you can also hold down the Ctrl key and move the mouse to help change your perspective) so that you can look around as if you were standing at the overlook point (in a perspective view of the planet). Once you've tilted all the way down, use the **Look** controls to examine the landscape. From here, use the **Move** controls to fly over the Badlands from this overlook point. When you're done cruising around the Badlands, answer Question 1.5.

> **Question 1.5** How does the terrain modeling (with the tilt function) aid in the visualization of the Badlands?

7. This ability to model the peaks and valleys of the landscape with aerial imagery "draped" or "stretched" over the terrain for a more realistic appearance is often used with many aspects of geospatial technology. Chapter 13 greatly expands on this by getting into how this is actually performed as well as doing some hands-on terrain analysis.

1.4 Using Google Earth Pro to Save Imagery

It's time to continue to the next leg of our South Dakota journey by heading to Mount Rushmore. Carved out of the side of a mountain in the Black Hills, Mount Rushmore National Memorial features the faces of presidents George Washington, Thomas Jefferson, Theodore Roosevelt, and Abraham Lincoln. For more information about Mount Rushmore visit **www.nps. gov/moru/index.htm.**

1. In GEP's **Search** box, type **Mount Rushmore**. GEP will zoom around to an overhead view of the monument. Like the Badlands, Mount Rushmore is overseen by the National Park Service—zoom out a little bit and you'll see the extent of the memorial's boundaries (still outlined in green).

2. You can also save an image of what's being shown in the view. GEP will take a "snapshot" and save it as a JPEG graphic file (.jpg), which is like a digital camera picture. Position the view to see the full outlined extent of Mount Rushmore, select the **File** pull-down menu, then select **Save**, and finally select **Save Image**. You'll see that a box called

Untitled Map has been added—click on that to give your image a name (like "Mount Rushmore") and then click on the **Save Image…** button at the top of the screen. Name the image **moru** (GEP will automatically add the .jpg file extension), and save it to your computer's hard drive.

3. Minimize GEP for now and go to the location on your computer where you saved the image and then open it to examine it (using a simple program such as Microsoft Office Picture Manager).

> **Question 1.6** Note that even though the graphic contains the locations of Mount Rushmore, the outline of the park, and information concerning location (including latitude, longitude, and elevation) at the bottom, it doesn't have any spatial reference for measurements. Why is this?

4. Close the image and return to Google Earth Pro. Even though the saved image doesn't have any spatial reference, Chapter 3 describes how to take unreferenced imagery (and other unreferenced data) and transform it to match referenced data. In Chapter 3's lab, you do this in a hands-on fashion. You can also turn off the **Parks/Recreation** layer for now.

1.5 Using Google Earth Pro to Make Placemarks and Measurements

While you're examining Mount Rushmore imagery, you can set up some points of reference to which you can return. GEP allows you to create points of reference as Placemarks. There are three points to mark on the map—the top of the mountain, the amphitheater, and the parking area (see the following graphic for their locations).

[Source: Google]

1. From the GEP toolbar, select the **Add Placemark** button:

2. A yellow pushpin (labeled "Untitled Placemark") will appear on the screen. Using your mouse, click on the **Pushpin** and drag it to the rear of the amphitheater so that the pin of the placemark is where the path meets the amphitheater.

3. In the Google Earth **New Placemark** dialog box, type **Mount Rushmore Amphitheater**.

4. By pressing the **Placemark** icon button next to where you typed the name, you can select an icon other than the yellow pushpin. Choose something more distinctive.

5. Click **OK** to close the dialog box.

6. Repeat the process by putting a new placemark at the top of the mountain where the memorial is. Name this new placemark **Top of the Memorial**.

7. Zoom GEP's view in tightly so that you can see the two placemarks clearly in the view (with the two placemarks in diagonally opposite corners of the view).

8. Next, select the **Show Ruler** button from the GEP toolbar:

9. In the **Ruler** dialog box that appears, select **Feet** from the **Length** pull-down menu.

10. Use the options under the **Line** tab. These will let you compute the distance between two points. The **Path** option will enable you to find out the total accumulated distance between more than two points.

11. Click the mouse on the point of the placemark in the amphitheater and drag it to the placemark on top of the mountain (your heading should be about 310 degrees or so). Click the mouse at the top of the mountain placemark.

12. Answer Question 1.7. When you're done, click on **Clear** in the **Ruler** dialog box to remove the drawn line from the screen, and then close the **Ruler** dialog box.

> **Question 1.7** What is the measured distance between the rear of the amphitheater and the top of the memorial? (Keep in mind this is the ground distance, not a straight line between the two points.)

13. This ability to create points of reference (as well as lines and area shapes) and then compute the distance between them are functions of a very basic GIS tool, which forms the basis for performing numerous tasks that will be discussed in upcoming chapters. Chapter 5 introduces GIS concepts, Chapter 6 expands on how spatial analysis can be performed with GIS, and Chapter 7 demonstrates how to take your data and create a professional quality map of the results.

14. Use the **Tilt** functions of GEP to get a perspective view on Mount Rushmore (as you did with the Badlands), and take a look at the mountain from a pseudo-3D view. Make sure that the option for **Terrain** in the **Layers** box is turned *on* and also make sure the **3D Buildings** option is turned *off*. Although you can see the height and

dimensions of the mountain, the famous four presidents' faces can't be seen, however you zoom or rotate the model.

> **Question 1.8** Even with **Terrain** turned on and the view switching over to a perspective, why can't the presidents' faces on the side of the memorial be seen?

15. Turn the **3D Buildings** option on. If you cannot see the separate Mount Rushmore 3D model (which lets you see the faces), select the **Tools** pull-down menu, select **Options**, then select the **3D View** tab, and make sure a checkmark is not in the **Use 3D imagery** (disable to use legacy 3D buildings) box.

1.6 Using Google Earth Pro to Examine Coordinates

1. You'll notice that next to the elevation value at the bottom of the GEP screen there's a set of coordinates for lat (Latitude) and lon (Longitude). Move the mouse around the screen and you'll see the coordinates change to reflect the latitude and longitude of whatever the mouse's current location is.

2. Zoom in closely on the road that enters the parking structure area of Mount Rushmore.

> **Question 1.9** What are the latitude and longitude coordinates of the entrance to the Mount Rushmore parking area?

3. You can also reference specific locations on Earth's surface by their coordinates instead of by name. In the **Search** box type the following coordinates: **43.836584, –103.623403**. GEP will rotate and zoom to this new location. Turn on **Photos** to obtain more information on what you've just flown to. You can also turn on the **3D Buildings** layer (if necessary, again using the legacy **3D Buildings** option, as you did with the Corn Palace and Mount Rushmore), to look at the location with a 3D version of it there. Answer Questions 1.10 and 1.11, and then turn off the **Photos** (and the **3D Buildings** option) when you're done.

> **Question 1.10** What is located at the following geographic coordinates: latitude 43.836584, longitude –103.623403?

> **Question 1.11** What is located at the following geographic coordinates: latitude 43.845709, longitude –103.563499?

Chapter 2 deals with numerous concepts related to coordinate systems, reference systems, and projections of a three-dimensional Earth onto a two-dimensional surface.

1.7 Other Features of Google Earth Pro

As noted before, you will be using GEP for several other labs in this book and exploring some of these applications of geospatial technology. GEP has a multitude of other features beyond those covered in this first lab. For instance, if you had a GPS receiver (say, one from Garmin or Magellan) you could import the data you've collected directly into GEP.

1. From the **Tools** pull-down menu, select **GPS**. You'll see several options for offloading information you collected in the field (such as waypoints) from the receiver into GEP.

   Chapter 4 will introduce you to several concepts related to GPS, as well as collecting data in the field and some pre-fieldwork-planning options.

2. Close the **GPS Import** dialog box now (unless you have a receiver you want to plug into a computer on which you can view your data in GEP).

   There are many other GEP layers and features to explore, especially in relation to how geospatial data is visualized. To look at one of these features, return to the Badlands once more and do the following:

3. In the **Layers** box, expand the options for **Gallery**, and then place a checkmark in the box next to **360 Cities**.

4. Fly back to the Badlands and scroll to the western edge of the park and you'll see another symbol on the map—a red circle marked "360."

Hover your cursor over it, and the text should read "Mako Sica aka The Badlands."

5. Double-click on the **360** circle. You'll zoom down and see a sphere on the surface, and then the view should dive inside the sphere. The view will change to a new view of the Badlands (use the mouse to move about and look around a full 360 degrees). In essence, this tool simulates what the Badlands would look like from that position, all around you. You can look in any direction, as well as zooming in or out of places. Answer Question 1.12. When you're done looking around, click on the **Exit Photo** button in **View** to return to regular GEP. Chapter 14 will present information about some more cool visualization techniques for geospatial data.

> **Question 1.12** How does the 360-degree image of the Badlands aid in visualizing the scenery? (Keep in mind that this image has been tied to a particular location.)

Geospatial technology is such a rapidly changing field that new advances come quickly. Chapter 15 also explores some of these current developments on the frontiers of geospatial technology.

Google Earth Pro changes regularly with new updates. Some of the other options you may want to explore are Google Mars, Google Sky, and Flight Simulator.

Google Mars

From the **View** pull-down menu, select **Explore**, and then select **Mars**. The view will shift to the familiar-looking globe, but this time covered with imagery (from NASA and the USGS) from the surface of Mars. The usual controls work the same way as they do with GEP, and there's a lot of Martian territory to be explored. When you're ready to return to GEP, from the **View** pull-down menu, select **Explore**, and then select **Earth**.

Google Sky

From the **View** pull-down menu, select **Explore**, and then select **Sky**. GEP's view will change: instead of looking down on Earth, you're looking up to the stars—and instead of remotely sensed aerial or satellite imagery, you're looking at space-telescope (including the Hubble) imagery. There's plenty of imagery to be explored in this new mode. When you want to get back "down to Earth" again, select the **View** pull-down menu, select **Explore**, and then select **Earth**.

Flight Simulator

From the **Tools** pull-down menu, select **Enter Flight Simulator**. A new dialog box will appear asking if you want to pilot an F-16 fighter jet or an SR-22 propeller airplane. Select whichever plane you want to fly; GEP will switch to the view as seen out the cockpit of your chosen plane, and you can fly across South Dakota (and the rest of GEP). Try not to crash (although you can easily reset if you do).

Closing Time

This introductory lab was pretty straightforward, but it served to introduce you to the basic functions of Google Earth Pro, as well as provided examples of many geospatial concepts you'll be working with over the course of this book. Chapter 2 starts looking at location concepts (like various coordinate systems) and returns to Google Earth Pro to start applying some of the topics you've worked with in this chapter. You're all set, and now you can exit GEP by selecting **Exit** from the **File** pull-down menu. There's no need to save any data (or Temporary Places) in this lab.

Where in the Geospatial World Are You?

Locations in a Digital World, Position Measurements, Datums, Coordinate Systems, GCS, Map Projections, UTM, and SPCS

Think about this. You're at the mall and your cell phone rings. It's your friend Sara calling you with a basic question: "Where are you?" For such a simple question, there's really a wide variety of answers. You could say "I'm at the mall," which gives Sara some basic location information—but only if she knows where the mall is. If she's from out of town and unfamiliar with the location of the mall, you haven't given her the information she needs to know where you are. If you tell her, "I'm in Columbus," then she'll have another type of information—but Columbus is a big city, and she still won't know where in the city you are. You could give her the name of the cross streets by the mall, the address of the mall, or the name of the store you're in at the mall, and even though all of the these contain different levels of information, your friend still won't have the information she needs to know precisely where you are.

"I'm at the mall" effectively conveys your location only if your friend has some sort of knowledge of the mall's location. If you give her a reference such as "The mall's on the north side of town," or "The mall is just off exit 5 on the freeway," then she'll have some idea of your location. For the location information to make sense, she needs to have something (perhaps a map of the city you're in) to which to reference the data. If you're asked to provide specific information about your exact location, it'll have to be much more detailed than "I'm at the mall." In a medical emergency, emergency medical technicians (EMTs) need to know your exact location (they can't afford to waste time searching the entire mall to find you). You need to provide precise location information so that the EMTs will be able to find you, no matter where you are.

Geospatial technology works the same way. First, since it deals with geospatial data (which is linked to non-spatial data), there has to be a way of assigning location information to the data being handled. There has to be some way of referencing every location on Earth. In other words, every location has to be clearly identifiable and measurable. Second, these measurements require some sort of reference system so that a location at one point on Earth can be compared with locations at other points. Third, since we're dealing with points on a three-dimensional (3D) Earth but only have two-dimensional (2D) surfaces to examine (i.e., maps of various kinds), there has to be a way of translating real-world data to an environment we can use. This chapter examines how these concepts are treated in all forms of geospatial technology. If you're going to be able to pin down the coordinates of a location precisely, you must first have a datum and a coordinate system.

 ## What Is a Datum?

datum a reference surface of Earth

A **datum** is a reference surface, or model, of Earth that is used for plotting locations anywhere on the actual surface of Earth. The datum represents the size and shape of Earth—which, contrary to popular belief, isn't perfectly round. Earth is actually an **ellipsoid** (or spheroid), meaning that it's larger at its center than it is at its poles. Think of taking a basketball and squeezing the top and bottom of the ball to compress it slightly and you'll have a basic spheroid shape. However, Earth isn't a smooth ellipsoid—gravitational forces affect different parts of Earth in different ways, causing some areas to be out of sync with the ellipsoid. Another model of Earth, called a **geoid**, places Earth's surface at mean sea level to try to account for these differences (see **Figure 2.1** to examine how the geoid and the ellipsoid stack up against each other and against the variable topography of Earth). The science of measuring Earth's shape to develop these kinds of models and reference surfaces is referred to as **geodesy**.

ellipsoid a model of the rounded shape of Earth

geoid a model of Earth using mean sea level as a base

geodesy the science of measuring Earth's shape

FIGURE 2.1

How the real-world Earth, the ellipsoid, and the geoid match up.

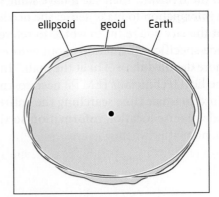

Developing a datum means creating a mathematical model against which to reference locations and coordinates. Thus, when you're plotting a point, you have a system for referencing this location. The difficulty in mapping with datums arises because there isn't just one datum to use for all measurements of Earth's locations. In fact, there are hundreds of datums in use. Some datums are global and measure the entire world, whereas some are localized to particular continents or regions. These are some of the more common datums you'll likely encounter with geospatial data:

➤ **NAD27** (the North American Datum of 1927): This datum was developed for measurements of the United States and the rest of North America. It has its center point positioned at Meades Ranch in Kansas.

NAD27 the North American Datum of 1927

➤ **NAD83** (the North American Datum of 1983): This datum was developed by the National Geodetic Survey, together with Canadian agencies, and is used as the datum for much data for the United States and the North American continent as a whole. While NAD83 is very commonly used, the U.S. National Geodetic Survey is currently preparing a new datum to take its place, but this won't be ready until 2022.

NAD83 the North American Datum of 1983

➤ **WGS84** (the World Geodetic System of 1984): This datum was developed by the U.S. Department of Defense, and is used by the Global Positioning System (GPS) for locating points worldwide on Earth's surface (see Chapter 4 for more information about GPS).

WGS84 the World Geodetic System of 1984 datum (used by the Global Positioning System)

Measurements made with one datum don't necessarily line up with measurements made from another datum. Problems arise if you take one set of data measured in one datum and your friend takes a second set of data measured from a different datum and you put both sets of data together. Both datasets refer to the same place, but they won't match up because you and your friend are using different forms of reference for measurement. For instance, data measured from the NAD27 datum could be off by a couple hundred meters from data measured from the NAD83 datum. This problem is compounded by the availability of several local datums, or reference surfaces, designed for smaller geographic areas. With the measurement differences that may arise, the best strategy when you're dealing with geospatial data is to keep everything in the same datum. A **datum transformation** is required to alter the measurements from one datum to another (for instance, changing the measurements made in NAD27 to NAD83). Datum transformation is a computational process, a standard one in many geospatial software packages. Datums can be used to set up geographic coordinate systems (GCS) for measuring coordinates.

datum transformation changing measurements from one datum to measurements in another datum

geographic coordinate system (GCS) a set of global latitude and longitude measurements used as a reference system for finding locations

latitude imaginary lines on a globe north and south of the Equator that serve as a basis of measurement in GCS

Equator the line of latitude that runs around the center of Earth and serves as the 0 degree line from which to make latitude measurements

longitude imaginary lines on a globe east and west of the Prime Meridian that serve as a basis of measurement in GCS

Prime Meridian the line of longitude that runs through Greenwich, England, and serves as the 0 degree line of longitude from which to base measurements

degrees, minutes, and seconds (DMS) the measurement system used in GCS

What Is a Geographic Coordinate System?

A **geographic coordinate system (GCS)** is a global reference system for determining the exact position of a point on Earth (see **Figure 2.2** for an example of GCS). GCS consists of lines of latitude and longitude that cover the entire planet and are used to find a position. Lines of **latitude** (also referred to as parallels) run in an east-to-west direction around the globe. The **Equator** serves as the starting point of 0, with measurements north of the Equator being numbers read as north latitude and measurements south of the Equator being numbers read as south latitude. Lines of **longitude** (also called meridians) run north-to-south from the North Pole to the South Pole. The **Prime Meridian** (the line of longitude that runs through the Royal Observatory in Greenwich, England) serves as the starting point of 0. Measurements made east of the Prime Meridian are numbers read as east longitude, while measurements made west of the Prime Meridian are numbers read as west longitude. The breaking point between east and west longitude is the 180th meridian.

Using this system (where measurements are made north or south of the Equator and east or west of the Prime Meridian), any point on the globe can be located. GCS measurements are not made in feet or meters or miles, but in **degrees, minutes, and seconds (DMS)**. A degree is a unit of measurement that can be broken into 60 subsections (each one referred to as a minute of distance). A minute can subsequently be broken down into 60 subsections (each one referred to as a second of distance). A single minute of latitude is equivalent to roughly 1 nautical mile (or about 1.15 miles). Lines of latitude are measured in angular units from the equator, while lines of longitude are measured in angular units from the Prime Meridian. It should be noted that

FIGURE 2.2

The geographic coordinate system (GCS) covers the globe, and locations are measured using latitude and longitude.

the distance between degrees of longitude varies at different places in the world—lines of longitude get closer together as they approach the poles and are furthest apart at the Equator.

When GCS coordinates are being written for a location, they are expressed in terms of the location's distance (that is, in terms of degrees, minutes, and seconds) from the Equator and the Prime Meridian. Latitude values run from 0 to 90 degrees north and south of the Equator, while longitude values run from 0 to 180 east and west of the Prime Meridian. For example, the GCS coordinates of a location at the Mount Rushmore National Memorial are 43 degrees, 52 minutes, 53.38 seconds north of the Equator (43°52'53.38" N latitude) and 103 degrees, 27 minutes, 56.46 seconds west of the Prime Meridian (103°27'56.46" W longitude).

Negative values can also be used when expressing coordinates using latitude and longitude. When distance measurements are made west of the Prime Meridian, they are often listed with a negative sign to note the direction. For instance, the location at Mount Rushmore, being west longitude, could also be written as –103°27'56.46" longitude. Negative values are used similarly in measurements of south latitude (below the Equator).

GCS coordinates can also be expressed in their decimal equivalent, referred to as **decimal degrees (DD)**. In the decimal degrees method, values for minutes and seconds are converted to their fractional number of degrees. For instance, 1 degree and 30 minutes of latitude (in DMS) equates to 1.5 degrees in DD (because 30 minutes is equal to 0.5 degrees). Using the Mount Rushmore coordinates as an example, the latitude measurement in decimal degrees would be 43.881494 north latitude (because 52 minutes and 53.38 seconds is equal to 0.881494 of one degree) and 103.465683 west longitude.

Using this system, any point on Earth's surface can be precisely measured and identified. Keep in mind that these locations are being made on a three-dimensional surface (in reference to a datum), therefore measurements between these locations have to take this into account. When making measurements on a sphere, the shortest distance between two points is referred to as the **great circle distance**. Thus, if you want to find the shortest distance between the cities of New York and Paris, you have to calculate the great circle distance between two sets of coordinates (see *Hands-on Application 2.1: Great Circle Distance Calculations* on page 48 for more information about using the great circle distance).

The latitude and longitude system of GCS also serves as the basis for breaking Earth up into multiple **time zones** to account for differences in time around the world. Time zones set up the concept of a 24-hour day, with each zone having a different time as it relates to the current time in Greenwich, England (also called Greenwich Mean Time or GMT). Earth is spanned by 360 degrees of longitude, and there are 24 hours in a day; 360 divided by 24 equals 15: there are 15 degrees of longitude for each hour. Thus, every 15 degrees away from the time zone containing the Prime Meridian marks the start of a new time zone (see **Figure 2.3** on page 46).

decimal degrees (DD) the fractional decimal equivalent to coordinates found using degrees, minutes, and seconds

great circle distance the shortest distance between two points on a spherical surface

time zones a method of measuring time around the world, created by dividing the world into subdivisions of longitude and relating the time in that subdivision to the time in Greenwich, England

FIGURE 2.3

The world's time zones are separated by lines of longitude (and geographic boundaries).

From 7.5 degrees west longitude to 7.5 degrees east longitude is one time zone, 7.5 to 22.5 degrees longitude is a second time zone, and so on. Following these sections of longitude around the world, the east coast of the United States is five time zones to the west of the Prime Meridian, which makes the time in New York and Boston five hours behind the time in Greenwich. The west coast of the United States is eight time zones to the west of the Prime Meridian (and three to the west of the east coast of the United States), which makes the time in San Francisco eight hours behind Greenwich (and three hours behind New York).

Hands-On Application 2.1

Great Circle Distance Calculations

To find the real-world shortest distance around the planet between two cities—New York and Paris—you'll need to calculate the great circle distance between them. To calculate great circle distances quickly, use the Surface Distance application online at **www.chemical-ecology.net/java/lat-long.htm**. Use the option for Great Circle Distance Between Cities to find the real-world shortest distance between New York and Paris. Use it to find the city closest to where you are, and calculate the great circle distances between various pairs of cities. Also, use a program such as Google Earth Pro to obtain the GCS coordinates (latitude and longitude) for a location close to you and find the surface distance from there to a nearby city.

Expansion Questions

➤ What is the great circle distance between New York and Paris?

➤ What is the great circle distance between the city closest to you and London, Paris, Rome, and Shanghai?

➤ Use Google Earth Pro to obtain the GCS (latitude and longitude) coordinates (in degrees, minutes, and seconds) for a location near you and GCS coordinates for a location at the White House in Washington, D.C., and calculate the surface distance between the two points. How many miles are you from the White House?

Thinking Critically with Geospatial Technology 2.1

Do You Really Need Printed Maps in a Digital World?

Let's face it, these days we're living in a digital world—road maps, directions, and property maps are all available in digital form with a few clicks of a mouse. With everything available digitally (and easily portable and accessible, thanks to laptops, tablets, and smartphones), is there still a need for printed maps? Do you really need to carry around a paper map of the road network in Los Angeles, California, or will a mapping app on your smartphone suffice to find your location and get you where you want to go? Are paper map collections at universities and libraries relics of the past, or are they still useful resources? When might it be necessary to have access to such a collection?

When you're driving a long distance and have GPS, satellite maps, and Internet mapping capabilities, is it necessary to have a printed map of the area you're traveling through? If your answer is yes, describe a situation in which it would be necessary. If your answer is no, why are printed maps still available?

Take a closer look at Figure 2.3, and you'll see that time zones don't exactly follow the lines of longitude. Political or geographic boundaries are often the reason why a particular area is entirely in one time zone even though that area may straddle two zones, and why lines of longitude shouldn't always be read as dividing lines between one time zone and the next. China, for instance, adopts one time measurement for the entire country, even though its geography crosses five time zones. The **International Date Line** marks the division between 24-hour periods and divides one day from another, but it also bends to accommodate considerations of geography, political boundaries, and convenience, and makes detours to include islands in one time zone or another, rather than strictly following the line of the 180th meridian.

International Date Line a line of longitude that uses the 180th meridian as a basis (but changes away from a straight line to accommodate geography)

 ## How Can Real-World Data Be Translated onto a Two-Dimensional Surface?

These measurements belong to a 3D world, but are all translated, for ease of use, into two dimensions (flat surfaces like pieces of paper and computer screens). A **map projection** is a translation of coordinates and locations from their places in the real world to a flat surface, and this type of translation is necessary if we want to be able to use this geospatial data. The biggest problem with a map projection is that not everything is going to translate perfectly from a 3D object to a 2D surface. In fact, the only representation of the world that can accurately capture every feature would be a globe. Because it's hard to keep a globe in the glove compartment of the car, we're going to have to make do with flat surface representations of the world (like paper maps and electronic screens), and realize that such representations will always have some sort of distortions built into them, simply because they're 2D versions of a 3D reality. One or more of the following will be distorted in a map projection: the shape of objects, the size of objects, the distance between objects, or the direction of objects. Some map projections may retain the shape of continents but misrepresent their land area. Other projections may present the land area accurately but throw the shapes out of place.

map projection the translation of locations on the three-dimensional (3D) Earth to a two-dimensional (2D) surface

An example of this translation of the world into two dimensions is the Mercator projection (see **Figure 2.4** on page 50), a map developed in 1569 by Flemish cartographer Gerardus Mercator for use as a navigation aid—it was designed so that every straight line on the map was a line of constant direction. The Mercator projection encompasses the whole globe, and although it keeps the shapes of areas intact (that is, the continents look the way they should), the sizes (that is, the areas of land masses) are very distorted, and become more distorted the further you go from the Equator. For example, take a look at Greenland and

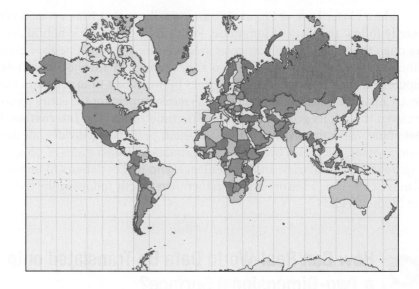

Africa—they look to be about the same size on the map, even though Africa is really more than 13 times larger in area than Greenland. Similar problems can be seen with Antarctica and Russia, which (although they have the right shape) look overwhelming in land area compared with their actual size.

Think of a projection this way. Imagine a see-through globe of Earth with the various landforms painted on it, and lines of latitude and longitude clearly marked. If you could place a lightbulb inside the globe and turn it on, it would cast shadows of the continents and the lines of latitude and longitude on the walls of the room. In essence, you've "projected" the 3D world on the surfaces of the walls. However, to make a map projection, you'd have to wrap a piece of paper around the globe and preserve or copy (i.e., project) the lines onto that paper. Then, when you unrolled the paper, you'd have the globe translated onto the map.

The real process of mapmaking is a much more complicated mathematical translation of coordinates, but the basic concept is similar. A map projection is created through the use of a developable surface, or a flat area on which the locations from Earth can be placed. In this way, the locations on Earth's surface get translated onto the flat surface. The three main developable surfaces (**Figure 2.5**) that are used in this fashion are a cylinder (which creates a cylindrical projection), a cone (which creates a conical projection), and a flat plane (which creates an azimuthal surface). The place where Earth "touches" the surface is called the point of tangency, which is where the map's distortion is at its least.

Like datums, there are numerous map projections available, each with its own specific uses and its own built-in distortions (see *Hands-on*

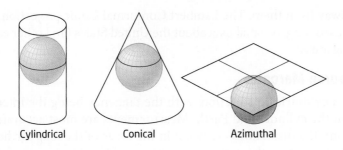

FIGURE 2.5

The three types of
developable surfaces in
map projections.

Cylindrical   Conical   Azimuthal

Application 2.2: Examining the Effects of Different Map Projections for
more information). Two projections that you'll often encounter with geospatial data are the Lambert Conformal Conic and the Transverse Mercator.

Lambert Conformal Conic

This is a conic projection, in which a cone intersects Earth at two parallels.
The choice of these tangency lines is important, as the distortion on the
map is reduced as you get closer to the lines and increases the farther you

Hands-On Application 2.2

Examining the Effects of Different Map Projections

Different map projections misrepresent Earth
in different ways, each one distorting certain
features and representing other features accurately. To examine the characteristics of different map projections interactively, go to **http://
indiemapper.io/app**. Indiemapper is a free online
application that enables you to make a variety of
different maps quickly and also examine how your
map will appear in different projections. When you
first start using Indiemapper, you'll have several
choices as to the source of the data you want to
map. For now, just choose **data library** and then
select a theme to map (such as Water Consump-
tion and Resources). By clicking on **Okay** your
theme will be chosen. Next, you'll have to choose
an attribute to map (such as **People**), a map
type (select **Choropleth**), and for now choose
No when asked about data standardization. In
order to see how the data looks in different projections, click on the **Projections** button in the

upper-right corner of the screen. You'll have 14
different choices of map projections that you can
select from—choose a projection from the available options and then click **Apply** to see your data
with a different map projection. Switch the projections to view the world map as Mercator, Lambert
Conformal Conic, Sinusoidal, and Transverse Mercator, and then examine each one. Note that the
Indiemapper controls allow you to zoom in and
out from the map.

Expansion Questions

➤ How does each of these map projections alter
the map as drawn? What kinds of distortions
(shape, area, distance, or direction) appear to
be built into each projection?

➤ Try some of the other map projection options
(such as Orthographic, Bonne, and Robinson).
What types of distortions and changes enter
the map with each projection?

move away from them. The Lambert Conformal Conic projection is commonly used for geospatial data about the United States (and other east–west trending areas).

Transverse Mercator

This is a cylindrical projection, with the tangency being the intersection between the cylinder and Earth. Measurements are most accurate at that point, and the distortions are worst in the areas of the map farthest away from that point, to the east or west. It's considered "transverse" because the cylinder is wrapped around the poles, rather than around the Equator. Transverse Mercator is used most for north–south trending areas.

GCS coordinates can find locations across the entire globe, down to precise measurements of degrees, minutes, and seconds. However, latitude and longitude is not the only coordinate system used today—there are several coordinate systems that use a map projection to translate locations on Earth's surface to a Cartesian grid system using x and y measurements. Two grid systems widely utilized with geospatial data are UTM and SPCS.

 ## What Is UTM?

Universal Transverse Mercator (UTM) the grid system of locating coordinates across the globe

The **Universal Transverse Mercator (UTM)** grid system works by dividing the world into a series of zones, and then determining the x and y coordinates for a location in that zone. The UTM grid is set up by translating real-world locations to where their corresponding places would be on a two-dimensional surface (using the Transverse Mercator projection). However, UTM only covers Earth from between 84° N latitude and 80° S latitude (and thus can't be used for mapping the polar regions—a different system, the Universal Polar Stereographic Grid System, is used to find coordinates in the Arctic and Antarctic).

UTM zone one of the 60 divisions of the world set up by the UTM system, each zone being 6 degrees of longitude wide

The first step in determining UTM coordinates is to find in which **UTM zone** the location is. UTM (**Figure 2.6**) divides the world into 60 zones, each one 6 degrees of longitude wide and set up as its own cylindrical projection. These zones are numbered 1 through 60, beginning at the 180th meridian and moving east. For instance, Mount Rushmore, outside of Keystone, South Dakota, is located in Zone 13. Beijing, China, is located in Zone 50. Sydney, Australia, is in Zone 56.

easting a measurement of so many units east (or west) of some principal meridian

northing a measurement of so many units north (or south) of a baseline

The next step is to determine the x and y grid coordinates of the point you want to locate. Unlike GCS, which uses degrees, minutes, and seconds, UTM uses meters as its unit of measurement. UTM's x and y coordinates are referred to as the **easting** and **northing** coordinates. The UTM northing is found by counting the number of meters the point is north of the Equator. For instance, a particular location at Mount Rushmore is located 4,859,580 meters to the north of the Equator.

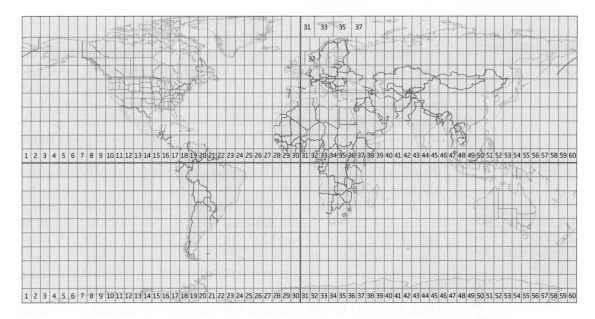

| 1 | 2 | 3 | 4 | 5 | 6 | 7 | 8 | 9 | 10 | 11 | 12 | 13 | 14 | 15 | 16 | 17 | 18 | 19 | 20 | 21 | 22 | 23 | 24 | 25 | 26 | 27 | 28 | 29 | 30 | 31 | 32 | 33 | 34 | 35 | 36 | 37 | 38 | 39 | 40 | 41 | 42 | 43 | 44 | 45 | 46 | 47 | 48 | 49 | 50 | 51 | 52 | 53 | 54 | 55 | 56 | 57 | 58 | 59 | 60 |

FIGURE 2.6

The UTM system divides Earth's surface between 80° S latitude and 84° N latitude into 60 zones.

Locations in the southern hemisphere handle northing a little differently because UTM doesn't use negative values. For southern hemisphere locations, measure the distance in meters (as a negative value) from the Equator to the particular location, then add 10,000,000 meters to the number. For example, a location at the Opera House in Sydney, Australia (in the southern hemisphere), has a northing of 6,252,268 meters. This indicates it would be approximately 3,747,732 meters south of the Equator (–3,747,732 + 10,000,000). Another way of conceptualizing this is to assume that (for southern hemisphere locations) rather than measuring from the Equator, you're measuring north of an imaginary line drawn 10,000,000 meters south of the Equator. This extra 10 million meters is referred to as a **false northing** because it's a value set up by the system's designers to avoid negative values.

The next step is to measure the point's easting. Unlike GCS, there is no one Prime Meridian from which to make measurements. Instead, each UTM zone has its own central meridian, which runs through the zone's center. This central meridian is given a value of 500,000 meters, and measurements are made to the east (adding meters) or west (subtracting meters) from this value. For example, a specific spot at Mount Rushmore is located 123,733.22 meters to the east of Zone 13's central meridian—therefore the easting for this location at Mount Rushmore is 623,733.22 meters (500,000 + 123,733.22). A location in the town of Buffalo, Wyoming, is also in Zone 13, and has an easting value of 365,568 meters. This indicates that this location is approximately 134,432 meters to the west of Zone 13's central meridian (500,000–134,432). No zone is more than 1,000,000 meters wide, therefore no negative values will be assigned to eastings in a zone. This value for

false northing a measurement made north (or south) of an imaginary line such as is used in measuring UTM northings in the southern hemisphere

false easting a measurement made east (or west) of an imaginary meridian set up for a particular zone

easting is referred to as a **false easting**, because the easting calculation is based on an imaginary central meridian in each zone.

When UTM coordinates are written, the false easting, northing, and zone must all be specified. If a false northing is used (that is, the coordinates are found in a location south of the Equator), then a notation is included, explaining that the coordinates are from the southern hemisphere. For example, the UTM coordinates of a location at Mount Rushmore would be written: 623733.22 E, 4859580.26 N, Zone 13. (See **Figure 2.7** for a simplified graphical example of finding the location at Mount Rushmore's coordinates with UTM. Also, see *Hands-on Application 2.3: Converting from Latitude/Longitude to UTM.*)

United States National Grid (USNG) a grid system of identifying locations in the United States

UTM values are also used as the basis for a related projected coordinate system, the **United States National Grid (USNG)**, also known as the military grid reference system (MGRS), which is used extensively by the military and by disaster relief organizations such as the Federal Emergency Management Agency (FEMA). USNG is particularly well-suited to emergency management and military applications because it uses a single identifier for the northing, easting, and zone information for every location—ensuring that the same coordinates are labeled and used in the same way by everyone. USNG measures coordinates in meters, using values derived from UTM northing and easting. It also uses UTM zone numbers in conjunction with smaller 100,000-meter gridded zones to derive location information. Depending on the number of digits computed for the position measurement, USNG accuracy ranges between 1000 meters and 1 meter.

FIGURE 2.7

Determining the UTM coordinates (easting, northing, and zone) of a location at Mount Rushmore (a simplified representation).

Hands-On Application 2.3

Converting from Latitude/Longitude to UTM

GCS and UTM are completely different systems, but there are utilities to make conversion from one system to another simple and easy. One of these is available at **www.earthpoint.us/Convert.aspx**. This Web resource can be used to convert from GCS to UTM. Input a coordinate for latitude and one for longitude, and click the **Calc** button. The corresponding UTM easting, northing, and zone will be computed for you and returned in a box below the inputs (along with other coordinate system conversions as well). You can also click **Fly To On Google Earth** and a marker will be placed in Google Earth at the coordinates you specify.

First, you're going to need GCS coordinates. A good source of GCS coordinates can be found at **www.lat-long.com**. This Website allows you to input the name of a location (try the Empire State Building) and the state (New York) or county, and it will return the latitude and longitude (in both

DD and DMS). Copy those coordinates into the converter, and click the **Calc** button to determine the UTM coordinates of the Empire State Building.

Expansion Questions

➤ What are the UTM coordinates for the latitude and longitude coordinate given for a location at the Empire State Building?

➤ What are the UTM coordinates for the latitude and longitude coordinate given for the Jefferson Memorial in the District of Columbia, the Space Needle in Washington, and the Transamerica Pyramid in California?

➤ Use the two Websites in conjunction with each other to determine the UTM coordinates of prominent locations around your local area. What locations did you identify, and what are their UTM coordinates?

For example, the UTM coordinates of our previously computed Mount Rushmore location were 623733.22 E, 4859580.26 N, Zone 13. In the USNG system, using 1-meter precision, the coordinates of the same location would be written as a single string: 13TFJ2373359580—"13T" represents UTM Zone 13, "FJ" identifies which one of smaller 100,000-meter grid zones the location is in, "23733" represents the value for easting (a shortened version of a UTM easting value of 623733), and "59580" represents the value for northing (a shortened version of a UTM northing value of 4859580).

What Is SPCS?

The **State Plane Coordinate System (SPCS)** is another grid-based system for measuring and determining coordinates. It was designed back in the 1930s, before the advent of computers. Today, SPCS is used as a coordinate system for United States data, especially city and county data and measurements. SPCS is a projected coordinate system that uses a translation method similar to UTM to set up a grid coordinate system to cover the United States. SPCS data was originally measured using the NAD27 datum, but now there are datasets using the NAD83 datum.

State Plane Coordinate System (SPCS) a grid-based system for determining coordinates of locations within the United States

Like UTM, SPCS requires you to specify a zone, an easting, and a northing in addition to the state in which the measurements are being made. SPCS divides the United States into a series of zones. Unlike UTM zones, which follow lines of longitude, **SPCS zones** are formed by following state or county boundaries. A state is often divided into multiple zones— for example, Ohio has two zones: Ohio-North and Ohio-South. Nevada has three zones: Nevada-West, Nevada-Central, and Nevada-East. Texas has five zones, and California has six, while all of Montana is represented by a single zone (**Figure 2.8**). To determine the coordinates for a point in SPCS, you must first know which state the point is in, and then you must identify the zone within that state. For example, to map data from Miami, the state is Florida and the zone is Florida-East, while Cleveland is in the state of Ohio, in the Ohio-North zone.

The next step is to determine the northing and easting. As with UTM, both of these are measured north and east from a predetermined baseline and meridian, but in SPCS, measurements are made in feet (using the NAD27 datum), or in feet or meters (using the NAD83 datum). Each of the SPCS zones has its own baseline and its own principal meridian. Two projected versions of SPCS are used—the Transverse Mercator version, which is usually used for states that have a north–south orientation (like Indiana), and the Lambert Conformal Conic version, which is usually used for states that have

SPCS zone one of the divisions of the United States set up by the SPCS

FIGURE 2.8

SPCS zones are formed by following state or county boundaries.

2478424.96 E,
531216.13 N,
Ohio, N

FIGURE 2.9

A simplified
representation of
SPCS measurements
and coordinates for
a location on the
YSU football field in
Youngstown, Ohio.

a more east–west orientation (like Ohio). Baselines and principal meridians will be positioned differently, depending on which version is being used.

Baselines are placed a certain distance under the zone's southern border and northings are measured from there (to ensure all positive values). In the NAD27 version, states using the Transverse Mercator format have their zone's origin 500,000 feet to the west of the principal meridian, while states using the Lambert Conformal Conic projection have their zone's origin 2,000,000 feet to the west of the meridian. Coordinates are determined by measuring distance using these false origin points, creating a false northing and a false easting. Thus, as each of the zones uses its own coordinate measurements, each zone has a different false northing and false easting from all other zones. Note that in the NAD83 version of SPCS, different values may be used for each zone's origin.

SPCS coordinates are referenced by listing the false easting, false northing, state, and zone. For example, the SPCS (NAD83 feet datum) coordinates of a location on the football field of Stambaugh Stadium in Youngstown, Ohio, are approximately: 2478424.96 E, 531216.31 N, Ohio, N (North Zone; see **Figure 2.9** for a simplified graphical representation of

Hands-On Application 2.4

Using the State Plane Coordinate System

Like the conversion tool in *Hands-on Application 2.3: Converting from Latitude/Longitude to UTM*, there are Web-based utilities to convert GCS coordinates to their SPCS counterparts. A good one is available free online at **www.earthpoint .us/StatePlane.aspx**, a Web resource similar to the one used in *Hands-On Application 2.3*. Enter the latitude and longitude coordinates of a location, click the **Calc** button, and it will calculate the SPCS coordinate for that location (it will also convert the other way, going from SPCS to GCS). Again, use the GCS coordinates you received from **www.lat-long.com** to compute the SPCS equivalents for them. Note that you also have to specify the SPCS zone when entering the latitude and longitude of the points. When calculating the SPCS coordinates, you may want to use the **Fly To On Google Earth** to verify your results.

Expansion Questions

➤ What are the SPCS coordinates for the latitude and longitude coordinate given for a location at the Empire State Building (New York Long Island Zone)?

➤ Working from their given latitude and longitude coordinates, what are the SPCS coordinates for the Jefferson Memorial in the District of Columbia (which uses Maryland's zone), the Space Needle in Washington (Washington North Zone), and the Transamerica Pyramid in California (California Zone 4)?

➤ Use the two Websites in conjunction to determine the SPCS coordinates of prominent locations around your local area (check Figure 2.8 to identify in which SPCS zone you are located). What locations did you choose, and what are their SPCS coordinates?

SPCS measurements—and see also *Hands-on Application 2.4: Using the State Plane Coordinate System* for a utility that will convert latitude and longitude coordinates into SPCS values).

Chapter Wrap-Up

This chapter provided a look at what goes into setting up coordinates and references for data used with geospatial technology. Whether you're reading coordinates from a GPS receiver or making measurements from a map, knowing how that geospatial information is put together and referenced has an impact on how you're able to make use of that data, especially if you're trying to fit several types of geospatial data together. Check out *Projection and Coordinate System Apps* for some downloadable apps and *Coordinate Systems and Projections in Social Media* for some YouTube videos that deal with this chapter's concepts.

In the next chapter, we're going to switch gears and look at how you can take your data (whether it's a remotely sensed image, some information created using GIS, or a paper map of the local fairgrounds) and get it to match up with all the other geospatial data you have. Before that, *Geospatial*

Lab Application 2.1: Coordinates and Position Measurements will have you start applying the coordinate and measurement concepts described in this chapter with some more work with Google Earth.

   Important note: The references for this chapter are part of the online companion for this book and can be found at **http://www.macmillanlearning.com/catalog/shellitoigt4e**.

Projection and Coordinate System Apps

Here's a sampling of available representative coordinate systems apps for your smartphone or tablet. Note that some apps are for Android, some are for Apple iOS, and some may be available for both.

➤ **Coordinates—Calculate and Convert a Position:** An app that calculates many types of coordinates (latitude and longitude, UTM, MGRS, etc.) for a location

➤ **Coordinate Tools Free:** An app that converts coordinates and plots them on a map

➤ **Distance Measure:** An app that allows you to calculate areas, distance, and great circle distance on a map

➤ **Globe Earth 3D Pro:** An app that shows a time zone map of the world on a virtual globe and also allows you to view flags and hear national anthems of countries from around the world

➤ **Latitude Longitude Convert:** An app that converts from DMS to decimal degrees and also plots the location using Google Maps

➤ **Map Projections—Ultimate Guide for Cartography:** An app that allows you to view a world map in many different map projections and also provides information about each projection

➤ **MyLocation:** An app that computes the latitude and longitude of your current location

➤ **World Clock:** An app that allows to you to select a location on a map and see its time zone as well as its current time

Coordinate Systems and Projections in Social Media

Here's a sampling of some representative YouTube videos related to this chapter's concepts.

You Tube **On YouTube, watch the following videos:**

➤ **Latitude and Longitude** (a video demonstrating concepts of latitude and longitude using Google Earth): **www.youtube.com/watch?v=swKBi6hHHMA**

➤ **Map Projections** (an Esri video showing map projection concepts): **www.youtube.com/watch?v=2LcyMemJ3dE**

➤ **Maps That Prove You Don't Really Know Earth** (a video about the distortions in the Mercator projection): **www.youtube.com/watch?v=KUF_Ckv8HbE**

Key Terms

datum (p. 42)

datum transformation (p. 43)

decimal degrees (DD) (p. 45)

degrees, minutes, and seconds (DMS) (p. 44)

easting (p. 52)

ellipsoid (p. 42)

Equator (p. 44)

false easting (p. 54)

false northing (p. 53)

geodesy (p. 42)

geographic coordinate system (GCS) (p. 44)

geoid (p. 42)

great circle distance (p. 45)

International Date Line (p. 49)

latitude (p. 44)

longitude (p. 44)

map projection (p. 49)

NAD27 (p. 43)

NAD83 (p. 43)

northing (p. 52)

Prime Meridian (p. 44)

SPCS zone (p. 56)

State Plane Coordinate System (SPCS) (p. 55)

time zones (p. 45)

United States National Grid (USNG) (p. 54)

Universal Transverse Mercator (UTM) (p. 52)

UTM zone (p. 52)

WGS84 (p. 43)

2.1 Geospatial Lab Application

Coordinates and Position Measurements

This chapter's lab will continue using Google Earth Pro (GEP), but only to examine coordinate systems and the relationships between various sets of coordinates and the objects they represent in the real world. In addition, you'll be making some measurements using GEP and using some Web resources for comparing measurements made by using different coordinate systems.

Note that this lab makes reference to such things as "coordinates for the White House" or "coordinates for Buckingham Palace"—these represent measuring a set of coordinates at one specific location at these places and are used as simplifications of things for lab purposes.

Objectives

The goals for you to take away from this lab are:

➤ To set up a graticule of lines in GEP

➤ To locate places and objects strictly by their coordinates

➤ To make measurements across long and short distances and then compare measurements with surface distance calculations

➤ To translate latitude/longitude coordinates into UTM

Using Geospatial Technologies

The concepts you'll be working with in this lab are used in a variety of real-world applications, including:

➤ Emergency rescue and relief operations, which need exact coordinates in order to locate incidents (like stranded civilians and downed power lines) and send the closest available emergency assistance

➤ Outdoor event planning groups, which need to know the coordinates of every entrance and exit of all buildings, temporary structures, and enclosures, as well as the locations of essential resources and amenities, such as first-aid facilities and other emergency services (on- and off-site), handicapped-accessible ramps, and restrooms (for both the disabled and the able-bodied)

[Greg Pease/The Image Bank/Getty Images]

Obtaining Software

The current version of Google Earth Pro (7.3) is available for free download at **www.google.com/earth/download/gep/agree.html.**

Important note: Software and online resources can change fast. This lab was designed with the most recently available version of the software at the time of writing. However, if the software or Websites have significantly changed between then and now, an updated version of this lab (using the newest versions) will be available online at **http://www.macmillanlearning.com/catalog/ shellitoigt4e.**

Lab Data

There is no data to copy in this lab. All data comes as part of the data that is installed with the software or is streamed across the Internet through GEP.

Localizing This Lab

This lab flies around the world to numerous locations, including Washington, D.C., and London. However, it can easily be changed to examine the coordinates of nearby areas. Rather than looking at the White House or Buckingham Palace, substitute the coordinates or locations for your own local government offices (like city halls or courthouses).

2.1 Examining Coordinates and Distance Measurements in Google Earth Pro

1. Start **Google Earth Pro.** Once Earth settles into view, scroll your mouse around Earth. You'll see a set of latitude and longitude coordinates appear at the bottom of the view—these represent the

coordinates assigned to your mouse's location. By default, GEP uses the GCS coordinate system and the WGS84 datum.

2. To examine the full graticule of latitude and longitude lines, select **Grid** from the **View** pull-down menu. Some key GCS lines will be highlighted in yellow amidst the web of lat/long lines—the Equator, the Prime Meridian, the Antimeridian (how Google Earth labels the 180th meridian), the Tropic of Cancer, and the Tropic of Capricorn.

3. We'll begin in Washington, D.C., so type **Washington, D.C.** into the **Search** box. GEP will rotate and zoom to the area. You'll also see the spaces between the lat/long lines grow smaller and new values appear as GEP zooms in.

4. Next, we'll go a specific location in Washington, D.C. Type these coordinates into the **Search** box: **38.897631, –77.036566**. These are the decimal degree lat/long coordinates of the White House.

5. Press the **Placemark** button on the toolbar (see *Geospatial Lab Application 1.1* for more information on placemarks). The yellow pushpin will automatically be placed at the lat/long coordinates you had GEP search for (so you don't have to place it anywhere manually). In the new placemark dialog box, type **The White House** for the name of the placemark and select another symbol for the yellow pushpin if you like.

6. The coordinates for the White House are in decimal degrees, but other methods of displaying coordinates are available in GEP. From the **Tools** pull-down menu, select **Options**.

7. In the **Show Lat/Long** options, select the radio button next to **Degrees, Minutes, Seconds**, click **Apply**, and then click **OK**.

8. From the **Places** box in GEP, right-click on **The White House** placemark, and then select **Properties**. The coordinates will be changed to degrees, minutes, and seconds (DMS). Answer Question 2.1. Close the White House placemark dialog box.

> **Question 2.1** What are the coordinates for the White House in degrees, minutes, and seconds?

9. Decimal degree coordinates for the Lincoln Memorial are: **38.889257, –77.050137**. Use the **Search** box to zoom to these coordinates. Once GEP arrives there, put a placemark at that spot, name it **Lincoln Memorial**, and answer Question 2.2.

> **Question 2.2** What are the coordinates for the Lincoln Memorial in degrees, minutes, and seconds?

10. Adjust the view so you can see both placemarks at the edges of the view. Select the **Ruler** tool and compute the distance between the White House and the Lincoln Memorial (see *Geospatial Lab Application 1.1* for the use of the **Ruler** tool in measurements of distance). Answer Question 2.3. Turn off the placemark for the Lincoln Memorial when you're done.

> **Question 2.3** According to GEP, what is the distance between the White House and the Lincoln Memorial?

11. It's such a short distance from the White House to the Lincoln Memorial that the differences between measurements should be very small, so let's look at some larger distances.

12. Buckingham Palace (in London, England) is located at these coordinates: **51.500821, –0.143089**. Use the **Search** box to zoom to these coordinates. Once GEP arrives there, put a placemark at that spot, name it Buckingham Palace, and then answer Question 2.4.

> **Question 2.4** What are the coordinates for Buckingham Palace in degrees, minutes, and seconds?

13. Zoom out and rotate the view so that you can see the placemarks for the White House and for Buckingham Palace in the same view. Use the **Ruler** to compute the distance between these two locations. Answer

Questions 2.5 and 2.6. Close the **Ruler** dialog box when you're done. Of course, this is a rough estimate because the scale of the view makes hitting the placemarks directly with the start point and the end point of the ruler difficult.

> **Question 2.5** According to your measurement in GEP, what is the computed distance (in miles) between the White House and Buckingham Palace?

> **Question 2.6** Why is the line curved rather than straight? What kind of distance is being computed here?

14. We'll now check your measurements using a Web utility. Go to **www .chemical-ecology.net/java/lat-long.htm**. This Website enables you to compute the surface (real-world) distance between two sets of lat/long coordinates (and also between two cities) in DMS format.

15. Use your answer to Question 2.1 as the degrees, minutes, and seconds for the latitude and longitude of point one (you are using North latitude and West longitude).

16. Use your answer to Question 2.4 as the degrees, minutes, and seconds for the latitude and longitude of point two (you are again using North latitude and West longitude).

17. Click the **Distance Between** button. The surface distance between the two points will be computed in kilometers, miles, and nautical miles. Answer Question 2.7. Your answers to Questions 2.5 and 2.7 should be relatively similar, given how they were both computed—if they're way off, redo your Question 2.5 measurement and make sure you're inputting the correct numbers for Question 2.7.

> **Question 2.7** According to this Website measurement, what is the computed surface distance (in miles) between the White House and Buckingham Palace?

2.2 Using UTM Coordinates and Measurements in Google Earth Pro

Universal Transverse Mercator (UTM) is a projected coordinate system, and GEP can track coordinates using this UTM system as well as lat/long.

1. From GEP's **Tools** pull-down menu, select **Options**. In the **Show Lat/Long** options, select the radio button for **Universal Transverse Mercator**. Click **Apply**, and then click **OK**.

2. Zoom out to see the whole world—you'll also see that the graticule has changed from a web of latitude/longitude lines to the UTM grid draped over Earth. You should see the UTM zones laid out across the world, numbered from 1 through 60.

> **Question 2.8** In what UTM zone is Buckingham Palace located? In what UTM zone is the White House located?

3. Double-click on **The White House** placemark (in the **Places** box) and GEP will rotate and zoom back to the White House.

4. Scroll the mouse around the White House area. You'll see the new coordinates appear at the bottom of the view, but this time they'll be the zone, easting, and northing measurements. Open the **Properties** of **The White House** placemark by right-clicking on it. The UTM easting, northing, and zone will appear.

> **Question 2.9** What are the UTM coordinates of the White House?

5. In the **Places** box, double-click on the Lincoln Memorial placemark, and GEP will zoom and rotate to it.

> **Question 2.10** What are the UTM coordinates of the Lincoln Memorial?

6. UTM coordinates are measured in meters rather than degrees of latitude or longitude—this enables an easier method of determining coordinates. If you move the mouse east, your easting will increase, and moving it west will decrease the easting. The same holds true for northing—moving the mouse north will increase the value of northing, while moving it south will decrease the northing value.

7. Change the view so that you can see both the White House and the Lincoln Memorial. Using your answers for Questions 2.9 and 2.10 (and perhaps the **Ruler** tool), answer Question 2.11.

> **Question 2.11** How many meters away to the north and east is the White House from the Lincoln Memorial?

8. UTM can also be used to determine the locations of other objects. Return to **The White House** placemark again. Answer Questions 2.12 and 2.13.

> **Question 2.12** What object is located approximately 912 meters south and 94.85 meters east of the White House?

> **Question 2.13** What are the full UTM coordinates of this object?

Closing Time

This lab served as an introduction for using and referencing coordinates and measurements, using Google Earth Pro. Chapter 3's lab will use a new software program, but we'll still briefly use Google Earth Pro. After that, we'll return to Google Earth Pro in Chapter 8, but we'll continue to use its functions in other chapters. You can now exit GEP by selecting **Exit** from the **File** pull-down menu. There's no need to save any data (or Temporary Places) in this lab.

Getting Your Data to Match the Map

Reprojecting, Georeferencing, Control Points, and Transformations

Because there are so many datums and map projections out there, it's important to know which one you're using when you're making measurements. Say, for instance, that you have two sets of geospatial data about your town that you need to bring together for analysis on one map (that is, a street map with building footprints). Each of these pieces of data is in a different projection and coordinate system, and each uses a different datum — the street map is in State Plane NAD27 feet and the buildings are in UTM NAD83 meters. When you put these things together, nothing is going to match up — the streets and buildings are going to be in different places in relation to each other because each one uses a different basis for its coordinates.

It's important to note when you're dealing with geospatial data that it can be constructed using a variety of different datums, coordinate systems, and projections. With all the different variations involved with these items, it's best to try to work with all geospatial data on the same basis (that is, to use the same datum, coordinate system, and projection). If you have multiple data layers in different map projections and you try to overlay them, those pieces of data won't properly align with each other (**Figure 3.1**).

How Can You Align Different Geospatial Datasets to Work Together?

Back in Chapter 2, we discussed the concept of having to transform data measured from different datums to a common datum so that everything would align correctly. It's best to **reproject** your data (or transform all of the datasets to conform to a common datum, coordinate system, and

reproject changing a dataset from one map projection (or measurement system) to another projection) before moving forward. Reprojecting data will translate the coordinates, map projections, and measurements from one system (for instance, SPCS) to another (such as UTM). Many geospatial software packages give you the capability to take your initial dataset (such as your street map in State Plane NAD27) and reproject it to create a new dataset with new coordinates (so now you have a new street map measured in UTM NAD83). This can also be done to change data from one map projection into another. In some cases, the software can "project on the fly" — calculate the necessary transformations without actually altering the base dataset. Whatever method you choose, you'll save yourself a lot of headaches later on by using a common basis for the various geospatial datasets you use. This is especially important when you're trying to get several different types of data to match up.

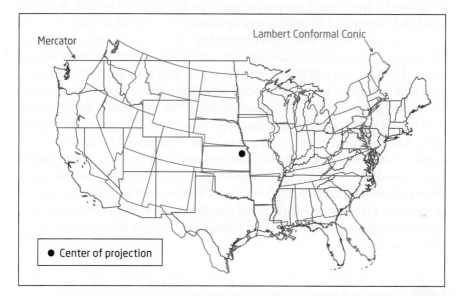

FIGURE 3.1

Measuring the same data with different projections produces maps that don't match.

Thinking Critically with Geospatial Technology 3.1

What Happens When Measurements Don't Match Up?

It's not uncommon to have multiple data sources as in the previous scenario, with different projections, coordinate systems, and datums that all need to be used together in a project. If everything doesn't align properly, you can end up with points being plotted in the wrong location with respect to the basemap on which they're being located, or street data that doesn't line up with a satellite image of the same area. What kind of effects could this type of simple data mismatch have on real-world companies or government agencies? What would happen if a road renovation crew is using sewer data and street data that don't align properly because they use different datums or projections? What could happen if the misalignment of data caused one dataset to be significantly off from its proper place on a map? Similarly, what effects could such misalignments have on projects such as shoreline construction, zoning maps, or species habitat mapping?

Problems are going to arise if one (or more) of your pieces of data doesn't have coordinates assigned to it, or if you don't have any information as to what projection or datum was used to construct it. Suppose you scan an old map of your town into your computer, and you want to add other layers to it (such as a road network or parcel layer) in a GIS to see how things have changed since that old map was made. You add the layers into the GIS, but the scanned historical map won't match up with any of your data. This is because that image doesn't contain any **spatial reference** (such as a real-world coordinate system). Although it might show such features as old roads or the dimensions of particular buildings, the only coordinate system the old map possesses is a grid that begins at a zero point at the lower left-hand corner and extends to the dimensions of the scanned image. For instance, coordinates would not be referenced by feet or degrees, but by whatever units you're measuring (such as 200 millimeters over from the left-hand corner and 40 millimeters up from the left-hand corner).

Items in the image can't be referenced by latitude and longitude because those types of coordinates haven't been assigned to anything in the image. Similarly, when you add this scanned image as a layer in a GIS, the software won't know how to match it up with other data layers that do have referenced coordinate systems. Without knowing "where" your scanned image matches up with the rest of the data, you won't be able to use this image as geospatial data or reproject it to match your other data. The same concept holds for other types of data used with geospatial technology: If all types of data automatically had the correct spatial reference when they were created, things would be a lot easier. So what you'll have to do is match the data to the map, or assign a real-world set of coordinates to your data so it can sync with other geospatial data. In other words, you're going to have to georeference your data.

spatial reference the use of a real-world coordinate system for identifying locations

 ## What Is Georeferencing?

georeferencing a process of aligning an unreferenced dataset with one that has spatial reference information

Georeferencing is the process of aligning an unreferenced dataset with one that has spatial reference information. Other terms used for similar processes include "image to map rectification," "geometric transformation," and "registration." Suppose you're hosting a meeting at your school and you want the invitation to include a map to direct people not only to the school, but also to your specific building on the campus. You go to your school's Website and download a map that shows the campus boundaries and specific building locations within those boundaries. So far so good — but you want to place that downloaded map in the larger context of a map of the city, so that people will be able to see where your campus is in relation to the surrounding roads and neighborhoods. The problem is that while the data you have of the surrounding area (whether aerial photos, satellite imagery, or GIS data layers) contains real-world spatial references, your downloaded map is just a graphic or PDF without any spatial reference at all. The campus map won't match up with the spatially referenced data, and your plan of a snappy geospatial-style invitation will be ruined. Or will it? If you could just georeference that campus map, everything would fit together. Well, the good news is, you can. The georeferencing procedure requires a few steps of work to proceed, but you'll be able to make your map (see *Hands-On Application 3.1: David Rumsey Historical Map Collection* and **Figure 3.2** for an example of an online georeferenced map collection).

Hands-On Application 3.1
David Rumsey Historical Map Collection

Information about this ongoing project for the collection and preservation of historical maps can be found at **http://www.davidrumsey.com**. The maps on the Website are part of a collection that stretches from the seventeenth century to the present. Although there are many ways of viewing these maps, some have been georeferenced to make them match up with current geospatial imagery. By looking at georeferenced maps in relation to current data, you can see how areas have changed over time.

On the Website, select the link for Google Maps under the Quicklinks section, then click on the map that appears in your Web browser. A new window will open (also available at **http://rumsey .geogarage.com**) showing you a map of the world, along with several symbols that indicate the presence of georeferenced maps. Clicking on a symbol will open a dialog box about its corresponding map and selecting the Open Image link in the box will show you the georeferenced map overlaid on the world map. A slider bar in the upper right-hand corner will allow you to adjust the transparency to see both the historical map and the current imagery together. Check out some of the map icons for a variety of places around the globe and see how this historical information syncs with modern boundaries and images.

The Rumsey Historical Map collection is also available as a layer in Google Earth Pro. To use the maps as another Google Earth Pro layer (like the roads or buildings), open the Gallery options, open the Rumsey Historical Maps option, and click on the Map Finder icon. You'll see a new icon representing the maps appear in Google Earth Pro that you can use to view the georeferenced maps. There's also an "About the Maps" pop-up where you can find more information about the map collection and how these maps were scanned and georeferenced for use in Google Earth Pro.

Expansion Questions
➤ From examining some of the historical maps in both the Web interface and on Google Earth Pro, how do the areas represented match up with their modern-day counterparts?

➤ Examine the "Chicago 1857" map. How does this historical map match up with modern-day Chicago? You can use Google Earth Pro to adjust the transparency of the map by using the tools in the Places Box (select the slider bar and move it about halfway over) — this will enable you to see current imagery of Chicago beneath the georeferenced layer. Are there any areas that have changed significantly between then and now?

➤ Examine the "United States 1833" map (the "Eagle" map with its icon located in Ohio), and make the layer semi-transparent. How does this map of the United States match up with current boundaries and coastlines?

FIGURE 3.2

A 1786 map of North America from the David Rumsey collection, georeferenced and overlaid in Google Earth.

How Can Data Be Georeferenced?

When data (such as an old scanned map or other unreferenced data) is georeferenced, each location in that data is aligned with real-world coordinates. To get started, you're going to need a source of real-world coordinates against which to align your unreferenced data. The choice of a source to use for the matching is important, because you have to be able to find the same locations on both the unreferenced image and the source. This is crucial to georeferencing because the process needs some common locations to "tie" the unreferenced data to the source data. In the georeferencing process, these spots (where you can find the coordinates on the source data) are referred to as **control points** (sometimes called ground control points or GCPs). Field work, such as surveying or making GPS measurements (see Chapter 4), can provide sources for these control points, but you can avoid having to go into the field by finding common locations on the unreferenced image and the source.

Clearly, the source should be something on which you can find the same locations as the ones you can see in your unreferenced data. For instance, a low-resolution satellite image isn't going to be a good match with the detailed map of your campus buildings. Trying to match individual locations such as buildings to an image like this isn't going

control points point locations where the coordinates are known; used to align the unreferenced image to the source

FIGURE 3.3

In the MapTiler program, the fully referenced aerial image of the Youngstown State University campus shown on the left is the potential source for the references that the unreferenced campus map on the right requires.

to work simply because the resolution is too low. A high-resolution image (where you can determine features of buildings) would be a more useful choice for this specific task. Datasets such as road networks (see Chapter 8) or orthophotos (see Chapter 9) are also good sources to use in finding these common points. In addition, your source data should be a similar projection to your unreferenced data. If the projections are different, some features will match up well and others will end up distorted. Whatever the source, the georeferencing process will align an unreferenced image against the projection and coordinates of the source map (**Figure 3.3**).

Where to select the control points is the key to the whole georeferencing procedure. Because control points are the objects that are going to connect the unreferenced data to the source data, they have to be placed in the same location in both sets of data; real-world coordinates for these points will be required. By finding these same locations on the source (that has real-world coordinates) and on the unreferenced data (that doesn't have real-world coordinates), we can start to match the two datasets together. Features that you can clearly find in the same spot in the unreferenced data and the source data will be good choices for the placement of control points. For instance, the intersections of roads and the corners of buildings make good control point locations, because they're well-defined and can be found on both sets of data. These are also locations that aren't likely to change over time, therefore even if the unreferenced data and the source aren't from the exact same time period, they're usually valid as control point selections.

Poor locations for control point selection spots would be those that can't be accurately determined between the two sets of data, either because they're poorly defined or because they might be in different places at different times — for example, the top of a building, the center of a field, the shoreline of a beach, or a shrub in a landscaping display. See **Figure 3.4** for some examples of good and poor control point location selections on the campus map.

Another consideration to keep in mind when you're selecting control point locations is not to bunch them all together in one area of the data and ignore the rest of the map. Placing all of the control points in one-quarter of the image is going to ensure that one section of the map is well-referenced, but the rest of the map may not fit very well. Make sure to spread out the control points to representative locations throughout the data so that all areas fit well in relation to their newly referenced spots. You'll want to use several control points in the process — some georeferencing processes may only need a few, whereas others may require many more.

FIGURE 3.4

Good (green stars) and poor (purple stars) locations for control point selection.

The minimum number of control points required to fit the unreferenced image to the source is three. Three control points are often enough to enable you to make a rough fit of the image to the source. With this initial fit, some programs (for instance, MapTiler, the program used in this chapter's Geospatial Lab Application) will attempt to match up the next control point location with a spot you select on the source. As you add more control points, the fit between the unreferenced data and the source should (hopefully) improve. Proper selection of control points is critical for georeferencing, because the procedure needs to have some sort of real-world coordinates to match to, and it's these points that will be used to calculate coordinates for the other locations in the image. Often, you'll find that deleting a control point, adding another one, and tinkering with the placement of others is necessary to get the best fit of the data (see *Hands-On Application 3.2: Online Georeferencing Resources* for several online examples of placing control points in relation to old maps). Once your data is properly aligned by using control points, you can then transform it to have the same coordinate system information as the source.

Hands-On Application 3.2

Online Georeferencing Resources

Georeferencing of maps has become a popular online activity. Several free georeferencing tools are available for anyone who wants to create spatially referenced versions of maps, especially old ones. You can select control points and transform the old maps to match up with current ones. For instance, the British Library is encouraging people to use online georeferencing tools to align and save old maps so that they can become part of the collection. Check out the following online resources to become part of the georeferencing effort (some sites will ask you to register for free before you can do any georeferencing, but you can usually view completed maps without registering):

1. The British Library Online Gallery: **http://www.bl.uk/maps**

2. David Rumsey Map Collection Georeferencer: **http://www.davidrumsey.com/view/georeferencer**

3. Georeferencer: **http://www.georeferencer.com**

4. The National Library of Scotland Map Georeferencer: **http://maps.nls.uk/projects/georeferencer**

5. The New York Public Library Map Warper: **http://maps.nypl.org/warper/**

6. WorldMap Warp at Harvard University: **http://warp.worldmap.harvard.edu**

Expansion Questions

➤ What kinds of historical georeferenced maps are available on these sites? How do the already georeferenced maps match up with modern maps?

➤ What kinds of difficulties did you find in trying to match unreferenced historical maps to modern maps (especially when it comes to selecting control points)?

How Is Data Transformed to a Georeferenced Format?

Take another look at Figure 3.3, which shows both the unreferenced campus map and the referenced source image. In order to make the unreferenced map align properly with the source image, there are a number of actions that have to be taken with the map. First, check out the football stadium (Stambaugh Stadium), which you'll find in the upper left quarter of both the map and the referenced image. In order to match the basic position and alignment of the stadium as it appears on the map with the stadium in the image, it must be rotated about 45 degrees clockwise. As you look at the other features, it becomes clear that the whole map must be rotated by some amount, because the image is oriented to the north and the map is oriented to the northeast. It also becomes apparent that the map might have to be stretched or skewed to match up with the image.

When unreferenced data is transformed or "rectified," it's warped to make it match up with the source. It can be **rotated** (turned, like changing the orientation of the map), **skewed** (by pulling the image, like at a slant), **scaled** (by altering the size and/or dimensions of the map to match the image), and/or **translated** (by altering the location and placement of the map). See **Figure 3.5** for examples of the effects of each of these transformations. This transformation of the data changes it to align with the new coordinate system. There are a number of transformation methods that are used, but a common one is the **affine transformation**, which will properly rotate, skew, scale, and translate the data.

The affine transformation calculates real-world X and Y coordinates for each unreferenced x and y coordinate. The formulae used for this are:

rotated when the unreferenced image is turned during the transformation

skewed when the unreferenced image is distorted or slanted during the transformation

scaled when the unreferenced image is altered or resized during the transformation

translated when the unreferenced image is shifted during the transformation

affine transformation a linear mathematical process by which data can be altered to align with another data source

FIGURE 3.5

Four ways that unreferenced data can be warped to align it with the source.

$$X = Ax + By + C$$

and

$$Y = Dx + Ey + F$$

where A, B, C, D, E, and F are the values calculated internally by the procedure and applied to the mathematical formulae as coefficients (and also control properties such as skewing and rotation). The affine transformation is a first-order transformation, which is the type of transformation that's usually utilized in this procedure. The end result is that unreferenced data will be turned into spatially referenced data.

Once you select some control points, the software runs the transformation and the data is georeferenced. Everything should work just fine. However, there are two questions that you should be asking yourself: "How do I know it worked?" and, more important, "How do I know the results are any good?" After all, just because the unreferenced data was transformed doesn't necessarily mean that it's a good match with the source. Since you know the coordinates of each place where you've put a control point, you should be able to match up those locations and see how closely your selected control points match the source. **Figure 3.6** shows a report from a software program that measures error values for each control point.

A transformation procedure uses a value called the **root mean square error (RMSE)** as an indicator of how well the now-referenced data matches with the source. After the initial set of control points has been chosen and the unreferenced data is aligned with the source, coordinates are computed for locations of the selected control points. The known coordinate location of each control point (in the source) is compared with the newly computed value for its corresponding location in the unreferenced data. If the control points are in exactly the same location in both the unreferenced data and the source, then the difference in their coordinates will be zero. However, the chances are that when placing coordinates — even if you select good locations, such as road intersections — the two points aren't going to match

root mean square error (RMSE) an error measure used in determining the accuracy of the overall transformation of the unreferenced data

	X	Y	Error [px]
1	-80.645	41.106	1.10
2	-80.646	41.104	2.45
3	-80.653	41.105	0.90
4	-80.651	41.109	1.02
5	-80.648	41.107	0.68
6	-80.643	41.108	1.56
7	-80.650	41.104	3.10

FIGURE 3.6

Some control point coordinate locations in the unreferenced map and the referenced image, and the computed error term for each one.

up exactly, though they might be very close. The software will examine all error values for all control points and report back a value for the RMSE.

The smaller the differences between where you put the control point on the unreferenced data (the estimated location) and its true location on the source data, the better the match between them will be. Thus, the lower the overall RMSE is, the better the transformation has worked. Usually, the georeferencing program will allow you to examine the difference between each point, so that you can see how closely your control point selections matched each other. If the RMSE is too high, it's a good idea to check the difference in coordinates for each point — if a few of them are significantly off, they can be adjusted, changed, or deleted, and new points can be selected to try to get a better fit. If some parts of the data match well and others don't, perhaps the control points need to be moved to other places to balance out the alignment better, or perhaps other control points need to be added or removed.

When you're georeferencing an image, new locations for the image's pixels will need to be calculated, and in some cases new values for the pixels will need to be generated as well. This process is referred to as *resampling*. The end result of a good georeferencing process will have your unreferenced data properly aligned with the source data (**Figure 3.7**). Maps that were not referenced can now match up with current maps and datasets (see *Hands-On Application 3.3: Georeferenced Historical Maps and the Spyglass* and *Hands-On Application 3.4: An Overview of the Georeferencing Process in ArcGIS* for examples).

FIGURE 3.7

Detail from the final georeferenced result of the campus map, overlaid and semi-transparent in Google Earth Pro.

Thinking Critically with Geospatial Technology 3.2

What Happens When the Georeferencing Is Wrong?

Think about this scenario. Surveying is being done for a new beachfront development, and a basemap of the area will be created (using GIS) from current high-resolution remotely sensed imagery. Because the basemap (and all layers created from it, such as property boundaries, utilities, and local land cover) will be derived from it, this initial layer is critical. However, something goes wrong in the georeferencing process of the image used to create the initial basemap, and it has incorrect coordinates — and then each layer created from it has incorrect coordinates. When on-site GPS coordinates and measurements are reprojected to match the new basemap, they'll also be inaccurate, because they use an incorrectly referenced basemap. What's going to be the end result for the land owners, developers, construction crews, and others involved in the project? What kind of long-term or widespread problems could be caused by this type of "cascading error" when the initial referencing is incorrect?

Hands-On Application 3.3

Georeferenced Historical Maps and the Spyglass

Esri Story Maps are great Web applications that have a variety of different layouts and tools (see Chapter 15 for more about Story Maps). One particularly useful tool is the Spyglass app, which allows you to view a portion of one map on top of another. In the case of historical maps, the Spyglass Story Map allows you to view an old georeferenced map on top of current satellite imagery, as viewed through a movable lens (the "spyglass"). Wherever you move the spyglass you'll see the historical map in the georeferenced location.

For example, check out the Spyglass Story Map of 1851 Washington, D.C., at **http://bit .ly/1ch2gb5**. When the Story Map opens, you'll see the spyglass placed over current satellite imagery of Washington, D.C. Within the spyglass you'll see a David Rumsey historical map of Washington, D.C., from 1851. By dragging the spyglass around the imagery, you'll see the historical georeferenced map underneath. You can also zoom in and out and the amount of area shown under the spyglass will change as well.

The Spyglass Story Map is a very neat way to look into the past by using a georeferenced historical map and current imagery. There are several other historical map examples that use the Spyglass available here: **http://storymaps.arcgis.com/en/app-list/ swipe-spyglass/gallery/#s=0&md=storymaps- apps:swipe-spyglass** (such as historical San Francisco, Chicago, Los Angeles, and New York City).

Expansion Questions

➤ From examining the 1851 Washington, D.C., map using the spyglass, what prominent items were to the immediate east and west of the U.S. Capitol in 1851 that are not there now?

➤ From examining the 1851 Washington, D.C., map using the spyglass, how was the White House referred to?

➤ From examining the 1851 Washington, D.C., map using the spyglass, were any of the following three things built in 1851: the Washington Monument, the Lincoln Memorial, or the Smithsonian Institution?

Hands-On Application 3.4

An Overview of the Georeferencing Process in ArcGIS

Although you'll use the free MapTiler program to align unreferenced data in the lab, georeferencing tools are common in other GIS software such as ArcGIS (which we'll discuss further in Chapter 5). In order to demonstrate georeferencing in ArcGIS, the company has provided a video showing the whole process, step by step, at **http://video.arcgis .com/watch/376/georeferencing-rasters-in-arcgis**. There's also a second video that demonstrates how georeferencing can be performed automatically,

without having to set up control points manually: **http://video.arcgis.com/watch/1553/automatic -georeferencing-in-arcgis-101**. Check out both videos.

Expansion Questions

➤ What other kinds of data besides imagery can be used as a source for georeferencing?

➤ What are the advantages and disadvantages of the automatic georeferencing option?

Georeferencing can also be performed on other data created without a spatial reference, such as unreferenced drawings and plans. A related procedure can also be used for remotely sensed images that need to be corrected.

Chapter Wrap-Up

Proper georeferencing of data is a key element of working with geospatial technology and allows for all types of maps and imagery to work side by side with other forms of geospatial data. This chapter's lab will have you using the MapTiler program. You'll start with an unreferenced campus map, and by the end it should be properly aligned with a real-world dataset. Also, check out *Georeferencing Apps* for a mobile app using georeferenced historical maps as well as *Georeferencing in Social Media* for more about historical maps online, the David Rumsey Historical Maps on Facebook and Twitter, and a YouTube video about online georeferencing tools.

In the next chapter, we'll be looking at the Global Positioning System, an extremely useful (and freely available) system for determining your position on Earth's surface and providing you with a set of coordinates so that you can find where you are.

Important note: The references for this chapter are part of the online companion for this book and can be found at **www.macmillanlearning .com/catalog/shellitoigt4e**.

Georeferencing Apps

Here's a sampling of available representative georeferencing-related apps for your smartphone or tablet. Note that some apps are for Android, some are for Apple iOS, and some may be available for both.

➤ **GeoEditor for MapTiler:** An app that allows you to display maps generated by MapTiler (as well as multiple other mapping and map editing functions).

➤ **Mapmatcher:** An app that allows you to compare current satellite imagery with georeferenced maps (including old historical ones) of different cities (and view the same position on up to four maps simultaneously). There are separate downloadable apps for each city and each app works the same way, for the following cities:

➤ **Mapmatcher Boston**

➤ **Mapmatcher Chicago**

➤ **Mapmatcher London**

➤ **Mapmatcher New York**

➤ **Mapmatcher San Francisco**

Georeferencing in Social Media

Here's a sampling of some representative Facebook and Twitter accounts, along with a YouTube video related to this chapter's concepts.

On Facebook, check out the following:

➤ **David Rumsey Map Collection:**
www.facebook.com/pages/David-Rumsey-Map-Collection/
216302804001

➤ **Klokan Technologies:**
(the designers of MapTiler): **https://www.facebook.com/klokantech**

Become a follower on Twitter of:

➤ **David Rumsey Map Collection:** @DavidRumseyMaps

➤ **Klokan Technologies:** @klokantech

On YouTube, watch the following video:

➤ **Georeferencer: David Rumsey Introduction** (a guide to using the online Georeferencer tools): **http://www.youtube.com/watch?v=0OrSVn8k8Mc**

Key Terms

affine transformation (p. 76)

control points (p. 72)

georeferencing (p. 70)

reproject (p. 68)

root mean square error (RMSE)
(p. 77)

rotated (p. 76)

scaled (p. 76)

skewed (p. 76)

spatial reference (p. 70)

translated (p. 76)

Georeferencing an Image

This chapter's lab will introduce you to some of the basics of working with georeferencing concepts by matching an unreferenced map image against a referenced imagery source. You'll be using the free version of the MapTiler software program for this exercise as well as briefly using Google Earth Pro.

Important note: MapTiler will take a map (in an image or PDF format) and align it to a source map and project it. The end result of the process will be a series of "tiles" used to examine your map and match it up with their data. While you'll be using georeferencing concepts from the chapter, the end result will be that the tiles of your map could be used later as part of a Web map.

Objectives

The goals for you to take away from this lab are:

➤ To familiarize yourself with the MapTiler operating environment

➤ To analyze both the unreferenced image and the source for appropriate control point locations

➤ To perform a transformation on the unreferenced data

➤ To evaluate the results of the georeferencing procedure

Using Geospatial Technologies

The concepts you'll be working with in this lab are used in a variety of real-world applications, including:

➤ The historical preservation of old maps, which utilizes georeferencing techniques with scanned versions of the maps, so that they may be matched up with current geography for comparison and analysis

➤ Engineering and real estate analysis, which utilizes scanned and geo-referenced versions of old railroad property valuation maps (often hand-drawn), to compare land parcel plans and values from the past with current maps and imagery

Obtaining Software

The current version of MapTiler is available for free download at **https://www.maptiler.com**. MapTiler comes in four versions — Free, Start, Plus, and Pro. For this Geospatial Lab Application, we will be using the Free version (version 7.2).

The current version of Google Earth Pro (7.3) is available for free download at **www.google.com/earth/download/gep/agree.html**.

Important note: Software and online resources can change fast. This lab was designed with the most recently available version of the software at the time of writing. However, if the software or Websites have changed significantly between then and now, an updated version of this lab (using the newest versions) will be available online at **www.macmillanlearning.com/catalog/shellitoigt4e**.

Lab Data

Copy the **Chapter 3** folder containing a PDF version of the map of the Youngstown State University (YSU) campus with no spatial reference attached to it (called CampusMap.pdf). Maps of YSU can also be downloaded from the campus map Web page at **http://www.ysu.edu/directions-to-campus**.

It also contains two empty folders, one called Output and one called GE. You will use these later in the lab to place the final rendered tiles of the map that you'll create through georeferencing — once for viewing the map in MapTiler and once again for viewing the map in Google Earth Pro.

Localizing This Lab

MapTiler is capable of georeferencing a graphic or PDF map of an area. If you have a graphic (such as a JPEG) or PDF of a map of your school, it may be substituted for the YSU campus map and you can match your own local data instead. The same steps will still apply; you'll just be finding appropriate local control points for your school rather than the ones for Youngstown, Ohio.

Alternatively, you could use a graphic or PDF of a map of a local park, hike or bike trail, fairgrounds, or local shopping district, and georeference that instead.

3.1 Getting Started with MapTiler

1. Start **MapTiler.** It will open with the default screen asking the location and format of your unreferenced image. Maximize the MapTiler window to give yourself plenty of room in which to work.

2. From the available options, click on **Standard Tiles**.

3. Either drag and drop the CampusMap.pdf file from the C:\Chapter3 folder into the MapTiler screen or click on **Open a file...** and navigate to the C:\Chapter3 folder and choose the CampusMap.pdf file.

4. When prompted about a Coordinate system, accept the default (WGS84) and click on **Set**.

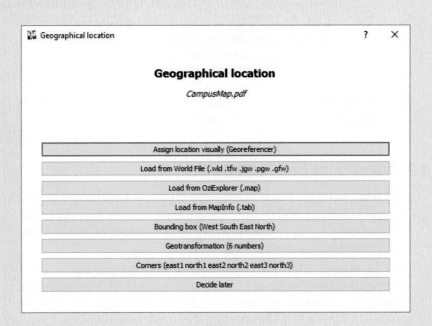

5. When prompted about Geographical location, select **Assign location visually (Georeferencer)**.

6. The campus map will be added to MapTiler. If it is not properly aligned vertically, click on the **Rotate** button on the right side of the screen.

7. In the **Search** box (next to the blue magnifying glass symbol), type **Youngstown State University** and then press the **Enter** key on the keyboard.

8. MapTiler will now display the campus in the referenced source (the left-hand view) and the unreferenced map (the right-hand view). If they are not shown side by side, click on the **Side by side** button at the top of the screen.

9. By clicking on the **blue Earth** symbol in the upper left-hand corner, you'll be able to choose other imagery and maps to use as the source you'll use for georeferencing. Try out a few of the options but when you're done, return to the default of **OSM Bright**. However, keep these other options in mind during the later steps of this lab if you want to use a different map when selecting your control points.

> **Question 3.1** Just by visual examination of the two maps, what types of transformative operations will have to be performed to the unreferenced map to make it match the source map?

3.2 Selecting the First Three Control Points

1. The next step is to choose the same positions on both images to use as control points. A good starting point will be the intersection of Wick Avenue and Lincoln Avenue (a major road intersection that borders the southeastern portion of the campus).

2. Use the **blue plus buttons** on both the source map and the unreferenced map to zoom in closer to this road intersection. You can also use the mouse to pan around each image until you can center this intersection on both the source and the unreferenced map.

Wick and Lincoln Intersection

3. You'll see that your cursor is now looking like black crosshairs. Starting with the source map, click the left mouse button in the intersection of Wick and Lincoln.

4. The cursor will change to larger crosshairs in an orange outline, with the space inside the crosshairs magnified for your benefit.

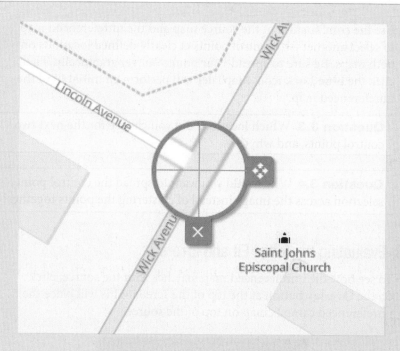

5. To select the location for the control point, position the tip of the crosshairs at the precise spot you want to place the control point. Get the exact location by using the crosshairs controls as follows:

a. By positioning the cursor inside the magnifying glass and holding down the left mouse button, you can drag the magnified version around the area.

b. By clicking and dragging the orange control with the four white arrows that is on the outer part of the crosshairs, you can move the entire crosshairs around the screen.

c. By clicking on the orange control with the white X you can remove the crosshairs from the screen entirely.

6. Now move over to the unreferenced map and click the left mouse button at the approximate place you want to place the control point that corresponds with the location you just identified on the source map. The same orange crosshairs will appear on the unreferenced map. Use the same controls as described above to place the crosshairs in the same place on the unreferenced map as you placed it on the source map.

7. When you have the two crosshairs positioned where you want them, click on the orange control with the four white arrows on one of the crosshairs and they will both disappear and be replaced by a blue point. This is the first control point.

Question 3.2 Why is this a good location for a control point?

8. Use the combination of the source map and the unreferenced map to select another two control points at clearly defined locations on both maps. Be sure to spread your points out geographically. Then click the **blue Lock** icon. MapTiler will perform an initial fit of the unreferenced map.

> **Question 3.3** Which locations did you choose for the next two control points, and why?

> **Question 3.4** Why would you want to spread the control point selection across the image instead of clustering the points together?

3.3 Evaluating the Initial Fit and Errors

1. To see how the unreferenced map matches with the source, click on the **Overlay** button at the top of the screen. This will place the unreferenced campus map on top of the source.

2. To make the unreferenced map semi-transparent so that you can see both layers at once (to see how they match up with each other), move the **blue control** on the **Slider bar** (next to the black eye icon on the upper-right corner of the screen). Moving it further to the left will make the unreferenced layer more transparent (and to the right makes it more opaque).

3. Pan around the unreferenced map — it's likely an okay fit with the source, but if you look closely, you'll see that things don't quite match up just yet. Click on the **Side by side** button so that we can add a few more control points and look at the overall fit.

4. Find a good location for a fourth control point and click on its location in the unreferenced map. You'll see that MapTiler will place the crosshairs in the unreferenced map where you clicked, but will also place the crosshairs in what it currently believes is the same location in the source map.

5. Use the controls to position the fourth control point into its location.

> **Question 3.5** Which location did you choose for the fourth control point, and why?

6. After it is set (i.e., the blue dot is where you placed the control point), click on the **Coordinates** button at the top of the screen. You'll see each of your control points listed, along with an error term that represents how many pixels the points are off by. The lower the error value, the better.

> **Question 3.6** Which point has the highest error term and what is that error value?

3.4 Editing Control Points

1. Return to the side by side view and add three more control points, so you have a total of seven control points.

> **Question 3.7** Which locations did you choose for the final three control points, and why?

2. Next, click on **Coordinates** to see the error values associated with each point. In order to reduce your error values, you'll need to edit a control point to try to get a better overall fit of the map. By editing a control point you'll be moving it to a different location — usually this will be a small adjustment, but sometimes you'll find yourself making a larger adjustment in the point's location. Other times you may find it necessary to delete a point and add a fresh one.

3. To zoom directly to one of the control points (for instance, the control point with the highest error value), double-click on its ID in the Coordinates box, and MapTiler will center on that control point.

4. Switch back to side by side mode. Hover the cursor over the control point and you'll see a gray outline appear around it. Click on that point and you'll see the orange crosshairs appear around the point in both the source map and the unreferenced map. Use the techniques previously outlined to adjust the point in either map, then click on the orange box with the white arrows on the crosshairs to re-fix that particular point.

5. Adjust your points as necessary to get low error values. You can always click on **Overlay** to see how the current georeferencing matches up with your chosen control points. You can also choose a different map to use as the source as well — for instance, you may want to compare

your unreferenced map to the imagery of the **MapBox Satellite** option.

> **Question 3.8** What are the final error values for each of your seven control points?

3.5 Completing the Georeferencing Process

When you have the unreferenced map set up the way you want it, it's time to complete the georeferencing process.

Important note: With the free version of MapTiler you're working with, you're unable to save your work and then reopen it later, so by completing the georeferencing process as outlined in this section, your results will be final (i.e., if you don't like how the final rendering of your map comes out, you'll have to start over from scratch rather than being able to return to this point — so make sure you're completely ready to proceed from here). Also, if you exit MapTiler anytime before the end of the entire lab, your work will be lost.

1. At the bottom of the MapTiler screen, click on **Save**.

2. A summary of the process will appear with a basic preview of georeferencing result. Click **Continue**.

3. A number of choices will appear for how you want to set up the final rendered map. Click on **Folder**.

4. Navigate to your **C:\Chapter3** folder and click on the **Output** folder, then click on **Select Folder**.

5. A new screen will appear showing the rendering progress. When it's completed, click on **View the Map**.

6. The georeferenced campus map will appear, overlaid onto the source map. You can zoom in on the view and use the blue **Slider bar** in the upper right-hand corner to adjust the transparency of the campus map layer to see how your georeferenced map lines up with the source.

> **Question 3.9** How does the map match up with the roads leading in and out of campus? Are there any locations where the roads don't match? How does the interior of campus match up?

7. When you're done examining the preview of the georeferenced map, click on **Close Preview**, then on each screen click on **Go Back** until you reach the main menu page again. *Important note*: Do not exit the MapTiler program yet.

3.6 Viewing Georeferencing Results in Google Earth

1. To see what your georeferenced map looks like in Google Earth Pro, on the main page click on the **Google Earth** option. What MapTiler will do in the next couple of steps is take your georeferenced map

and convert it into a format that you can then open in Google Earth Pro.

2. On the next screen, click **Continue**.

3. On the next screen, click **Render**.

4. When prompted, navigate to your **C:\Chapter3** folder, choose the **GE** folder, and then click on **Select Folder**.

5. After the rendering process completes again, you'll be back to the screen that says "View this map." However, you can now click on **Quit MapTiler**.

6. Start **Google Earth Pro**.

7. From the **File** pull-down menu, click **Open**, then navigate to your **C:\ Chapter3\GE** folder, then click on the file called **doc.KML** and then click **Open**.

8. Your georeferenced map will open in Google Earth Pro. You can zoom and tilt the map as necessary to get a good look at the georeferenced map overlaid on the Google Earth Pro imagery and terrain.

9. Under **Temporary Places** in Google Earth Pro, you'll see a layer called **GE**. Click once on this layer to select it (so it's highlighted in blue). Next, click on the **gray square icon** (this is the transparency icon), and then the **Slider bar** next to it can be used to adjust the transparency of the georeferenced map again, so that you can get a better look at how your georeferencing results match up with Google Earth Pro.

> **Question 3.10** How does the map match up with Google Earth Pro? How does showing the georeferenced map overlaid on Google Earth Pro's terrain aid in viewing the campus map?

Closing Time

MapTiler is a highly versatile tool that is used for making data (such as unreferenced images) match the map. Any type of map that is drawn to scale (such as a park trail map, a fairgrounds map, a theme park diagram, or an unreferenced image) can be aligned using MapTiler. You can now exit Google Earth Pro by selecting **Exit** from the **File** pull-down menu.

Finding Your Location with the Global Positioning System

4

GPS Origins, Position Measurement, Errors, Accuracy, GNSS around the World, Applications, and Geocaching

Here's something for you to think about: You're out hiking in a forest. A storm comes up, and you quickly take shelter well off the trail. By the time things are all clear, you've become completely disoriented, and you can't pinpoint your location: You're well and truly lost. Or consider this: You're driving in unfamiliar territory. It's late at night, you're miles from anywhere, you missed an important turn a few miles back, and none of the names or route numbers on the signs mean anything to you. In both cases, you're out of cell phone range, and even though you have a printed map, you really have no idea where you are. How do you start navigating yourself out of this mess?

A few years ago, situations like these could have been potentially disastrous, but today there's a geospatial technology tool that's ready and available to help get you out of these situations. Using a small, hand-held, relatively inexpensive device, you can determine the coordinates of your position on Earth's surface to within several feet. Simply stand outside, turn the device on, and shortly your position will be found and perhaps even plotted on a map stored on the device. Start walking around, and it'll keep reporting where you are and even take into account some other factors like your elevation, speed, and the compass direction in which you're moving. Even though it all sounds like magic, it's far from it—there's an extensive network of equipment across Earth and in space continuously working in conjunction with your hand-held receiver to locate your position. This, in a nutshell, is the operability of the Global Positioning System (**GPS**).

GPS receivers have become so widespread that they seem to be everywhere. In-car navigation systems have become standard items on the shelves of electronics stores. GPS receivers are also on sale at sporting goods stores because they're becoming standard equipment for campers, hikers, fishing enthusiasts, and golfers. Runners can purchase a watch containing a GPS

GPS the Global Positioning System, a technology using signals broadcast from satellites for navigation and position determination on Earth receiver that will tell them their location, the distance they run, and their average speed. Smartphones will often have GPS capability as a standard function. Use of the GPS is free for everyone and available worldwide, no matter how remote the area in which you find yourself. So the first questions to ask are—who built this type of geospatial technology, and why?

 ## Who Made GPS?

NAVSTAR GPS the U.S. Global Positioning System

The acronym "GPS" is commonly used to describe any number of devices that use Global Positioning System technology to tell us where we are, but the term is often used inaccurately. The term "GPS" originated with the initial designers and developers of the setup, the U.S. Department of Defense, and is part of the official name of the system called **NAVSTAR GPS**. Other countries besides the United States have developed or are currently engaged in developing systems similar to NAVSTAR GPS, so strictly speaking we should only use the term "GPS" when we're referring to the United States system. Because the United States isn't the only nation with this technology, it would be more accurate to describe GPS as one type of global navigation satellite system (**GNSS**).

GNSS the global navigation satellite system, an overall term for the technologies that use signals from satellites to find locations on Earth's surface

GPS isn't the first satellite navigation system. During the 1960s, the U.S. Navy used a system called Transit, which used satellites to determine the location of sea-going vessels. The drawback of Transit was that it didn't provide continuous location information—you'd have to wait a long time to get a fix on your position rather than always knowing where you were. Another early satellite navigation program of the 1960s, the Naval Research Laboratory's Timation program, used accurate clock timings to determine locations from orbit. The first GPS satellite was launched in 1978, and the 24th GPS satellite was launched in 1993, completing an initial full operational capability of the satellite system for use. Today, GPS is overseen and maintained by the 50th Space Wing, a division of the U.S. Air Force headquartered at Schriever Air Force Base in Colorado.

 ## What Does the Global Positioning System Consist Of?

Finding your position on the ground with GPS relies on three separate components, all operating together: a space segment, a control segment, and a user segment.

Space Segment

space segment one of the three segments of GPS, consisting of the satellites and the signals they broadcast from space

The **space segment** is made up of numerous GPS satellites (also referred to as "SVs" or "space vehicles") orbiting Earth in fixed paths. GPS satellites make two orbits around Earth every day (their orbit time is

FIGURE 4.1

The constellation of GPS satellites in orbit around Earth.

actually about 11 hours and 58 minutes) at an altitude of 20,200 kilometers (12,552 miles).

The satellites are set in a pattern of specific orbits called a **constellation**, in which the satellites are specifically arranged for maximum coverage over Earth (**Figure 4.1**). The way the GPS constellation is designed allows for a person to be able to receive enough signals to find their location wherever they are on the planet. Twenty-four satellites is the minimum for a full constellation, and there are currently several additional operational GPS satellites in orbit to improve global coverage.

The job of GPS satellites is to broadcast a set of signals down to Earth from orbit. These signals (discussed later) contain information about the position of the satellite and the precise time at which the signal was transmitted from the satellite (GPS satellites measure time using extremely accurate atomic clocks carried onboard). The signals are sent on carrier frequencies L1 (broadcast at 1575.42 MHz) and L2 (broadcast at 1227.60 MHz). Additional carrier frequencies are planned for future GPS upgrades—L2C intended as a second civilian frequency; L5, intended to be used for safety-of-life functions; and L1C, intended to have GPS operate with other countries—and GNSS. You need to have a direct line of sight to the satellites to receive these frequencies (which means that you have to be in the open air, not indoors).

constellation the full complement of satellites comprising a GNSS

Control Segment

control segment one of the three segments of GPS, consisting of the control stations that monitor the signals from the GPS satellites

The **control segment** of GPS represents a series of worldwide ground stations that track and monitor the signals being transmitted by the satellites. These ground stations, also known as control stations, are spread out to enable continuous monitoring of the satellites. The control stations collect the satellite data and transmit it to the master control station at Schriever Air Force Base. In the control segment, corrections and new data are uploaded to the satellites so that the satellites will be broadcasting correct data. The control stations also monitor the satellites' positions and relay updated orbit information to the satellites (**Figure 4.2**).

User Segment

user segment one of the three segments of GPS, consisting of the GPS receivers on the ground that pick up the signals from the satellites

channels the number of satellite signals a GPS unit can receive

single frequency a GPS receiver that can pick up only the L1 frequency

dual frequency a GPS receiver that can pick up both the L1 and L2 frequencies

The **user segment** represents a GPS unit somewhere on the Earth that is receiving satellite signals. The receiver can then use the information in these signals to compute its position on Earth. A key component of a GPS receiver is how many **channels** the receiver has—the number of channels reflects the number of satellites from which the receiver can obtain signals at one time. Thus, a 12-channel receiver can potentially pick up signals from up to 12 satellites. The type of unit you're using is also important—a **single frequency** receiver can only pick up the L1 frequency, but a **dual frequency** receiver can pick up both L1 and L2. One thing to keep in mind about the receiver is that no matter what size, capabilities, or number of

FIGURE 4.2

A control station monitors the signals from the satellites and sends correction data back to the satellites.

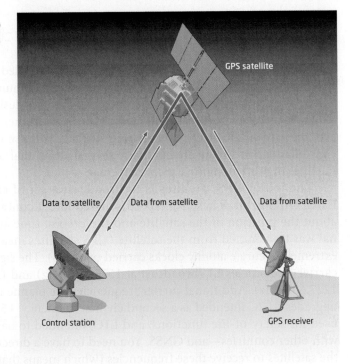

GPS satellite

Data to satellite   Data from satellite   Data from satellite

Control station   GPS receiver

channels it has, a receiver (like the name implies) can only receive satellite data—it can't transmit data back up to the satellites. This is why GPS is referred to as a one-way ranging system: satellites broadcast information, and receivers receive it.

 # How Does GPS Find Your Position?

At this point, the three segments of GPS have satellites broadcasting signals, ground stations monitoring and correcting those signals, and users receiving those signals—and all three of these segments work together to determine your position on Earth's surface. The signals being sent from the satellites are the key to the whole process—the signals contain information about the satellite's status, the orbit, and the location (referred to as the **almanac**) that's sending them, as well as more precise data about satellite location (this information is referred to as the **ephemeris**).

The information is sent in one of two digital pseudo-random codes—the **C/A code** (coarse acquisition code) and the **P code** (precise code). The C/A code is broadcast on the L1 carrier frequency and is the information that civilian receivers can pick up. The P code is broadcast on the L1 and L2 carrier frequencies and contains more precise information, but a military receiver is required to pick up this signal. The **Y code** is an encrypted version of the P code and is used to prevent false P code information from being sent to a receiver by hostile forces. This "anti-spoofing" technique is commonly used to make sure that only the correct data is being received. Basically, the satellites are transmitting information in the codes about the location of the satellite, and they can also be used to determine the time when the signal was sent. By using this data, the receiver can find its position relative to that one satellite.

The signals are transmitted from space using high-frequency radio waves (the L1 and L2 carrier frequencies). These radio waves are forms of electromagnetic energy (discussed in Chapter 10) and will therefore be moving at the speed of light. Your receiver can compute the time it took for the signal to arrive from the satellite. If you know these high-frequency radio waves move at the speed of light (c), and you also know how long the transmission time (t) was, then you can calculate the distance between you and that one satellite by multiplying t by c. This result will give you the **pseudorange** (or distance) between your receiver and the satellite transmitting the signal.

Unfortunately, this still doesn't tell us much—you know where you are on the ground in relation to the position of one satellite over 12,000 miles away, but that by itself gives you very little information about where you are on Earth's surface. It's like waking up in an unknown location and being told by a passerby that you're 500 miles from Boston—it gives you very little to go on in determining your real-world location.

almanac data concerning the status of a GPS satellite, which is included in the information being transmitted by the satellite

ephemeris data referring to the GPS satellite's position in orbit

C/A code the digital code broadcast on the L1 frequency, which is accessible by all GPS receivers

P code the digital code broadcast on the L1 and L2 frequencies, which is accessible by the military

Y code an encrypted version of the P code

pseudorange the calculated distance between a GPS satellite and a GPS receiver

trilateration finding a location in relation to three points of reference

The process of determining your position is referred to as **trilateration**, which means using three points of reference to find where you are. Trilateration in two dimensions is commonly used when plotting a location on a map. Let's say, for example, that you're on a road trip to Michigan—you've been driving all night, and you've arrived in an unknown location with no identifying information to tell you where you are. The first person you bump into tells you (somewhat unhelpfully) that you're 50 miles away from Ann Arbor, Michigan (see **Figure 4.3**). That puts you somewhere on a circle with Ann Arbor at its center 50 miles away, and gives you a choice of far too many possibilities to be of any use to you.

The second person you run into tells you (again, not being overly helpful) that you're located 150 miles away from Traverse City, Michigan. This again puts your location somewhere on a circle 150 miles from Traverse City—but when you combine this with the information that places you 50 miles away from Ann Arbor, you can cut your options down considerably. There are only two cities in Michigan that are 50 miles from Ann Arbor and 150 miles away from Traverse City—you're either in Lansing or in Flint (see **Figure 4.4**).

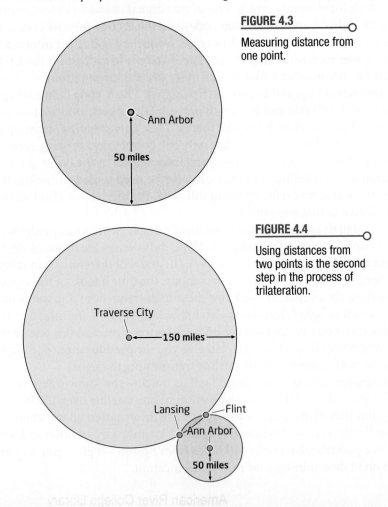

FIGURE 4.3

Measuring distance from one point.

FIGURE 4.4

Using distances from two points is the second step in the process of trilateration.

Luckily, the third person you see tells you that you're 60 miles from Kalamazoo. Your first thought should be "Who are all these geographically minded people I keep running into?" but your second thought is that you now know exactly where you are. Lansing is the only option that fits all three of the distances from your reference points, therefore you can disregard Flint and be satisfied that you've found your location (see **Figure 4.5** and *Hands-On Application 4.1: Trilateration Concepts*).

The same concept applies to finding your location using GPS, but rather than locating yourself relative to three points on a map, your GPS receiver is finding its position on Earth's surface relative to three satellites. Also, because a position on the three-dimensional (3D) Earth is being found with reference to positions surrounding it, a spherical distance is calculated rather than a flat circular distance. This process is referred to as trilateration in three dimensions (or **3D trilateration**). The concept is similar, though finding the receiver's position relative to one satellite means it's located somewhere on one sphere (similar to only having the information about being 50 miles from Ann Arbor). By finding its location in relation to two satellites, the receiver is finding a location on a common boundary between two spheres. Finally, by finding the location relative to three satellites, there are only two places where those three spheres intersect, and the way the geometry works out, one of them is in outer space and can be disregarded. That leaves only one location where all three spheres intersect, and thus one location on Earth's surface relative to all three satellites. That position will be the location of the GPS receiver (**Figure 4.6**).

3D trilateration finding a location on Earth's surface in relation to the positions of three satellites

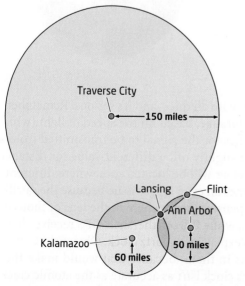

FIGURE 4.5

The measurement of distances from the third point completes the process of trilateration by finding a location.

Hands-On Application 4.1

Trilateration Concepts

Trilateration can be used to determine your location on a map. Go to **http://electronics .howstuffworks.com/gadgets/travel/gps1.htm** to view the article by Marshall Brain and Tom Harris at How Stuff Works. Check out how three measurements are made to determine a single location (as in the previous Michigan example), then start Google Earth Pro and zoom to your location. Use Google Earth Pro and its measuring tools to set up a similar scenario — find your location relative to three other cities and figure out the measurements to do a similar

trilateration. Then zoom in closer to identify three local locations and calculate the measurements from each one that you need to trilaterate your position.

Expansion Questions

➤ What three cities did you choose, and what measurements did you use to trilaterate your city?

➤ What three local locations did you choose, and what measurements did you use to trilaterate your current position?

FIGURE 4.6

Finding a location using GPS: the results of 3D trilateration.

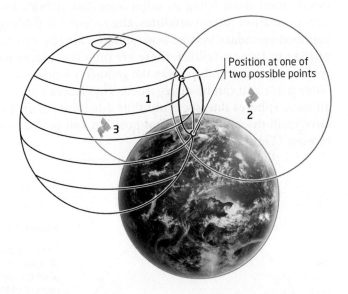

There's just one problem with all this—and it's serious. Remember, what you're calculating is a pseudorange, based on the speed of light (which is a constant) and the time it takes for the signal to be transmitted from space to Earth. If that time is just slightly off, a different value for distance will be calculated, and your receiver will be located somewhere different from where it really is. This becomes an even bigger issue because the satellite has a super-precise and very expensive (somewhere in the tens of thousands of dollars range) atomic clock, while the off-the-shelf GPS receiver has a less precise (and very much less expensive) quartz clock. Obviously, GPS receivers can't have atomic clocks in them, because that would make them far too expensive. The receiver's clock isn't as accurate as the atomic clock, and timing errors can cause inaccuracies in calculating distances when it comes to finding a position. Fortunately, there's an easy solution to this mess: your receiver picks up additional information from a fourth satellite. By using a fourth measurement, clock errors can be corrected and your position can be determined accurately. The position you'll receive will be measured using the WGS84 datum.

Why Isn't GPS Perfectly Accurate?

If GPS can be used to find a location on Earth's surface, the next question to ask is: "Just how accurate is GPS?" Through the use of pseudorange measurements from four satellites, you might reasonably expect the system to give the user one accurately measured location. However, in normal everyday use with an ordinary off-the-shelf civilian receiver, accuracy is usually somewhere within 10–15 meters horizontally. This doesn't mean that the location is always off by 15 meters—it could be off by 5, 10, or 15 meters. This is due to a number of factors that can interfere with or delay the transmission or reception of the signal—and these delays in time can cause errors in position.

One of the biggest sources of error was actually worked into the system intentionally to make GPS less accurate. The United States introduced **selective availability (SA)** into GPS with the goal of intentionally making the C/A signal less accurate, presumably so that enemy forces couldn't use GPS as a tool against the U.S. military. Selective availability introduced two errors into the signals being transmitted by the satellites—a delta error (which contained incorrect clock timing information) and an epsilon error (which contained incorrect satellite ephemeris information). The net effect of this was to make the accuracy of the C/A to within about 100 meters. As you can imagine, this level of accuracy certainly limited GPS in the civilian sector—after all, who wants to try to land an airplane with GPS when the accuracy of the runway location could be hundreds of feet off? In the year 2000, the U.S. government turned selective availability off and has no announced plans to turn it back on (in fact, the U.S. Department of Defense has announced that new generations of GPS satellites will not have the capability for selective availability).

Although selective availability was an intentional error introduced into the signals broadcast by the satellites, other errors can occur naturally at the satellite level. Ephemeris errors can occur when a satellite broadcasts incorrect information about orbit position. These types of errors indicate that the satellite is out of position, and the ephemeris data needs to be corrected or updated by a control station. Typically, these errors can introduce 2 meters or so of error.

Even the arrangement of satellites in space (that is, how they are distributed) can have an effect on the determination of position. The

selective availability (SA) the intentional degradation of the timing and position information transmitted by a GPS satellite further the satellites are spread out from one another and the wider the angle between them, the more accurate a measurement obtained by the receiver will be. The error introduced by a poor geometric arrangement of the satellites is referred to as the position dilution of precision, or **PDOP**. Some receivers will calculate a value for PDOP for you to indicate what relative error could be introduced because of the arrangement of the satellites being used (basically, the lower the value the better).

PDOP the position dilution of precision; describes the amount of error due to the geometric position of the GPS satellites

You should also keep in mind that the signals are being broadcast from space and have to pass through Earth's atmosphere before reaching the receiver. Though some of these atmospheric effects can be modeled, and attempts can be made to compensate for them, atmospheric problems can still cause errors in accuracy by making changes to the speed of the signals (**Figure 4.7**). Atmospheric interference, especially in the ionosphere and troposphere, can cause delays in the signal on its way to the ground. The ionosphere can alter the propagation speed of a signal, which can cause an inaccurate timing measurement, while the amount of water vapor in the troposphere can also interfere with the signals passing through it and also cause delays. It's estimated that the problems caused by conditions in the ionosphere can cause about 5 meters (or more) of position inaccuracy, while conditions in the troposphere can add about another 0.5 meters of measurement error.

The atmosphere is not the only thing that can interfere with the signal reaching the receiver. The receiver needs a good view of the sky to get an initial fix, therefore things that would block this view (such as a heavy tree canopy) may interfere with the signal reception. In addition, the signals may be reflected off objects (such as tall buildings or bodies of water) before reaching the receiver rather than taking a straight path from the satellite (**Figure 4.8**). This **multipath** effect can introduce additional delays into the reception of the signal because the signal is reaching the receiver later than it should (and will add further error to your position determination).

multipath an error caused by a delay in the signal due to reflecting from surfaces before reaching the receiver

FIGURE 4.7

The effects of the ionosphere and troposphere on GPS signals.

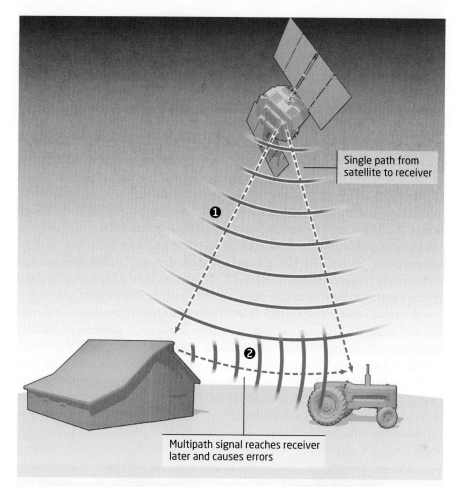

FIGURE 4.8

The effect of multipathing on the GPS signal.

Single path from satellite to receiver

❶

❷

Multipath signal reaches receiver later and causes errors

It's quite possible for any or all of these various types of error measurements to combine together to contribute several meters worth of errors into your position determination. When the ephemeris introduces a few meters of error, and the ionosphere and troposphere add a few more, and then multipathing and PDOP add a few more on top of that, before you know it they're starting to add up to a pretty significant difference between your position as plotted and where you actually are. Therefore there should be some manner of solutions to improve the accuracy of GPS and get a better fix on your position.

How Can You Get Better Accuracy from GPS?

A few methods have been developed to improve the accuracy of GPS position measurements, and to deal with the problems that cause the kinds of errors described above. The first of these is a process referred to as differential GPS (or **DGPS**). This method uses one or more base stations at locations

DGPS differential GPS; a method of using a ground-based correction in addition to the satellite signals to determine position on the ground to provide a correction for GPS position determination. The base station is positioned at a place where the coordinates are known, and it receives signals from the satellites. Because the base station knows its own coordinates, it can calculate a correction to any error that causes the position measurement from the satellites to locate it in the wrong place. The base station can then broadcast this correction out to receivers equipped to collect this type of differential correction. Thus, when you're using DGPS, your receiver is picking up the usual four satellite signals plus an additional correction from a nearby location on the ground (see **Figure 4.9** for an example of the operation of DGPS). As a result, the position measurement can be reduced to 5 meters or less (and often less than 1 meter).

Today, there are DGPS locations all over the world. In the United States, the U.S. Coast Guard has set up several reference stations along the coasts and waterways to enable ships to find their exact location and to improve navigation. The U.S. Coast Guard Navigation Center operates the Nationwide Differential GPS (**NDGPS**) system and aims to provide coastal regions with the most accurate position information that's available.

A related program, **CORS** (Continuously Operating Reference Stations), is operated by the National Geodetic Survey and consists of numerous base stations located across the United States and throughout the world. CORS collects GPS information and makes it available to users of GPS data (such as surveyors and engineers) to make more accurate position measurements. CORS data provides a powerful tool for location information for a variety of applications (such as the U.S. Army establishing CORS in Iraq as an aid in the rebuilding efforts).

NDGPS Nationwide Differential GPS, which consists of ground-based DGPS locations around the United States

CORS Continuously Operating Reference Stations; a system operated by the National Geodetic Survey to provide a ground-based method of obtaining more accurate GPS positioning

FIGURE 4.9

The operation of DGPS.

DGPS

Satellite-based augmentation systems (**SBAS**) are additional resources for improving the accuracy of position measurements. SBAS works in a similar way to DGPS, in that a correction is calculated and picked up by your receiver in addition to signals from the regular four satellites. However, in this case, the correction is sent from an additional new satellite, not from an Earth-bound station. A good example of an SBAS is the Wide Area Augmentation System (**WAAS**), which was developed by the Federal Aviation Administration (FAA) to obtain more accurate position information for aircraft. WAAS has developed widespread uses beyond the world of aviation as a quick and easy method of improving overall GPS position accuracy.

WAAS operates through a series of base stations spread throughout the United States that collect and monitor the signals sent by GPS satellites. These base stations calculate position correction information (similar to the operation of DGPS) and transmit this correction to geostationary WAAS satellites. These WAAS satellites then broadcast this correction signal back to Earth, and if your receiver can pick up this signal, it can use the information to improve its position calculation (see **Figure 4.10**). WAAS reduces the position error to 3 meters or less. One drawback to WAAS is that it will only function within the United States (including Alaska and Hawaii) and nearby portions of North America (although WAAS base stations have been added in Mexico and Canada, greatly expanding the availability of WAAS advantages). There are other systems that improve the accuracy of position calculation, but they all operate on the same principle as WAAS and DGPS (some sort of correction is calculated and transmitted to a receiver, which uses it to determine its position).

SBAS satellite-based augmentation system; a method of using correction information sent from an additional satellite to improve GPS position determination

WAAS Wide Area Augmentation System; a satellite-based augmentation system that covers the United States and other portions of North America

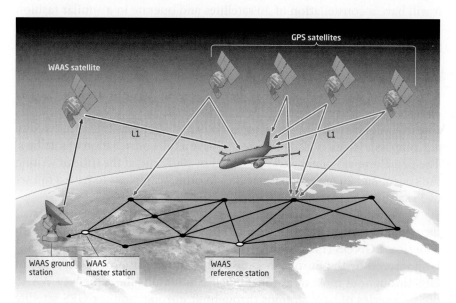

FIGURE 4.10

The setup and operation of the WAAS system.

What Other GNSS Are There Beyond GPS?

This chapter has been devoted to GPS: the way it operates, the types of errors it's liable to make, and how its accuracy issues can be resolved. However, as we touched on earlier, "GPS" properly refers to the U.S. NAVSTAR GPS. Other countries have developed (or are in the process of developing) their own versions of GPS or GPS augmentation systems, presumably so as not to be completely dependent on U.S. technology. There are other full constellations of GPS-style satellites being developed and put into orbit, and numerous other enhancement systems are either in operation or in the planning stage all around the globe.

GLONASS the former USSR's (now Russia's) GNSS

GLONASS was the name of the USSR's counterpart to GPS and operated in a similar fashion. The GLONASS equivalent to the C/A code operated at horizontal accuracies of about 100 meters (or better) with an equivalent of the P code for more accurate use. GLONASS consisted of a full constellation of satellites with an orbital setup similar to GPS. By 2001, however, there were only a few GLONASS satellites in operation. In recent years Russia has renewed its GLONASS program—by the end of 2011 a fully operational constellation was back in place, and at the time of writing there are plans to launch additional satellites in the near future. In 2007, Russia announced that the civilian signal from GLONASS would be made freely available for use, and at the time of writing GLONASS accuracy was reported to be approximately 5 meters (or less).

Galileo the European Union's GNSS, currently in development

Galileo is the European Union's version of GPS. When it's completed, it will have a constellation of 30 satellites and operate in a similar fashion to GPS. The first of the GIOVE (Galileo in Orbit Validation Element) satellites was launched in 2005, and since then the program has continued to develop, with several satellites now orbiting as part of its constellation (at the time of writing it was partially completed with 18 satellites in orbit and projected to have a full constellation by 2020). Galileo promises four different types of signals, including information available to civilians as well as restricted signals. Also in development is China's **Beidou-2** (also referred to as Compass), which is also projected to have a full constellation of satellites. The first Compass satellite was launched in 2007, and at the time of writing the system had 20 operational satellites in orbit, with plans to have a full set of 35 satellites in place by 2020.

Beidou-2 China's GNSS, currently in development

EGNOS an SBAS that covers Europe

Other satellite-based augmentation systems have been developed across the world to operate like WAAS but to function in regions outside of North America. **EGNOS** (European Geostationary Navigation Overlay System) is sponsored by the European Union and provides improved accuracy of GPS throughout most of Europe. EGNOS functions the same way WAAS does—a series of base stations throughout Europe monitor GPS satellite data, calculate corrections, then transmit these corrections to geostationary EGNOS satellites over Europe, which in turn broadcast this data to Earth.

An EGNOS-enabled receiver is necessary to utilize this correction data, and the system is capable of accuracies of about 1.5 meters. Japan also operates its own SBAS, known as **MSAS** (Multifunctional Satellite Augmentation System), which covers Japan and portions of the Pacific Rim. MSAS operates in a similar way to WAAS and EGNOS, but it covers a different region of the world and uses just two satellites to provide coverage.

MSAS an SBAS that covers Japan and nearby regions

Many other GPS enhancement systems exist or are in development throughout the world, including the following:

➤ Beidou-1: an SBAS operating in China

➤ GAGAN: an SBAS in development for India

➤ IRNSS (also referred to as NAVIC): an Indian satellite navigation system, consisting of seven satellites.

➤ QZSS: an expanded SBAS based in Japan, currently in development

 # What Are Some Applications of GPS?

GPS is used in a wide variety of settings. Any application that requires location-based data to be collected firsthand can apply GPS to the task. For instance, geologists use GPS to mark the location of water wells when testing groundwater, while civil engineers use GPS to determine the coordinates of items such as culverts, road signs, and utility hole covers. Coordinates measured with a GPS receiver are commonly used with other geospatial technologies in **GPX** format, a standard for storing and mapping collected GPS data. In the following sections you'll see just a handful of the ways GPS is being used in industry, jobs, and the public sector.

GPX a standard file format for working with GPS data collected by GPS receivers

Emergency Response

When you make a 911 call, the call center is able to fix your location and route emergency services (police, firefighters, or emergency medical technicians) to where you are. When you make a 911 call from a cell phone, the Enhanced 911 (E-911) system can determine your position—either using cell towers to determine where you are, or using a GPS receiver in your phone to find your location.

Farming

Precision farming methods rely on getting accurate location data about factors such as watering levels, crops, and soils. Having the ability to map this data gives farmers information about where crops are thriving and what areas need attention. GPS serves as a data-collection tool to quickly (and accurately) obtain location information for farmers, which can then be combined with other geospatial technologies, such as using Geographic Information Systems (GIS) to place the GPS data on a map.

Forensics

Recovery of human remains has been greatly assisted through the use of GPS. Rather than having to rely solely on surveying equipment, GPS can provide quick data and measurements in addition to other measurements made in the field. The same concepts have been applied to identifying the locations of artifacts and remains at archeological sites.

Public Utilities

City or county data for items that need to be measured manually (such as culverts, road signs, or utility hole locations) can be done quickly and easily with GPS. When field-based data is necessary to create or update utility data, workers equipped with GPS can collect this type of data quickly and accurately.

Transportation

GPS provides a means of continuously tracking travel, whether on the road, in the water, or in the air. Since GPS receivers can receive signals constantly from satellites, your travel progress can be monitored to make sure that you stay on track. When your GPS-determined position is plotted in real time on a map, you can see your exact point in your journey and measure your progress. Delivery and shipping companies can use this type of data to keep tabs on their fleets of vehicles and to check if they deviate from their routes.

Wildlife Management

In much the same way you track vehicles, wildlife and endangered species can be tagged with a GPS receiver to monitor their location and transmit it to a source capable of tracking those movements. In this way, migration patterns can be determined and other large-scale movements of wildlife can be observed.

Beyond its uses in the private and public sectors, GPS plays a big part in personal recreation. If you're a runner, GPS can track your position and measure your speed. If you're a hiker, GPS can keep you from getting lost when you're deep into unfamiliar territory. If you're into fishing, GPS can help locate places reputed to be well stocked with fish. If you're a golfer, you can use GPS to measure your ball's location and the distance to the pin. In fact, there's a whole new form of recreation that's grown in popularity along with GPS, called **geocaching**.

geocaching using GPS and a set of coordinates to find locations of hidden objects

In geocaching, certain small objects (referred to as "geocaches") are hidden in an area and the coordinates of the object are listed on the Web (a log is maintained). From there, you can use a GPS receiver to track down those

Hands-On Application 4.2

Things to Do Before You Go Geocaching

Geocaching is a widespread and popular recreational activity today. As long as you have a GPS receiver and some coordinates to look for, you can get started with geocaching. This chapter's lab will have you investigating some specific geocaching resources on the Web, but before you start looking for caches, there are some online utilities to explore first. Open your Web browser and go to **http://www.geocaching.com.** You'll first have to create a free account (with a username and a password) and log in with it to use the Website's functions. There are several functions of the Website to use free of charge, but for many additional features you'll have to pay for a premium subscription service.

Even just using the free features, the main page of the Website will allow you to search for locations of geocaches near a particular location, address, city, or zip code, as well as create a Web map of search results. In the Web map, you can click on caches to obtain more information about each cache (such as when the cache was planted, the difficulty rating of the cache, and so on).

Expansion Questions

➤ What geocaches are located around your own local area or zip code?

➤ Select the cache closest to your current location on the Web map. When was that cache hidden, what is the size of the cache, and what is the difficulty rating of locating that cache?

coordinates (which usually involve hiking or walking to a location) and locate the cache. In essence, geocaching is a high-tech outdoor treasure hunt (see *Hands-On Application 4.2: Things to Do Before You Go Geocaching*). This type of activity (using a GPS receiver to track down pre-recorded coordinates to find objects or locations) is also used in educational activities, in and out of the classroom, to teach students the relationship between a set of map coordinates and a real-world location, as well as give them practice in basic land navigation and GPS usage.

An example of the kind of item that's used in geocaching is a **geocoin**, a small object about the size of a casino chip. A geocoin is "trackable" in that the unique ID number on the coin can be logged into Geocaching.com, and the location of where the coin is taken from and the location it's placed in can then be mapped (**Figure 4.11**). Geocoins are primarily for recreational geocaching but can also be used in several educational activities involving geographic studies. For example, VirginiaView (one of the StateView programs of the AmericaView consortium—see Chapter 15 for more information) has put together a set of lesson plans that involve students tracking geocoin movements and relating these activities to larger projects such as the tracking of invasive species and watershed characteristics. See Virginia-View's Website, **http://virginiaview.cnre.vt.edu**, for further information on combining trackable geocoins with geographic education.

geocoin a small coin used in geocaching with a unique ID number that allows its changing location to be tracked and mapped

FIGURE 4.11

A VirginiaView geocoin and a map showing how it can be tracked as it moves from location to location. *[VirginiaView/Virginia Geospatial Extension Program]*

Thinking Critically with Geospatial Technology 4.1

What Happens If GPS Stops Working?

What would happen if the Global Positioning System stopped working—or at least working at 100% capability? If the satellites stopped operating and weren't replaced, GPS would essentially stop functioning. If the number of satellites dropped below full operational capacity, GPS would only work some of the time and its accuracy would likely be greatly reduced. An operational failure of GPS forms the scenario of this article from GPS World: **http://www.gpsworld.com/gps-risk-doomsday-2010**.

If something like this came to pass, how would the lack of GPS affect life on Earth? What industries would be impacted by sudden deprivation of access to GPS? What would be the effect on navigation at sea and in the air, and how would other kinds of travel be disrupted? What would be the military implications if the world were suddenly deprived of GPS? How would all of these factors affect your own life?

Is society too reliant on GPS? Since selective availability was turned off in 2000, there's been an overwhelming increase in GPS usage in many aspects of everyday life. Is the availability of GPS taken for granted? If the system could fail or be eradicated, what impact would this have on people who believe that this technology would always be available to them?

In a similar vein, there's been a growing awareness of the vulnerability of GPS as so many functions in today's society rely on it. Everything from military drones to shipping to delivery systems has become reliant on GPS. The U.S. Department of Homeland Security recently published a report on the risks posed to the United States from disruptions in GPS (see a summary here: **http://www.gps.gov/news/2013/06/2013-06-NRE-public-summary.pdf**). What kinds of potential wrongdoing or dangers could occur through jamming, spoofing, or misusing GPS in relation to a nation's infrastructure?

Chapter Wrap-Up

This chapter provided an introduction to how the Global Positioning System operates as well as how GPS is able to find where a receiver is located on the ground. An increasingly popular GPS application is the kind of in-car navigation system manufactured by companies such as Garmin, Magellan, and TomTom (many of these also operate on smartphones). These types of systems not only plot your location on a map using GPS, but can also show surrounding roads and nearby locations (such as restaurants or gas stations), locate an address, and show you the shortest route from your current position to any location. The only one of these features that's actually using GPS is the one that finds your current location—the rest are all functions that are covered in Chapter 8. Until then, starting in the next chapter, we'll begin to focus on the actual computerized mapping and analysis that underlies those types of in-car systems, all of which are concepts of GIS.

This chapter's lab will start putting some of these GPS concepts to work, even if you don't have a GPS receiver. Before that, check out *GPS Apps* for information on some available apps that use the GPS capabilities of your mobile device, as well as *GPS and GNSS in Social Media* for Facebook, Twitter, and Instagram accounts, along with a blog related to GPS.

Important note: The references for this chapter are part of the online companion for this book and can be found at **www.macmillanlearning.com/catalog/shellitoigt4e.**

GPS Apps

Here's a sampling of available representative GPS apps for your smartphone or tablet. Note that some apps are for Android, some are for Apple iOS, and some may be available for both.

➤ **Commander Compass Lite:** An app that determines your position and then turns your mobile device into a compass for navigation

➤ **Compass GPS:** An app that uses your device's GPS in conjunction with a compass for navigation

➤ **Cyclemeter GPS:** An app designed for cyclists that finds your position and keeps track of your distance, riding time, and speed

➤ **Golf GPS Rangefinder and Scoring:** An app designed for golfers which determines location and gives a view of golf courses and their features through aerial imagery

➤ **GPS Essentials:** An app that contains several GPS utilities (and bills itself as the "Swiss army knife" of GPS navigation)

- ➤ **Trails—Outdoor GPS Logbook:** An app that uses GPS to create a GPX layer of where you walk or drive
- ➤ **Travel Altimeter Lite:** An app that determines your position and elevation and also serves as a compass
- ➤ **Wikiloc outdoor navigation GPS:** An app that uses GPS to find your position for use in outdoor activities such as hiking

GPS and GNSS in Social Media

On Facebook, check out the following:

- ➤ European GNSS Agency:
    http://www.facebook.com/EuropeanGnssAgency
- ➤ Garmin:
    http://www.facebook.com/Garmin
- ➤ Geocaching.com:
    http://www.facebook.com/geocaching
- ➤ GPS World Magazine:
    http://www.facebook.com/GPSWorldMag
- ➤ Magellan GPS:
    http://www.facebook.com/MagellanGPS
- ➤ TomTom:
    http://www.facebook.com/TomTom

Become a Twitter follower of:

- ➤ EGNOS Portal: @EGNOSPortal
- ➤ European GNSS Agency: @EU_GNSS
- ➤ Garmin Blog Team: @Garmin
- ➤ Galileo GNSS: @GalileoGNSS
- ➤ Geocaching.com: @GoGeocaching
- ➤ GPS World Magazine: @GPSWorld
- ➤ Magellan GPS: @magellangps
- ➤ TomTom: @TomTom
- ➤ Trimble: @trimbleGIS

Become an Instagram follower of:

➤ **Garmin:** @garminpics
➤ **Geocaching:** @geocaching

For further up-to-date info, read up on this blog:

➤ **Garmin Blog** (the official blog from Garmin with news and updates about GPS products and applications): **http://garmin.blogs.com**
➤ **Geocaching Blog** (the official blog from Geocaching.com): **https://www.geocaching.com/blog**

Key Terms

4.1 Geospatial Lab Application

GNSS Applications

This chapter's lab will examine the use of the Global Positioning System (and other global navigation satellite systems) in a variety of activities. The big caveat is that the lab can't assume that you have access to a GPS receiver unit and that you're able to use it in your immediate area (and unfortunately, we can't issue you a receiver with this book). It would also be useless to describe exercises involving you running around (for instance) Boston, Massachusetts, collecting GPS data, when you may not live anywhere near Boston.

This lab will therefore use the free Trimble GNSS Planning Online software for examining GNSS satellite positions and other planning factors for the use of GNSS for data collection. In addition, some sample Web resources will be used to examine related GNSS concepts for your local area.

Important note: Though this lab is short, it can be significantly expanded if you have access to GPS receivers. If you do, some of the geocaching exercises described in Section 4.5 of this lab can be implemented. Alternatively, rather than just examining cache locations on the Web (as in Section 4.4), you can use the GPS equipment to hunt for the caches themselves.

Objectives

The goals for you to take away from this lab are:

➤ To examine satellite visibility charts to determine how many satellites are available in a particular geographic location and to find out when during the day these satellites are available

➤ To read a DOP chart to determine what times of day have higher and lower values for DOP for a particular geographic location

➤ To explore some Web resources to find publicly available geocaches or permanent marker sites that you can track down with a GPS receiver

Using Geospatial Technologies

The concepts you'll be working with in this lab are used in a variety of real-world applications, including:

➤ Precision agriculture, in which farmers utilize GNSS to find more efficient means of maintaining crops through having access to very accurate crop locations

➤ Geology, which uses GNSS for collecting location data while in the field, such as the precise coordinates of groundwater wells

[© Richard Hamilton Smith/Corbis Documentary/Getty Images]

Obtaining Software

No software or data is necessary to download for this lab—everything you will be working with can be done through a Web browser. However, the Trimble GNSS Planning Online software requires you to have Microsoft Silverlight installed on your computer in order to work properly. You can download and install Silverlight (for free) from here: **http://www.microsoft.com/getsilverlight**. Note that you'll also need to be using a browser that is compatible with Silverlight—at the time of writing, this included browsers such as Safari (with enabled plug-ins), Mozilla Firefox (Extended Support Release), and Internet Explorer (but not Microsoft Edge). Microsoft has indicated it will continue to support Silverlight and Internet Explorer until 2021—this lab was designed using the most recent available version of Internet Explorer.

Important note: Software and online resources can change fast. This lab was designed with the most recently available version of the software at the time of writing. However, if the software or Websites have significantly changed between then and now, an updated version of this lab (using the newest versions) will be available online at **www.macmillanlearning.com/catalog/shellitoigt4e**.

Lab Data

There is no data to copy in this lab. All data comes as part of the Trimble GNSS Planning Online software.

Localizing This Lab

Note: This lab uses one location and one date to keep information consistent. Although this lab looks at GNSS sky conditions in Orlando, Florida, there are many more geographic locations that can be selected using Trimble GNSS Planning Online software. Its purpose is to be able to evaluate local conditions for GNSS planning—so rather than selecting Orlando, use the default latitude and longitude for you own location and examine that data instead. You can also use today's date rather than the one used in the lab.

Similarly, in Section 4.4 of the lab, you'll be examining Web resources for geocaching locations in Orlando, Florida. Use your own zip code rather than the one for Orlando and see what's near your home or school instead.

4.1 Getting Started with Trimble GNSS Planning Online

1. Start your Web browser and go to **http://www.gnssplanningonline .com/**. Note that it may take a minute for Microsoft Silverlight to load the Web resources for the program.

2. The visibility of GNSS satellites will change with the time of day and your position on Earth. The Trimble Planning program will allow you to examine data from stations around Earth to determine GNSS information. The first thing we'll do is select a location from which to examine satellite visibility. By default, Trimble Planning will use your current location. You could also manually select your location for this lab, but for our purposes we'll just enter the latitude and longitude of Orlando, FL. In the **Latitude** box type: N **28.5635** and in the **Longitude** box type: W **81.3724**.

3. Next, we'll choose the date to examine GNSS satellite availability. Click on the small **Calendar** button to the left of the button marked **Today**. This will bring up a calendar from which you can select a specific date. For purposes of this lab, choose **April 12, 2017**. For **Time Zone**, choose the **Eastern Time (US and Canada)** option. Note: If you wanted to use today's date, pressing the **Today** button would automatically adjust the desired date to the current one.

4. Lastly, to examine an entire day's worth of satellite information, select **24** from the pull-down menu next to **Time Span [hours]**.

5. Click the **Apply** button again now that the location, date, and time are set.

4.2 GNSS Satellite Availability Information

1. Trimble GNSS Planning Online allows you to examine satellite visibility (and other factors) from a local area at various times throughout the day. To choose which group of satellites you're examining, press the **Satellite Library** button. Answer Question 4.1.

> **Question 4.1** How many GNSS satellites are currently being tracked by the GNSS Planning Online? How many of them are considered healthy?

2. You'll see in the window all of the various satellites being tracked by Trimble GNSS Planning Online. By clicking on the button that corresponds with the name of the GNSS you'll see the active and inactive satellites that correspond with the particular GNSS. Click on **Galileo** and answer Question 4.2.

> **Question 4.2** How many Galileo satellites are currently active?

3. Next, click on **BeiDou** (this is the name that Trimble GNSS Planning Online uses for Beidou-2/Compass) and answer Question 4.3.

> **Question 4.3** How many Compass satellites are currently active?

4. For this lab application, we will only be working with GPS satellites. Remove the checkmarks from the Glonass, Galileo, BeiDou, and QZSS boxes, and then click on the GPS box so that only the GPS satellites are selected (and the GPS button is indented).

4.3 GNSS Satellite Visibility Information

1. With only the GPS satellites chosen, we'll now look at the number of satellites available for the chosen location on the chosen date. Press the **Number of Satellites** button and the view will change. A graph will appear, showing the number of visible satellites on the y-axis and a 24-hour time frame on the x-axis. Answer Questions 4.4 and 4.5.

> **Question 4.4** At what specific times of day are the maximum number of GPS satellites visible? What is this maximum number? *Hint*: You can use the time slider bar in the lower-right corner of the Number of Satellites screen to give you the time that corresponds with a particular stack on the chart. These times are available in 10-minute intervals.

> **Question 4.5** At what specific times of day are the minimum number of GPS satellites visible? What is this minimum number?

2. To determine which satellites are visible at specific times, press the **Visibility** button. A new chart will appear with the different satellites on the **y**-axis represented as colored bars stretching across the chart and the times those satellites are visible on the x-axis. Answer Questions 4.6 and 4.7.

Question 4.6 Which satellites are visible from Orlando at 9 a.m.? *Hint*: You can again use the time slider bar to move the horizontal line to a particular time. You can also hover your cursor over a particular colored bar to obtain the satellite number that goes along with that bar.

Question 4.7 Which satellites are visible from Orlando at 4:00 p.m.?

3. To examine the position dilution of precision (PDOP) at a particular time of day, press the **DOPs** button. A new chart will appear showing the values of DOP on the y-axis and the times of day on the x-axis. Different types of DOP will be charted, each with their own different colored line. Answer Questions 4.8 and 4.9.

Question 4.8 At what time(s) of day will the maximum values for DOP be encountered? What is this maximum value? *Hint*: You can again use the time slider bar to move the horizontal line to a particular time. You can also hover your cursor over a particular colored line to determine which type of DOP that line represents.

Question 4.9 At what time(s) of day will the minimum values for geometric DOP be encountered? What is this minimum DOP value?

4. To examine the ionospheric effects for a particular time of day, press the **Iono Information** button. A new chart will appear. It has two *y*-axes showing the Ionosphere Index values (on the right, with values from 1 to 10, with 1 being the least amount of ionospheric interference and 10 being the highest, based on the total electron count [TEC] and the ionospheric scintillation) while the *x*-axis shows the times of day at which this is occurring. Answer Questions 4.10 and 4.11.

Question 4.10 At what time(s) of day will the highest value of the ionospheric index be encountered? What is this value? *Hint*: You can again use the time slider bar to move the horizontal line to a particular time.

Question 4.11 At what time(s) of day will the lowest value of the ionospheric index be encountered? What is this value?

5. Take another look at the three graphs you've used and answer Questions 4.12 and 4.13.

> **Question 4.12** Just based on the number of visible satellites and the ionospheric and DOP calculations, what would be the "best" (i.e., peak) time range during which to perform GPS field measurements in Orlando and why?

> **Question 4.13** Just based on number of visible satellites and the ionospheric and DOP calculations, what would be the "worst" time range during which to perform GPS field measurements in Orlando and why?

4.4 GPS and Geocaching Web Resources

With some preliminary planning information available from Trimble GNSS Planning Online, you can get ready to head out with a GPS receiver. For some starting points for GPS usage, try the following:

1. Go to **http://www.geocaching.com**. This Website is home to several hundred thousand geocaches (small hidden items) at locations throughout the world and lists their coordinates so that you can use a GPS to track them down.

2. In order to access the features of Geocaching.com, you will have to create an account. The basic account is free—click on **Sign In**, then follow the steps to create your account. The Website may send you an email with a link to use to verify your account information. Once your account is ready, log in to it using the username and password you just created and you'll be moved to the main Web page.

3. On the main page of the Website, enter **32801** (one of Orlando's many zip codes) in the box that asks for a zip code. Several potential geocaches should appear on the next Web page.

4. You can also search for benchmarks (permanent survey markers set into the ground) by going to **http://www.geocaching.com/mark**. On this benchmark page, input **32801** for the postal code. Answer Questions 4.14 and 4.15.

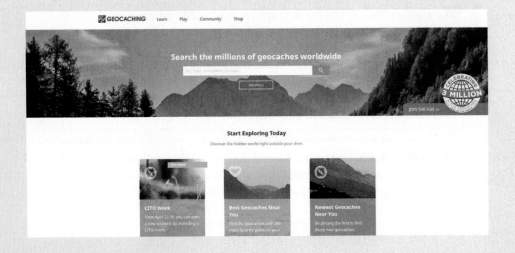

Question 4.14 How many benchmarks are available in the database for this postal code? How many of them are located within 1 mile or less of the Orlando 32801 postal code?

Question 4.15 What are the coordinates of the closest benchmark to the origin of the postal code?

4.5 GPS and Geocaching Applications

As we touched on back in the introduction, there's a lot more that can be done application-wise if you have access to a GPS receiver. For starters, you can find the positions of some nearby caches or benchmarks by visiting **http://www.geocaching.com** and then tracking them down using the receiver and your land navigation skills. However, there are a number of different ways that geocaching concepts can be adapted so that they can be explored in a classroom setting.

The first of these is based on material developed by Dr. Mandy Munro-Stasiuk and published in her article "Introducing Teachers and University Students to GPS Technology and Its Importance in Remote Sensing Through GPS Treasure Hunts" (see the online chapter references for the full citation of the article). Like geocaching, the coordinates of locations have been determined ahead of time by the instructor and given to the participants (such as university students or K–12 teachers attending a workshop) along with a GPS receiver. Having only sets of latitude/longitude or UTM coordinates, the participants break into groups and set out to find the items (such as statues, monuments, building entrances, or other objects around their campus or local area). The students are required to take a photo of each object they find, and all participants must be present in the photo (this usually results in some highly creative picture taking). The last twist is that the "treasure hunt" is a timed competition, with the team that returns first (and having found all the correct items) earning some sort of reward, such as extra credit or—at the very least—bragging rights. In this way, participants learn how to use the GPS receivers, how to tie coordinates to real-world locations, and they also get to improve their land navigation skills by reinforcing their familiarity with concepts such as how northings and eastings work in UTM.

A similar version of this GPS activity is utilized during the OhioView SATELLITES Teacher Institutes (see Chapter 15 for more information on this program). Again, participants break up into groups with GPS receivers, but this time each group is given a set of small trinkets (such as small toys or plastic animal figures) and instructed to hide each one, registering the coordinates of the hiding place with their GPS receiver and writing them on a sheet of paper. When the groups reconvene, the papers with the coordinates are switched with another group, and each group now has to find the other's hidden objects. In this way, each group sets up its own "geocaches" and then gets challenged to find another group's caches. Like before, this is a timed lab with a reward awaiting the team of participants that successfully finds all of the hidden items at the coordinates they've been given. This activity helps to reinforce not only GPS usage and land navigation, but also the ties between real-world locations and the coordinates being recorded and read by a GPS receiver. Both of these activities have proven highly successful in reinforcing these concepts in the participants involved.

However you get involved with geocaching, there are numerous GPS apps available for your smartphone or tablet that can be used in the field to obtain your coordinates and plot your location on a map. See pages 115–116 for some examples of free apps to check out.

Closing Time

This lab illustrated some initial planning concepts and some directions in which to take GPS fieldwork (as well as some caches and benchmarks that are out there waiting for you to find) to help demonstrate some of the concepts delineated in the chapter. The geocaching field exercises described in Section 4.5 can also be adapted for classroom use to help expand the computer portion of this lab with some GPS fieldwork as well.

Starting with Chapter 5, you'll begin to integrate some new concepts of geographic information systems into your geospatial repertoire.

Working with Digital Geospatial Data and GIS

5

Geographic Information Systems, Modeling the Real World, Vector Data and Raster Data, Attribute Data, Joining Tables, Metadata, Esri, ArcGIS, and QGIS

Once you've measured and collected your geospatial data, it's time to do something useful with it. The nature of geospatial data lends itself to such projects as analyzing where one location lies in relation to another or the appeals of one location versus another. When a new library is to be built, it should be placed at the optimal location to serve the greatest number of people without duplicating the function of other libraries in the area. When a company is deciding where to open a new retail store, they'll need information about the location—the city, the town, the neighborhood, the block, how many people live there and why. When a family is planning a vacation to Florida, they'll want to know the best way to get there and the best places to stay, not too far from the theme parks they're planning to visit. All of these concepts involve the nature of geospatial data and the ability to analyze or compare particular locations and their attributes. Geospatial technology is used to address these ideas and solve these types of problems (and many, many more). Whenever examination, manipulation, or analysis of geospatial data is involved (for instance, if you need to tie non-spatial data to a location or if you want to use this information to create a map of your analysis), **geographic information systems (GIS)** are essential to getting these tasks done.

GIS is a powerful tool for analyzing geospatial data (the "geographic information" of the title). This ability to utilize geospatial data explicitly is what separates GIS from other information systems or methods of analysis. For instance, a spreadsheet enables you to tabulate housing values and the amount of money paid for a house at a particular address, but even though you're using data that has something to do with a location, there's nothing in the spreadsheet that allows you to examine that data in a spatial context (such as translating the housing values into a map of where these houses are).

geographic information system (GIS) a computer-based set of hardware and software used to capture, analyze, manipulate, and visualize geospatial information

The ability to map the real-world locations of the houses, to examine trends in housing values in spatial terms, and to search for significant patterns or clusters in the geospatial data are what makes GIS unique. Using the coordinate and measurement concepts discussed in the previous chapters allows us to create, map, and analyze digital geospatial data in a wide variety of geospatial technology applications (see **Figure 5.1** and *Hands-On Application 5.1: Using GIS Online* for an example of GIS data used by a county government).

GIS is by its nature 100% computer-based—whether it involves software running on a desktop, laptop, mobile device, or served off the Web, the "system" being referred to is a computer system. The information handled by GIS is geospatial information—something that has direct ties to a location on Earth's surface. GIS can also utilize non-spatial data and link it directly to a location. For instance, if you have a set of data points that represent the locations of local car washes (that is, geospatial data) and a separate spreadsheet of data of how many cars use a particular car wash (that is, non-spatial data), you can link the usage data in the spreadsheet to the car wash location so that the location contains that data. From there, you can do analyses of local population demographics or road conditions about how they relate to the usage statistics of the car wash at that particular location. This wouldn't be possible without GIS being able to link non-spatial data to a location. Lastly, GIS can also perform multiple types of operations related to geospatial data—it can be used to analyze, capture, create, manipulate, store, or visualize all manner of geospatial information, not just simply design maps (though it can do that too).

FIGURE 5.1

The online GIS utility for Mahoning County, Ohio.

Hands-On Application 5.1

Using GIS Online

There are numerous online GIS resources for viewing and analyzing geospatial data. An example of a local GIS is from Mahoning County, Ohio. Open your Web browser and go to **http://gisapp. mahoningcountyoh.gov/mahgis**. In the default Layer Search Tools pop-up select the By Value tab. Use Properties as the Search layer, PARCEL ID or Owner Name as the Search alias, enter 29-042-0-002.00-0 (this is the parcel number for the Boardman Township Park), then click Search. The map will zoom to the parcel of the park and a table will appear at the bottom of the screen with attribute information about that parcel (such as the sale date, acreage, and land value).

Next, in the upper-right corner under the Layers icon (the globe's) pull-down menu, choose the option for Map Layers. Here, you can select which GIS layers are displayed on the map. Make sure there is a checkmark in the box of the most recent date for Aerial imagery to look at overhead aerial photography of the area. You can also choose the option for Zoning Data to overlay a zoning map on the aerial photos. With the Zoning Data layer turned on, you can click on a particular parcel to see how it has been zoned. You can use the mouse wheel or the options under the Navigation icon to zoom in and out of the view and get a closer look at the map and the surroundings of the park.

Expansion Questions

➤ From the tabular information provided about the park's parcel of land, how large is it in terms of acreage and what is the market value of the land?

➤ From examining the aerial photos and the zoning information, what types of land uses are adjacent to the park boundaries (commercial, residential, etc.)?

➤ Check your local state or county government's Websites to see if a similar kind of online GIS data viewer has been set up for your area (it may be part of your county assessor's Website or a separate GIS department for the state or county). If a GIS is available, what types of layers are available for you to view and interact with?

GIS isn't a development that just popped up over the last couple of years. Large-scale mapping efforts were done using computer-based cartography in the 1950s and 1960s by a number of different government agencies and other labs (including Harvard Laboratory). The term "GIS" first appeared in the early 1960s with the implementation of **CGIS** (the acronym for the Canadian Geographic Information System), which was designed to provide large-scale mapping of land use in Canada. The development of CGIS is attributed to the late Roger Tomlinson, who has become known as "the father of GIS."

GIS developments have continued over the years through the efforts of government agencies and the private sector, and these days GIS technology and concepts are extremely widespread (see *Hands-On Application 5.2: GIS Current Events Maps* for another example of using GIS concepts online).

CGIS the Canadian Geographic Information System—a large land inventory system developed in Canada, and the first system of this type to use the term "GIS"

How Does GIS Represent Real-World Items?

GIS has a multitude of applications, ranging from mapping natural gas pipelines to determining which parts of the natural landscape are under the greatest pressure to be developed. No matter the application, the data (such as the locations and lengths of pipelines or the area and extent of land cover) needs to be represented somehow. GIS provides a means of representing (or modeling) this data in a computer environment so that it can be easily analyzed or manipulated. A model is a kind of representation that simplifies or generalizes real-world information. For instance, if you were tasked to make a map of all of the stop signs in your city, you'd need to be able to represent the locations of all the signs on a map at the scale of the entire city. By laying out a map of the city and placing a dot at each sign location, you would have a good representation of them. The same holds true if you had to come up with a map of all the roads in your town. You could attempt to draw the roads at their correct width, but the roads' locations can be represented simply and accurately by lines drawn on the map.

Whenever you're considering how to model real-world phenomena with a GIS, you first have to decide on the nature of the items you want to represent. In the GIS, there are two ways of viewing the world. The first is called the **discrete object view**. In the discrete object view, the world is made up of a series of objects, each of which having a fixed location, or a fixed starting and stopping point, or some sort of fixed boundary. A street starts at one point and ends at another, or there is a definite boundary between your property and your neighbor's. Around a university campus, there will be numerous objects representing buildings, parking lots, stop signs, benches, roads, trees, and sidewalks. By viewing the world in this way, real-world items can be modeled in the GIS as a series of objects (such as a collection of lines to represent a road system, or a series of points to represent fire hydrant locations).

When adapting the discrete object view of the world to a data model, items in the real world are represented in the GIS by one of three objects (**Figure 5.2**):

➤ **Points**: These are zero-dimensional objects, a simple set of coordinate locations.

➤ **Lines**: These are one-dimensional objects, created from connecting starting and ending points (and any points in between that give the line its shape).

➤ **Polygons**: These are two-dimensional objects that form an area from a set of lines (or having an area defined by a line forming a boundary).

discrete object view a conceptualization of the world in which all reality can be represented by a series of separate objects

points zero-dimensional vector objects

lines one-dimensional vector objects

polygons two-dimensional vector objects

Hands-On Application 5.2

GIS Current Events Maps

There are a lot of GIS applications available online that are used for mapping and analysis of geospatial data. For some interactive examples, go to **www.esri.com/esri-news/maps**. The Website will show you different categories of Map Galleries related to current events, such as public information or public policy issues. These interactive Web maps and applications were created using GIS software from a company called Esri. We'll talk a lot more about Esri and its software platform, ArcGIS, later in this chapter, but for now we'll be examining some of its online GIS applications.

On the main page, select one of the categories of Web maps to examine and you'll be presented with several different maps and applications to view. When you open a Web map, you'll be placed into a GIS environment where you can pan around the map, zoom in and out, and access different features of the maps. Some Web maps will allow you to measure distances and areas and get information back (as a pop-up) about layers on the map, while others will use pre-made templates to convey geospatial and attribute information to you.

Expansion Questions

➤ What kinds of features are being mapped in the Web maps available online? (You may want to choose one of the available categories and investigate several available Web maps.)

➤ How are the Web maps able to convey non-spatial information to the user? (*Hint*: click on some of the features in a Web map itself.)

Points   Lines   Polygons

FIGURE 5.2

Discrete objects in GIS: basic points, lines, and polygons.

These three items are referred to as **vector objects**, as they make up the basis of the GIS **vector data model**. Basically, the vector data model is a means of representing and handling real-world spatial information as a series of vector objects—items are realized in the GIS as points, lines, or polygons. For instance, locations of fire hydrants, crime incidents, or trailheads can be represented as points; roads, streams, or power conduits are represented as lines; and land parcels, building footprints, and county boundaries are represented as polygons. When dealing with a much larger geographic area (for instance, a map of the entire United States), cities may be represented as points, interstates as lines, and state boundaries as polygons (**Figure 5.3**).

A great deal of GIS data has already been created (such as road networks, boundaries, utilities, etc.), but in order to update or develop your own data,

vector objects points, lines, and polygons that are used to model real-world phenomena using the vector data model

vector data model a conceptualization of the world that represents spatial data as a series of vector objects (points, lines, and polygons)

FIGURE 5.3

Points (city locations),
lines (interstates), and
polygons (state borders)
as represented in GIS.

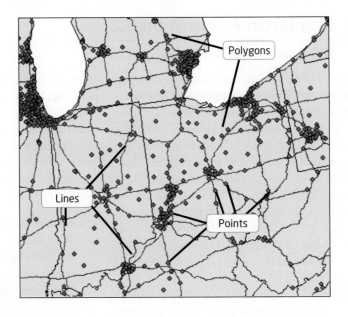

FIGURE 5.3

Points (city locations),
lines (interstates), and
polygons (state borders)
as represented in GIS.

digitizing the creation
of vector objects through
sketching or tracing
representations from a map
or image source

there are a few steps that need to be taken. **Digitizing** is a common way to create the points, lines, and polygons of vector data. With digitizing, you are in essence "tracing" or "sketching" over the top of areas on a map or other image to model features in that map or image in the GIS. For instance, if you have an aerial photograph of your house being shown on the screen of your GIS software, you could create a polygon by using the mouse to sketch the outline of your house and save it in the GIS as a representation of your house. If you sketch outlines of four of your neighbors' houses as well, you'll have several polygons stored in a single layer in the GIS. When you examine the data, you'll find you have five polygon entries in that data layer (your house and four others). Similarly, you could sketch each house's driveway as a line object by tracing the mouse over the appropriate place in the image, which would result in a second data layer that's comprised of five lines. With digitizing, each of your sketches can be translated into a vector object.

Digitizing is commonly done through a process referred to as "heads-up digitizing" or "on-screen digitizing," which is similar to the sketching just described. In heads-up digitizing, a map or other image (usually an aerial photograph or satellite image) is displayed on the screen as a backdrop. Decide if you'll be sketching points, lines, or polygons and create an empty data layer of the appropriate type (to hold the sketches you'll make). From there, you use the mouse to sketch over the objects, using as much detail as you see fit (**Figure 5.4**). Keep in mind that despite your best efforts, it may likely be impossible or unfeasible to sketch every detail in the image. For instance, when you're tracing a river, it may not be possible to capture each bend or curve by digitizing straight lines.

FIGURE 5.4 ───────○

An example of heads-up digitizing—tracing the boundary of the football field on Youngstown State University's (YSU's) campus to create a polygon object.

When objects are created in GIS, they can be set up with coordinates, but the GIS still needs to have knowledge of how these objects connect to each other and what relation each object has to the others. For instance, when two lines that represent crossing streets are digitized, the GIS needs to have some sort of information that an intersection should be placed where the two lines cross. Similarly, if you have digitized two residential parcels, the GIS would require some information that the two parcels are adjacent to one another. This notion of the GIS being able to understand how objects connect to one another independent of their coordinates is referred to as **topology**. With topology, geometric characteristics aren't changed, no matter how the dataset may be altered (by projecting or transforming the data). Topology establishes adjacency (how one polygon relates to another polygon, in that they share a common boundary), connectivity (how lines can intersect with one another), and containment (how locations are situated inside of a polygon boundary).

topology how vector objects relate to each other (in terms of their adjacency, connectivity, and containment) independently of the objects' coordinates

The layers of the National Hydrography Dataset (**NHD**) are examples of vector data applications in GIS. The NHD is a set of GIS data compiled and distributed by the USGS that maps out various bodies of water across the United States. Lakes, rivers, streams, canals, ponds, and coastlines are among the features accessible in GIS through NHD. These layers are represented by line objects (such as rivers) or polygon objects (such as lakes) with numerous attributes attached to each feature about these water features (**Figure 5.5**). NHD serves as the basis for water resources within The National Map (see Chapter 1) and can be freely downloaded from the Web for use in GIS.

NHD The National Hydrography Dataset—a collection of GIS data of water resources for the United States

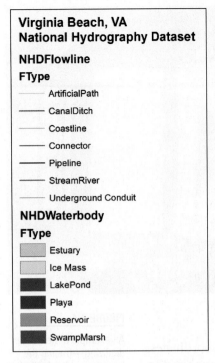

**Virginia Beach, VA
National Hydrography Dataset**

NHDFlowline
FType

ArtificialPath
CanalDitch
Coastline
Connector
Pipeline
StreamRiver
Underground Conduit

NHDWaterbody
FType

Estuary
Ice Mass
LakePond
Playa
Reservoir
SwampMarsh

FIGURE 5.5

Several different features from the National Hydrography Dataset for Virginia Beach, Virginia. *[Data from NHD/USGS]*

continuous field view a conceptualization of the world in which all items vary across Earth's surface as constant fields, and values are available at all locations along the field

There's also a second way of viewing the world, in which not everything has a fixed boundary or is an object. For instance, things such as temperature, atmospheric pressure, and elevation vary across Earth and are not best represented in GIS as a set of objects. Instead, they can be thought of as a surface that's made up of a near-infinite set of points. This way of viewing the world is called the **continuous field view**. This view implies that real-world phenomena continuously vary; rather than a set of objects, a surface filled with values is used to represent things. For instance, elevation gradually varies across a landscape from high elevations to low elevations filled with hills, valleys, and flatlands, and at every location we can measure the height of the land. Similarly, if you're standing at a ranger station in a park and preparing a rescue plan for stranded hikers, you'll want to know the distance from the station to every location in the park. Thus, you could have two layers in your GIS—one showing the elevation at every place and another showing the distance to the ranger station at each location.

How Can You Represent the Real World as Continuous Fields?

When these kinds of continuous fields are represented in GIS, the three vector objects (and thus, the vector data model) are often not the best way of representing data. Instead, a different conceptualization called the **raster data model** is usually used. In the raster data model, data is represented using a set of evenly distributed square **grid cells**, with each square cell representing the same area on Earth's surface (see **Figure 5.6** for the layout of the grid cells). For instance, to represent an entire section of the landscape, each grid cell might represent 10 square feet, or 30 square meters, or 1 square kilometer, depending on the grid cell size being used for the model. Regardless of the grid cell size, each cell has the same resolution (that is, the same grid dataset cannot mix cells of 30 square feet together with cells of 5 square feet). Also, each grid cell contains a single value representing the data being modeled. For instance, in an elevation raster, each grid cell could contain a value for the elevation at that area on the ground. Datasets such as land cover, soils, or elevation are all commonly represented using the raster data model.

The National Land Cover Database (**NLCD**) is an example of a raster dataset used in GIS. Developed by a consortium of United States agencies (including the USGS and the Environmental Protection Agency), NLCD provides 30-meter raster data covering the entire United States, wherein each grid cell is coded with a value that corresponds to a specific land cover type at that location. NLCD designations include categories such as "Open Water," "Developed, High Intensity," "Deciduous Forest," "Pasture/Hay," and "Woody Wetlands." NLCD provides a means of broad land-cover

raster data model a way of representing spatial data that utilizes a series of equally spaced and sized grid cells

grid cell a square unit, representing some real-world size, which contains a single value

NLCD the National Land Cover Database is a raster-based GIS dataset that maps the land-cover types for the entire United States at 30-meter resolution

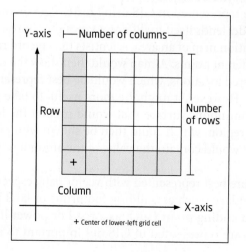

FIGURE 5.6

Grid cells of raster data model.

FIGURE 5.7

A section of the 2011 NLCD showing land cover in Columbiana County, Ohio, modeled with raster grid cells. *[Data from National Land Cover Database]*

classification at a state, regional, or national scale (**Figure 5.7**). NLCD products are available for land cover circa 1992, 2001, 2006, and 2011. A separate dataset is available that details the change in land-cover types between the 1992 and 2001 datasets (and datasets showing change between the other dates as well), enabling researchers not only to measure land-cover types, but also to see how the landscape is changing (such as where agricultural fields or forests are converting over to urban land uses). NLCD data is distributed free of charge via The National Map in a format that can be easily read or converted by GIS software. See *Hands-On Application 5.3: The National Map Advanced Viewer for* a Web application that will allow you to view NLCD data in relation to other GIS layers (such as the NHD).

Real-world items can be represented in GIS using raster data or vector data, although each data model lends itself to different types of features. For instance, to create a precipitation map of an area, scientists may record rainfall amounts at dozens of different points. A map would then show the rainfall levels only at these scattered locations, which would be best represented with points (vector objects). However, it might be more useful to take that data and create an entire continuous surface that would predict the level of rainfall at all areas in the region, which could then be shown with raster data in which every grid cell would contain the value of the amount of predicted rainfall.

However, some objects are best represented with vector data, especially those having a definite boundary (such as building footprints or land parcels) or a fixed starting and ending point (such as a road or power line). No matter how these items are represented in GIS, it's important to note

Hands-On Application 5.3

The National Map Advanced Viewer

We've already discussed how to download GIS data from The National Map back in Chapter 1 (see *Hands-On Application 1.3: The National Map* on page 11) but you can also use the features of The National Map to view different GIS layers (including both NHD or NLCD data) online rather than just downloading the data. To get started, go to **https://viewer.nationalmap.gov/advanced-viewer** to open The National Map Advanced Viewer. To begin, in the Search box, type: Pittsburgh, PA, and the Viewer will change to show you downtown Pittsburgh at the intersection of the three rivers shown on the basemap: the Ohio River, the Allegheny River, and the Monongahela River.

To examine various layers from The National Map on this basemap, choose the Layers icon (the one with the three squares stacked on each other) and a new pane will appear on the right-hand side of the screen showing available layers. From the options, place a checkmark in the Hydrography–Cached box (this will display the NHD features on the map). Next, place a checkmark for the option for

National Land Cover Database (NLCD) and expand the NLCD options. You'll see that by default the NLCD 2011 layer will appear but the past dates of NLCD are also available. If you click on the Legend icon (the one with the three symbols with vertical lines next to them) the pane will change to a legend describing what land-cover types each of the raster grid cell colors refer to. The other layers that can be downloaded from The National Map (including vector layers such as structures, transportation, boundaries, and watersheds) can be viewed here as well.

Expansion Questions

➤ For the Pittsburgh area, how do the vector NHD representations of the three main rivers match up with the raster NLCD representations of them? Are there any significant NHD areas not represented in NLCD?

➤ Search for your own local area and examine the NHD and NLCD layers for it. How do the vector NHD and raster NLCD match up with each other?

that both vector and raster data structures are only modeling the spatial features of real-world items—but GIS is also capable of handling all of the non-spatial data that goes along with them.

How Is Non-Spatial Data Handled by GIS?

Whether you're mapping schools ranked by average SAT scores or the locations of the top tourist destinations in the United States, all of this numeric data needs somehow to be represented in GIS. For instance (going back to the digitizing example), if you have two data layers (one containing polygons representing housing footprints and one containing lines representing the driveways), each would contain multiple objects. The GIS will have information about the spatial properties of the objects in that layer, but what it does not have is any other information about those objects (such as the tax-assessed value of the house, the house's address, the name of the owner, the number of people living in the house, the material the driveway is made of, or how old the driveway is). All of these **attributes** represent other non-spatial information associated with each of the objects. When representing attribute data in GIS, these values can take one of four forms: nominal, ordinal, interval, or ratio data.

Nominal data are values that represent some sort of unique identifier. Your Social Security number or telephone number would both be examples of nominal data—both of these values are unique to you. Names or descriptive information that are associated with a location would be nominal data, as they're unique values, the same as a value from a land-cover classification scheme. Also, the difference between numerical nominal values is not significant—you can't add your Social Security number to a friend's number and come up with a relative's number—in the same way you can't subtract your phone number from one friend's number and come up with another friend's phone number.

Ordinal data is used to represent a ranking system of data. If you have data that is placed in a hierarchy where one item is first, another is second, and another is third, that data is considered ordinal. For instance, if you have a map of the city and you're going to identify the locations of the homes of the most successful competitors in a local car-racing event, the points on the map will be tagged as to the location of the individual who took first place, the one who came in second, and so on. Ordinal data deals solely with the rankings themselves, not with the numbers associated with these ranks. For instance, the only values mapped for the car-race winners is their placement in the race, not their winning times, the cars' speeds, or any other values. The only thing ordinal data represents is a clear value of what item is first, which is second, and so on. No measurements can be made of how much better the winning driver's time was than the time of the driver who took second place, only that one driver was first and one driver was second.

Interval data is used when the difference between numbers is significant, but there is no fixed zero point. With interval data, the value of zero is just another number used on the scale. For instance, temperature measured in degrees Celsius would be considered interval data because values can fall below zero—the value of zero only represents the freezing point of water, not the bottom of the Celsius temperature scale. However, because there is no fixed zero point, we can make differences between values (for instance, if it was 15 degrees yesterday and 30 degrees today, we can say it was 15 degrees warmer), but dividing numbers wouldn't work (because a temperature of 30 degrees is not twice as warm as a temperature of 15 degrees).

Ratio data are values with a fixed and non-arbitrary zero point. For instance, a person's age or weight would be considered ratio data because a

attributes the non-spatial data that can be associated with a spatial location

nominal data a type of data that is a unique identifier of some kind—if numerical, the differences between numbers are not significant

ordinal data a type of data that refers solely to a ranking of some kind

interval data a type of numerical data in which the difference between numbers is significant, but there is no fixed non-arbitrary zero point associated with the data

ratio data a type of numerical data in which the difference between numbers is significant, but there is a fixed non-arbitrary zero point associated with the data person cannot be less than zero years in age or weigh less than zero pounds. With ratio data, the values can be meaningfully divided and subtracted. If we want to know how much time separated the car-racing winners, we could subtract the winning driver's time from the second-place driver's time and get the necessary data. Similarly, a textbook that costs $100 is twice as expensive as a textbook that costs $50 (as we divide one number by the other). Ratio data is used when comparing multiple sets of numbers and looking for distinctions between them. For instance, when mapping the locations of car wash centers, the data associated with the location is the number of cars that use the service. By comparing the values, we can see how much one car wash outsells the others, or calculate the profit generated by each location.

Each layer in the GIS has an associated **attribute table** that stores additional information about the features making up that layer. The attribute table is like a spreadsheet where objects are stored as **records** (rows) and the information associated with the records—the attributes—is stored as **fields** (columns). (See **Figure 5.8** for an example of an attribute table in conjunction with its GIS objects.) To use the housing example again, the houses' attribute table would consist of five records, each representing a house polygon. New fields could be created, having headings such as "Owner" and "Appraised Value" so that this type of descriptive, non-spatial data could be added for each of the records. These new attributes can be one of the four data types (for instance, "Owner" would be nominal data, while "Appraised Value" would be ratio data). In this way, non-spatial data is associated with a spatial location.

attribute table a spreadsheet-style form where the rows consist of individual objects and the columns are the attributes associated with those objects

records the rows of an attribute table

fields the columns of an attribute table

FIGURE 5.8

Attribute table associated with a GIS data layer.

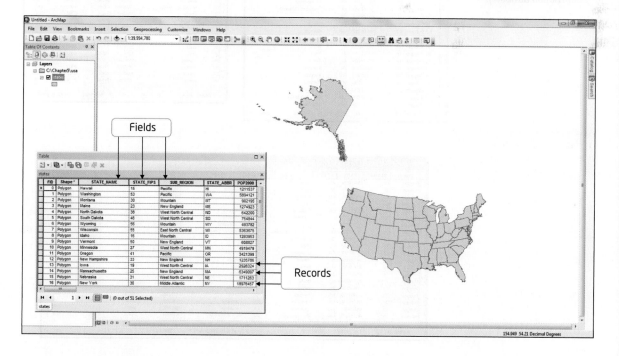

A raster attribute table can't be handled the same way as one for vector objects because a raster dataset is composed of multiple grid cells (for instance, a 30-meter resolution grid of a single county could take millions of cells to model). Rather than having separate records for each grid cell, raster data will often be set up in a table with each record featuring the value for a raster cell and an attribute showing the count of how many cells comprise the dataset. For instance, a county-land-use dataset may contain millions of cells, but only consist of seven values. Thus, the raster would have seven records (one for each value) and another field whose value would be a summation of how many cells have that value. New attributes would be defined by the raster values (for instance, the type of land use that value represents).

Besides creating new fields and populating them with attributes by hand, a **join** is another way GIS allows non-spatial data to be connected to spatial locations. This works by linking the information for records in one table to corresponding records in another table. This operation is managed by both tables having a field in common (this field is also referred to as a **key**). For instance, **Figure 5.9** shows two tables, the first from a shapefile's attribute table that has geospatial information about a set of points representing the states of the United States, while the second table has non-spatial

join a method of linking two (or more) tables together

key the field that two tables have to have in common with each other in order for the tables to be joined

FIGURE 5.9

A joining of two attribute tables, based on their common field (the state's abbreviation attribute).

Census information about housing stock for those states. Because both tables have a common field (that is, the state's abbreviation), a join can be performed on the basis of that key.

After the join, the non-spatial Census information from the second table will be associated with the geospatial information from the states' table. For instance, you can now select the record for Alaska and have access to all of its housing information because the tables have been related to each other. Joining tables is a simple method for linking sets of data together, and especially for linking non-spatial spreadsheet data to geospatial locations. With all of this other data available (and linked to a geospatial location), new maps can be made. For instance, you could now create maps of the United States showing any of the attributes (total number of houses, number of vacation homes, or percentage of the total houses that are vacation homes) linked to the geospatial features.

What Other Kind of Information Do You Need to Use GIS Data?

Once you have your geospatial and attribute data within the GIS, there's one more important piece of data you're going to need—descriptive information about your data. When creating data or obtaining it from another source, you should have access to a separate file (usually a "readme" text file or an XML document) that fully describes the dataset. Useful things to know would be information about the coordinate system, projection, and datum of the data; when the data was created; what sources were used to create it; how accurate the data is; and what each of the attributes represents. All of this kind of information (and more) is critical to understanding the data in order to use it in GIS properly. For instance, if you have to reproject the data, you're going to have to know what projection it was originally in; if you're going to be mapping certain attributes, then you'll have to know what each one of those strangely named fields in the attribute table actually represents. In GIS, this "data about your data" is referred to as **metadata**.

The Federal Geographic Data Committee (FGDC) has developed some standards for the content and format that metadata should have. These include:

metadata descriptive information about geospatial data

➤ Identification of the dataset (a description of the data)

➤ Data quality (information about the accuracy of the data)

➤ Spatial data organization information (how the data is represented—whether it's vector or raster, for instance)

➤ Spatial reference information (the coordinate system, datum, and projection used)

Thinking Critically with Geospatial Technology 5.1

What Happens When You Don't Have Metadata?

Say you're trying to develop some GIS resources for your local community, and you need data about the area's sewer system. You get the data in a format that your GIS software will read, but it doesn't come with any metadata—meaning you have no information on how the sewer data was created, its accuracy, what the various attributes indicate, when it was created, or the datum, coordinate system, or projection the data is in. How useful is this dataset for your GIS work? When metadata for a GIS layer is unavailable (or incomplete), what sorts of issues can occur when you're trying to work with that data?

> ➤ Entity and attribute information (what each of the attributes means)
>
> ➤ Distribution information (how you can obtain this data)
>
> ➤ Metadata reference information (how current the metadata is)
>
> ➤ Citation information (if you want to cite this data, how it should be done)
>
> ➤ Time period information (what dates the data covers)
>
> ➤ Contact information (whom to get in touch with for further information)

What Kinds of GIS Are Available?

There are many different types of GIS software packages available, everything from free lightweight "data viewers," to open source programs, to expensive, commercially available software packages. There are numerous companies producing a wide variety of GIS software. According to a 2011 GIS Salary Survey conducted by the Urban and Regional Information Systems Association (URISA), GIS products from the company Esri were found to be among the most popular and widely used.

Esri a key developer and industry leader of GIS products

Located in Redlands, California, **Esri** was founded by Jack and Laura Dangermond in 1969. By 1982, Esri had released the first of its "Arc" products, the initial version of Arc/Info, a key GIS software package. Esri's GIS software has gone through many changes and evolutions over the years to reach its current form: **ArcGIS**, an entire GIS software platform. The concept of a platform means multiple pieces are brought together—software on your desktop computer, cloud-based software, mobile apps, and more that all use the same common techniques and tie together. This is the best way to think of ArcGIS: a GIS platform that you can use in a variety of different ways, whether on a desktop computer, a Web browser, an app on a tablet, or a toolkit for developers.

ArcGIS Esri's GIS platform

ArcGIS Desktop is the part of the platform that runs on a personal computer. At the time of this writing, the most recent iteration of the software was ArcGIS Desktop 10.5. ArcGIS Desktop is available in three different varieties: ArcGIS Basic, ArcGIS Standard, and ArcGIS Advanced. These can best be thought of as different levels of the same software package—all of them are considered ArcGIS Desktop, but the Basic version has the smallest number of functions, the Standard version has the mid-range of functions, and the Advanced version has the largest number of functions.

ArcGIS Desktop the component of ArcGIS that runs on a desktop computer

One of the main components of ArcGIS Desktop is **ArcMap**, which is used for viewing and analyzing data (see **Figure 5.10**). Within ArcMap, GIS datasets can be added and treated as different map layers. ArcMap offers multiple tools for analysis and data manipulation and enable you to make maps from your data (things you'll be doing in the labs for Chapters 6 and 7). Within ArcMap, you'll find the **Catalog**, which is used for managing your data and moving it from place to place (such as from the hard drive to a flash drive).

ArcMap the component of ArcGIS Desktop used for viewing and analyzing data

Catalog the component of ArcMap used for managing data

The other part of ArcGIS Desktop is a new 64-bit stand-alone GIS program called **ArcGIS Pro**, which runs alongside ArcMap. ArcGIS Pro (see **Figure 5.11**) is the latest GIS from Esri, which features a new software interface, greater compatibility with Web-based GIS maps and data, 3D visualization, and new tools and enhancements to working with GIS.

ArcGIS Pro a new 64-bit stand-alone GIS program that comes as part of ArcGIS Desktop

ArcGIS Online is the cloud-based part of the ArcGIS platform. ArcGIS Online allows you to build Web maps and applications using only your Web browser as well as transfer your data and maps from ArcMap or ArcGIS Pro to the cloud. You'll work with ArcGIS Online's capabilities later, in Chapter 15. See *Hands-On Application 5.4: Esri News and Videos* for more information about Esri software and applications.

ArcGIS Online a cloud-based resource for creating and sharing GIS Web maps and applications

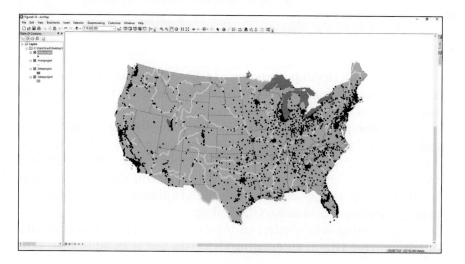

FIGURE 5.10

The basic layout of ArcMap.

FIGURE 5.11

The basic layout of Esri's ArcGIS Pro.

Hands-On Application 5.4

Esri News and Videos

Esri data and software products are used in numerous applications across the globe. To keep informed about new Esri updates, software, and real-world uses of GIS, Esri publishes a pair of print publications (and makes articles and content available free online) as well as an online magazine. Check the following sites for its publications and look for articles related to fields of interest to you:

1. ArcNews: **www.esri.com/esri-news/arcnews**

2. ArcUser: **www.esri.com/esri-news/arcuser**

3. ArcWatch: **www.esri.com/esri-news/arcwatch**

In addition, Esri also produces a series of short informative videos about GIS. They pack a lot of rich content into a short amount of time. Check them out at **http://www.esri.com/videos.**

Expansion Questions

➤ How is Esri software enabling current developments in environmental monitoring, national security, urban planning, and law enforcement?

➤ What new developments are currently underway in the GIS field?

Python a free computer programming language used for writing scripts in ArcGIS

Script a short piece of computer code

While ArcGIS Desktop is a powerful GIS program that contains a wide variety of tools and features, computer programming can be used to create customizations of the software or help automate tasks or perform analysis. The **Python** programming language is used for writing computer code (called a **script**) for working with ArcGIS (see **Figure 5.12**). Python (named after the Monty Python comedy team) is commonly used with ArcGIS and is simple to use for programming tasks. You could write short scripts within ArcGIS Desktop itself or use free tools found online to write your scripts separately then run them in ArcGIS Desktop.

Programming is becoming an increasingly important facet of GIS for customizing GIS to automating GIS tasks and is an additional skill

```
import arcpy

# input data is in NAD 1983 UTM Zone 11N coordinate system
input_features = r"C:/data/Redlands.shp"

# output data
output_feature_class = r"C:/data/Redlands_Project.shp"

# create a spatial reference object for the output coordinate system
out_coordinate_system = arcpy.SpatialReference('NAD 1983 StatePlane California V FIPS 0405 (US Feet)')

# run the tool
arcpy.Project_management(input_features, output_feature_class, out_coordinate_system)
```

FIGURE 5.12

An example of using Python in ArcMap to project a layer from one coordinate system to another.

set in demand by employers in the geospatial world. While Python is used regularly with ArcGIS Desktop applications, there are a variety of other programming languages that play a part in working with GIS. For instance, Javascript is used with many web-based geospatial applications, along with HTML 5 and CSS, while SQL is used for programming with databases.

With the proliferation of numerous Esri products on the market and in use today, it's no surprise that a lot of the available GIS data is in an Esri data format. ArcGIS can also import data from other formats into an Esri format to use. For example, CAD (computer-aided design) data produced by programs such as AutoCAD can be imported into ArcGIS to use alongside other layers. Raster data is widely available in (or can be easily imported into) Esri Grid format, which can be directly read by Esri software. Current vector data can usually be found in one of the following two different varieties: shapefile or geodatabase.

A common GIS file type is the **shapefile**. It can hold only one type of vector object, thus there are point shapefiles, line shapefiles, and polygon shapefiles. Despite the name, a shapefile actually consists of multiple files that have to be present together to represent the data properly. The files all start with the same prefix (that is, if the shapefile is called "roads," then all the files that it consists of will be called "roads") but will have different extensions, including .shp, .shx, .dbf, and others. When copying a shapefile, you need to move all files with the same prefix to their new location for the shapefile to function properly.

A **geodatabase** is the format developed for ArcGIS Desktop. It consists of a single item that contains all of the geospatial information for a dataset. Its object-oriented structure is set up so that multiple layers can be stored in a single geodatabase, each layer being its own **feature class**. For example, a geodatabase called "Orange County" could store the county boundaries as a polygon feature class, highways in the county as a line feature class, and the city streets as another line feature class. Geodatabases can also contain **feature datasets**, or subgroups of feature classes organized together. For instance, in the "Orange County" geodatabase, if you have multiple feature classes, each highlighting different road types and planning features, you could have a feature dataset called "Transportation" to group all of

shapefile a series of files (with extensions such as .shp, .shx, and .dbf) that make up one vector data layer

geodatabase a single item that can contain multiple layers, each as its own feature class

feature class a single data layer in a geodatabase

feature dataset a grouping of multiple feature classes together within a geodatabase

Geodatabase

Feature Dataset

Feature Classes Within Feature Dataset

Feature Classes

Orange County.gdb
- Transportation
  - Highways
  - Railroads
  - Streets
- Boundary
- FireHydrants
- Parcels

the transportation-related features classes together all under one heading (see **Figure 5.13**). In addition, geodatabases can also store other data used in GIS, such as tables and rasters, and they can have GIS data in other formats (such as shapefiles or CAD data) imported into them. In this way, a geodatabase serves to organize your GIS data and hold all of the various layers you're using in a project together in one place.

Geodatabases come in two varieties: the first is a **file geodatabase**, which is stored as a folder or directory filled with files. A file geodatabase is essentially unlimited storage—each feature class can hold a maximum of 2 TB of data, but there is no limit on how many feature classes a file geodatabase can contain. Second is a **personal geodatabase**, which is stored as a single file that can hold a maximum of 2 GB of data. It is also the same file format (and thus directly compatible) as Microsoft Access.

Esri isn't the only company making GIS software—there are several other software packages and GIS companies providing a wide range of products that are used throughout the industry. No list of such things can hope to be comprehensive, but here are some of the more notable commercial GIS software products (and their vendors):

➤ Geomedia (from Intergraph/Hexagon Geospatial)

➤ Manifold System (from Manifold)

➤ MapInfo Pro (from Pitney Bowes Business Insight)

➤ Maptitude (from Caliper)

➤ TerrSet (from Clark Labs)

There are also several open source GIS products available for use and download via the Internet. An example is GRASS (Geographic Resources Analysis Support System), a long-running free (and open source) GIS software that can handle vector, raster, and image data with many different types of analysis. **QGIS** is another prominent open source GIS program that you'll have the opportunity to use in Geospatial Lab Application 5.1. QGIS is freely available for download, can use many types of GIS data

file geodatabase a single folder that can hold multiple files, with nearly unlimited storage space

personal geodatabase a single file that can contain a maximum of 2 GB of data

QGIS an open source GIS program

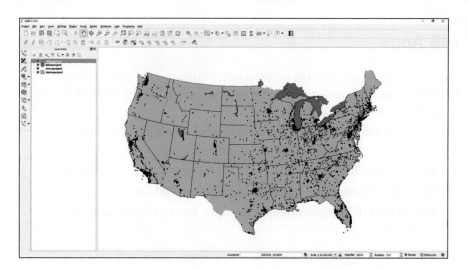

FIGURE 5.14

The basic layout of QGIS.

formats (including file geodatabases, shapefiles, and rasters), and has a layout similar to that of ArcMap (see **Figure 5.14**). QGIS features multiple types of analytical capabilities as well as mapping and data creation tools.

Chapter Wrap-Up

No matter the provider, the software, or the data representation format, GIS has a wide variety of uses, ranging from public utility mapping to law enforcement analysis, fire tracking, landscape planning, and many, many more. The next chapter will examine a number of different types of GIS analysis and address how to take this geospatial data and start doing things with it. Check out GIS Apps for some downloadable apps that use GIS concepts (including one for ArcGIS). Also take a look at GIS in Social Media for information from some blogs, Facebook pages, Twitter accounts, and Esri's YouTube channel.

The lab for this chapter provides an introduction to GIS and some basic handling of geospatial data. Two versions of the lab have been designed—the first one, *Geospatial Lab Application 5.1: GIS Introduction: QGIS Version* uses the free QGIS software package, and the second one, *Geospatial Lab Application 5.2: GIS Introduction: ArcGIS Pro Version* uses ArcGIS Pro. A third version of the Lab, which uses ArcMap, *Geospatial Lab Application 5.3: GIS Introduction: ArcMap Version,* is available online. All labs use the same data and concepts, but implement them differently depending on which software package you have available.

Important note: The references for this chapter are part of the online companion for this book and can be found at **www.macmillanlearning .com/catalog/shellitoigt4e.**

GIS Apps

Here's a sampling of available representative GIS apps for your smartphone or tablet. Note that some apps are for Android, some are for Apple iOS, and some may be available for both.

> **Collector for ArcGIS:** An app designed for obtaining field data as a geodatabase and then mapping via ArcGIS Online on a mobile device
> **iGIS:** An app for working with GIS (and shapefiles) on a mobile device
> **QGIS Experimental:** A version of QGIS designed for mobile devices
> **QField for QGIS Experimental:** An open source GIS application for mobile devices

GIS in Social Media

Here's a sampling of some representative Facebook, Twitter, and Instagram accounts, along with some YouTube videos and blogs related to this chapter's concepts.

On Facebook, check out the following:

> **Esri:**
>> **www.facebook.com/esrigis**
> **GIS users:**
>> **www.facebook.com/GISuser**
> **QGIS:**
>> **www.facebook.com/pages/QGIS-Quantum-GIS-/298112000 235096**

Become a Twitter follower of:

> **ArcGIS Pro:** @ArcGISPro
> **Esri:** @Esri
> **Esri Apps:** @EsriApps
> **Esri Globe:** @EsriGlobe
> **GIS Geospatial News:** @gisuser
> **Python Software Foundation:** @ThePSF
> **QGIS:** @qgis
> **What is GIS:** @GISdotcom

Become an Instagram follower of:

➤ **Esri:** @esrigram

You Tube **On YouTube, watch the following video:**

➤ **Esri TV** (Esri's YouTube channel featuring numerous videos of ArcGIS applications and resources): **www.youtube.com/user/esritv**

For further up-to-date info, read up on these blogs:

➤ **AnyGeo Blog** (a blog about GIS, maps, and location-based technologies): **http://blog.gisuser.com**

➤ **ArcGIS blog: http://blogs.esri.com/esri/arcgis**

➤ **ArcGIS blog—ArcGIS Pro Category: https://blogs.esri.com/esri/ arcgis/category/arcgis-pro**

➤ **ArcGIS blog—Python Category** (blog posts concerning using Python and GIS): **https://blogs.esri.com/esri/arcgis/category/subject-python**

➤ **Esri Insider** (a blog about multiple geospatial and Esri software applications): **http://blogs.esri.com/esri/esri-insider**

➤ **QGIS Planet** (a blog about QGIS developments and applications): **http://plugins.qgis.org/planet**

Key Terms

ArcGIS (p. 144)
ArcGIS Desktop (p. 145)
ArcGIS Online (p. 145)
ArcGIS Pro (p. 145)
ArcMap (p. 145)
attribute table (p. 141)
attributes (p. 140)
Catalog (p. 145)
CGIS (p. 131)
continuous field view (p. 136)
digitizing (p. 134)
discrete object view (p. 132)
Esri (p. 144)

feature class (p. 147)
feature dataset (p. 147)
fields (p. 141)
file geodatabase (p. 148)
geodatabase (p. 147)
geographic information systems (p. 129)
grid cells (p. 137)
interval data (p. 140)
join (p. 142)
key (p. 142)
lines (p. 132)
metadata (p. 143)

GIS Introduction: QGIS Version

This lab will introduce you to some of the basic features of GIS. You will be using a free open source program to navigate a GIS environment and begin working with geospatial data. The labs in Chapters 6 and 7 will utilize several more GIS features; the aim of this lab is to familiarize you with the basic GIS functions of the software. This lab uses the free QGIS software package.

Objectives

The goals for you to take away from this lab are:

➤ To familiarize yourself with the QGIS software environment, including basic navigation and tool use with both the Map and Browser

➤ To examine characteristics of geospatial data, such as their coordinate system, datum, and projection information

➤ To familiarize yourself with adding data and manipulating data layer properties, including changing the symbology and appearance of geospatial data

➤ To familiarize yourself with data attribute tables in QGIS

➤ To join a table of non-spatial data to a layer's attribute table in QGIS

➤ To make measurements between objects in QGIS

Using Geospatial Technologies

The concepts you'll be working with in this lab are used in a variety of real-world applications, including:

➤ Public works, where GIS is used for mapping features of the infrastructure such as sewer lines, water mains, and streetlight locations

➤ Archeology, where GIS is used for mapping boundaries and the spatial characteristics of dig sites, as well as the locations of artifacts found there

Obtaining Software

The version of QGIS used in this lab is 2.18, and available for free download at **http://qgis.org/downloads**.

Important note: Software and online resources can change fast. This lab was designed with the most recently available version of the software at the time of writing. However, if the software or Websites have significantly changed between then and now, an updated version of this lab (using the newest versions) will be available online at **www.macmillanlearning.com/catalog/shellitoigt4e.**

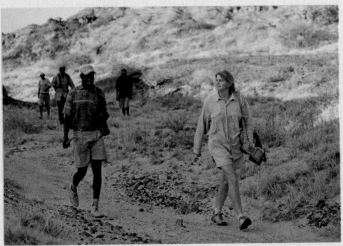

[Kenneth Garrett/National Geographic/Getty Images]

Lab Data

Copy the folder **Chapter5QGIS**—it contains a CSV file called CAPop which has population information as well as a folder called "CalData" which is the file geodatabase (see page 148) you'll be using in this lab. A file geodatabase is a single folder that contains many individual layers—if you look at the contents of the CalData folder in Windows it will contain numerous computer files that look meaningless—but to the GIS software these multiple files make up the GIS data layers we will be using. This lab's data was downloaded free of charge from The National Map (see page 11) and the population information was taken from the U.S. Census Bureau. The GIS layers in the geodatabase have already been projected for you to the U.S. National Atlas Equal Area projection.

Localizing This Lab

The data used in this lab is a variety of features of attributes of the state of California drawn from The National Map and the U.S. Census Bureau. However, you can easily work with the same data from your own state as The National Map makes the same layers available for all 50 states. From The National Map, download the Boundaries and Transportation geo-databases for your own state, which contain the layers used for California in this lab. If you are in, for instance, Michigan, you can download county boundaries, the structures layer, the trails, railroads, and airports for Michigan from The National Map. Similarly, the U.S. Census Bureau has population values for each county, which can be placed into a CSV table like the one of California used in this lab (to make a CSV table, open the table in Microsoft Excel and then use Save As to save the file as CSV Comma Delimited format).

5.1 An Introduction to QGIS

1. Start **QGIS** and it will open in its initial mode. The left-hand column (where you'll see the word Layers) is referred to as the Map Legend—it's where you'll have a list of available data layers. The blank screen that takes up most of the interface is the Map View, where data will be displayed.

Important note: If you don't see the Map Legend on the screen (or if other panels are open where the Map Legend should be), you can control which panels are available to view. From the **View** pull-down menu, select **Panels**, and then you can place an x in whichever panels you want to see on the screen. To begin, make sure that the only panel with an x next to its name is **Layers Panel**.

Important note: Before you begin adding and viewing data, the first thing to do is examine the data you have available to you. To do so in QGIS, you'll use the QGIS Browser—a utility designed to enable you to organize and manage GIS data.

2. Right click on the QGIS toolbar and a new menu of options will appear. From this menu, place a checkmark in the **Browser Panel** option. The Browser dialog box will open in a new window above the

Map Legend. Double-click on **Browser Panel** and the dialog box will detach itself and become a new floating window that you can move and resize.

3. The Browser can be used to manage your data and to get information about it. In the Browser tree (the section going down the left-hand side of the dialog box), navigate to the **C:** drive, and open the **Chapter5QGIS** folder. Verify that you have the CalData.gdb folder (this is the file geodatabase) and the CAPop.csv file (this is the table with population data).

Important note: This lab assumes that the Chapter5QGIS folder is in this location on the C: drive. If you have copied the data to another location (or another drive), navigate to that one instead.

4. Close the **Browser** window.

5.2 Adding Data to QGIS and Working with the Map Legend

1. Back in the main window of QGIS, by default there should be a column of buttons running down the left-hand side of the screen, to the left of the Map Legend. Click on the **button** icon showing four points connected in a "V" shape with a plus sign to start adding the data you've examined in the Browser to the map. This button is the tool to Add Vector Layers to QGIS (for this lab you'll use this option because all of your shapefiles represent vector data).

2. In the **Add Vector Layer** dialog box, click on the radio button for **Directory**, then from the **Type** pull-down menu choose **OpenFileGDB**. This will allow you to choose individual files from within the file geodatabase.

3. Next, select **Browse** and navigate to the **C:\Chapter5QGIS folder**. In that folder click once on the **CalData** folder and then click on **Select Folder**.

4. Next, click on **Open**. You'll see a list of the different layers that are within the file geodatabase and from which you can choose.

5. To begin, choose **CalBounds** and click **OK**.

6. You'll now see the CalBounds feature class listed in the Map Legend, and its content (a set of polygons displaying each of the counties in California) will be displayed in the Map View.

7. In the Map Legend, you'll see an "x" in the box next to the CalBounds feature class. When the checkmark is displayed, the layer will be shown in the Map View, and when the checkmark is removed (by clicking on it), the layer will not be displayed.

8. Now add two more layers from the CalData geodatabase: **CalAirports** (a point layer), and **CalTrails** (a line layer). All three of your layers will now be in the Map Legend.

9. You can manipulate the "drawing order" of items in the Map Legend by grabbing a layer with the mouse and moving it up or down in the Map Legend. Whatever layer is at the bottom is drawn first, and the layers **above** it in the Map Legend are drawn on top of it. Thus, if you move the CalBounds layer to the top of the Map Legend, the other layers will not be visible in the Map View because the CalBounds' polygons are being drawn over them.

10. Next, we'll get some basic information about your GIS layers. Right-click on the **CalBounds** layer and choose the option for **Properties**. In the Properties dialog box, select the Metadata tab. Information about your layer (also known as the "data about the data" discussed in Chapter 5) will be shown. At the bottom of the dialog box will be a separate box called Properties—scrolling through it will give a variety of information including information about the number of features in the layer. If each feature in the CalBounds layer represents a county in California then this is the number of counties in the dataset.

11. Do the same for the CalAirports and CalTrails layers and examine their information.

> **Question 5.1** How many airports are being represented as points in the CalAirports dataset? (Remember, each feature is a separate polygon.) How many trails are being represented as lines in the CalTrails dataset?

5.3 Symbology of Features in QGIS

1. You'll also notice that the symbology generated for each of the three objects is simple—points, lines, and polygons have been assigned a random color. You can significantly alter the appearance of the objects in the Map View to customize your maps.

2. Right-click on the **CalBounds** layer and select **Properties**.

3. In the **Properties** dialog box, select the **Style** tab. Here, you can alter the appearance of the states by changing their Style and Color, and then elements of how the polygons are outlined.

4. You can select one of the suggested appearances for your polygons, and by first clicking on **Fill**, and then clicking on the solid color bar beside the **Color** pull-down menu you can select a different color for your layer. Clicking once on **Simple fill** will bring up more options for

adjusting the color, border color, border thickness, and so on. These options allow you to alter the CalBounds layer into something more appealing to you. Once you've selected new colors click **Apply** to make the changes, and then click **OK** to close the dialog box.

5. Also change the appearances for the CalAirports and CalTrails feature classes. Note that you can also alter the points for the airports into shapes such as stars, triangles, or a default airport symbol, as well as changing the size and color of the airports. Several different styles are available for the lines of the trails as well.

5.4 Obtaining Projection Information

1. Chapter 2 discussed the importance of projections, coordinate systems, and datums. All of the information associated with these concepts can be accessed using QGIS.

2. Right-click on the **CalBounds** shapefile and select **Properties.** Click on the **Metadata** tab in the dialog box. At the bottom of the dialog box will be a separate box called Properties. Scrolling through it will give a variety of information including some shorthand about the Layer Spatial Reference System. Carefully read through the abbreviated information to find what units of measurement are being used for the CalBounds layer.

3. Click **OK** to exit the CalBounds Properties dialog box.

> **Question 5.2** What units of measurement (feet, meters, degrees, etc.) are being used in the CalTrails dataset?

> **Question 5.3** What units of measurement (feet, meters, degrees, etc.) are being used in the CalAirports dataset?

4. QGIS also allows you to specify the projection information used by the entire project. Because your layers are in the U.S. National Atlas Equal Area projection, you'll want to tell QGIS to use this projection information while you're using this data in this lab. To do this, select the **Project** pull-down menu, and then choose **Project Properties**.

5. In the **Project Properties** dialog box, select the **CRS** tab (this stands for Coordinate Reference System).

6. Place a checkmark in the box labeled **Enable "on the fly" CRS transformation OTF.**

7. In the filter box, type **U.S. National Atlas**—of the hundreds of supported projections, this will limit your search to find the U.S. National Atlas Equal Area projection.

8. In the bottom panel, select the **U.S. National Atlas Equal Area** option.

> **Question 5.4** To which "family" of projections does the U.S. National Atlas Equal Area projection belong?

9. Click **Apply** to make the changes, and then click **OK** to close the dialog box.

5.5 Examining Attribute Tables

1. Each of your three layers is represented by points (airports), lines (trails), and polygons (the counties), but each layer also has a corresponding set of attributes that accompanies their spatial features. As noted in the chapter, these non-spatial attributes are stored in an attribute table for each layer. To get started working with this non-spatial data, in the Layers Panel, right-click on the CalBounds layer, then choose **Open Attribute Table**.

2. A new window with the CalBounds attribute table will open. Each of the rows of the table is a record representing an individual feature (in this case, a single polygon of a California county) while the columns of the table are the fields representing the attributes for each of those records. By perusing the attribute table you can find information about a particular record. However, with 58 records it could take awhile to find some specific information. One way to organize the table is to sort it—scroll across the CalBounds attribute table until you find the field called County_Name. Click once on the County_Name table header and you'll see that the table will sort itself into alphabetical order by the name of the county. Click on the County_Name table header a second time and you'll see that the table sorts itself into reverse alphabetical order. Click it a third time to return the sorting of the table to alphabetical order.

3. With the table sorted, scroll through it to find the record with the County_Name attribute of "San Diego." Each county has a FIPS code assigned to it—this is a numeric value that is a government designation to uniquely identify things such as states, counties, and townships. This FIPS code information for each California county is located in this attribute table.

> **Question 5.5** What is the county FIPS code for San Diego County?

5.6 Joining a Table to an Attribute Table

1. While there are a lot of records in the CalBounds attribute table, QGIS will allow you to add further attributes to them. This can be done by

taking another table that contains additional non-spatial information and joining it to the CalBounds attribute table.

2. To begin, the first thing you'll need to do is add a table of non-spatial data to QGIS with which to work.

3. From the initial set of buttons on the left-hand side (from which you added the initial vector data), choose the button for **Add Delimited Text Layer.**

4. In the Create a Layer from a Delimited Text File dialog box, first click **Browse** and navigate to the **C:\Chapter5QGIS** folder and choose the **CAPop.csv** file.

5. Next, under **File Format**, select the radio button for **CSV** (comma separated values).

6. Next, under Geometry definition, select the radio button for **No geometry** (attribute only table) because this is only a table of values, not something that will be a spatial layer.

7. Leave the other settings as their defaults. At the bottom of the dialog box you'll see a preview of what your added table will look like. If everything looks okay, click **OK**.

8. A new item called CAPop will be added to the Layers Panel—this is the table containing a new set of attributes. Right-click on **CAPop** and choose **Open Attribute Table**. You'll see the new non-spatial table open in QGIS—it contains a field with the county's FIPS code, the name of that county, and the population of that county. In order to join two tables together, they need to contain the same field—in this case, the CAPop table contains a field called N that has the name of the county, while your CalBounds layer's attribute table contains a field called County_Name that has the same information. Because these two tables (the spatial data from CalBounds and the non-spatial data from CAPop) have a field in common (i.e., they both have a field containing the name of each county), they can be joined together.

9. To perform the join, right-click on the layer you want to join attributes to (in this case, we want to join information to the **CalBounds** layer), and select **Properties**, then select the **Joins** tab.

10. At the bottom of the Joins dialog box, press the **green "plus" button** which indicates you want to join a table to the layer with which you're working.

11. In the Add vector join dialog, under Join layer select **CAPop** (this is the table containing the information you want to join to the spatial layer).

12. Under Join field, select **N** (this is what the county name field is called in the CAPop table).

13. Under Target field, select **County_Name** (this is what the county name field is called in the CalBounds layer).

14. Put an "x" in the **Choose which fields are joined** box, then put an "x" next to the **CAFIPS, Pop2010,** and **N** options. This will join the data from the Pop2010 field from CAPop to CalBounds.

15. Put an "x" in the **Custom field name prefix** box. This means that any of the joined fields will start with the prefix CAPop_ so that you can keep them separate.

16. Leave the other options alone and click **OK** to perform the join. Then back in the Layer Properties dialog, click **OK** to close it.

17. Open the CalBounds layer's attribute table and scroll across to the far right. You'll now see the population values from the CAPop table joined for each record in the attribute table (i.e., each county now has a value for population joined to it).

18. Now that CAPop_Pop2010 is a field in the attribute table, you can sort it from highest to lowest populations (or vice versa) by clicking on the field header, just like you can do to sort any field (in the same way you sorted the counties alphabetically by name). Keeping this ability in mind, answer the following questions.

> **Question 5.6** What is the 2010 population of Los Angeles county?

> **Question 5.7** What is the 2010 population of the California county with FIPS code 065?

> **Question 5.8** What county in California has the lowest 2010 population and what is that population value?

> **Question 5.9** What county in California has the third highest 2010 population and what is that population value?

5.7 Navigating the Map View

1. We'll now just focus on one area of California—San Diego County. In the CalBounds attribute table, scroll through until you find San Diego County, then click once on the header of the record on the far-left of the attribute table.

2. Back in the View, you'll see San Diego County now highlighted in yellow.
3. QGIS provides a number of tools for navigating around the data layers in the View. Clicking on the **magnifying glass with the four arrows** icon will zoom the Map View to the full extent of all the layers. It's good to use if you've zoomed too far in or out in the Map View or need to restart.
4. The other magnifying glass icons allow you to zoom in (the **plus** icon) or out (the **minus** icon). You can zoom by clicking in the window or clicking and dragging a box around the area you want to zoom into.
5. The **hand** icon is the **Pan** tool that allows you to "grab" the Map View by clicking on it and dragging the map around the screen for navigation.

6. Use the **Zoom** and **Pan** tools to center the Map View on San Diego County so that you're able to see the entirety of the state and its airports and trails.

7. With San Diego County found, return to the CalBounds attribute table and hold down the **Ctrl** key on the keyboard, then click the header record for the county again to remove the blue highlighting, then close the attribute table.

5.8 Interactively Obtaining Information

1. Even with the Map View centered on San Diego County, there are an awful lot of point symbols there, representing the airports. We're going to want to identify and use only a couple of them. QGIS has a searching tool that allows you to locate specific objects from a dataset.

2. Right-click on the **CalAirports** layer and select **Open Attribute Table.** The attribute table for the layer will appear as a new window.

3. What we'll first want to do is find out which one of these 1163 records represents San Diego International airport. To begin searching the attribute table, click on the button at the bottom that says **Show All Features** and choose the option for **Field Filter**. A list of all of the attributes will appear; choose the one called **Name**.

4. You will now be able to search through the attribute table on the basis of the Name field. At the bottom of the dialog box, type **San Diego International**.

5. Click on the **Enter key** on the keyboard. The record where the Name attribute is "San Diego International" will appear as the only record in the attribute table. Click on the record header next to the Name field (the tab should be labeled 269) and you'll see the record highlighted in dark blue, just like when you selected San Diego County previously. This indicates that you've "selected" the record.

6. Drag the dialog box out of the way so that you can see both it and the Map View at the same time.

7. You'll also see that the point symbol for the San Diego International airport has changed to a yellow color. This means it has been "selected" by QGIS—any actions performed on the CalAirports layer will only affect the selected items, not all of the CalAirports.

However, all that you've done so far is locate and select an object. To obtain information about it, you can use the **Identify Features** tool (the icon on the toolbar is the white "i" and a white cursor in a blue circle). Note that Identify Features will only work with the layer that is selected in the Legend, so make sure that **Cal Airports** (i.e., the layer in which you want to identify items) is highlighted in the Legend before choosing the tool.

8. Select the **Identify Features** tool, and then click on the point representing San Diego International airport (zooming in as necessary). A new panel will appear called **Identify Results** (you may have to grab the top of the panel and drag it up to expand it so you can see what is inside the panel), listing all of the attributes for the CalAirports layer. By clicking on the **View feature form** button, you'll be able to see all of the field attribute information that goes along with the record/point representing San Diego International airport.

> **Question 5.10** What is the 3-letter FAA Airport Classification Code for San Diego International airport?

9. Close the **Identify Results** box after you've answered the question.

5.9 Labeling Features

Rather than dealing with several points on a map and trying to remember which airport is which, it's easier to simply label each airport so that its name appears in the Map View. QGIS gives you the ability to do this by creating a label for each record and allowing you to choose the field with which to create the label.

1. Right-click on the **CalAirports** layer in the Map Legend and select **Properties**.

2. Click on the **Labels** tab.

3. From the pull-down menu at the top of the dialog box, choose **Show labels for this layer**.

4. From the pull-down menu next to **Label with**, choose **Name**.

5. For now, accept all the defaults and click **Apply** and then **OK**. Back in the Map View, you'll see that labels of all the airport names have been created and placed near each point.

6. Change around any of the options—font size, color, label placement, and even the angle—to set up the labels however you feel is best for examining the map data.

5.10 Measurements on a Map

With all the airports now labeled, it's easier to keep track of all of them. With this in mind, your next task is to make a series of measurements between points to see what the Euclidian (straight line) distance is between airports around San Diego.

1. Zoom in tightly so that your Map View contains the airports San Diego International and North Island Naval Air Station.

2. Select the **Measure Line** tool from the toolbar—it's the one that resembles a gray ruler with a line positioned over the top.

3. In the Measure dialog box, choose **miles** from the pull-down menu so that your measurements will be computed in miles rather than the default of meters.

4. You'll see that the cursor has turned into a crosshairs. Place the crosshairs on the point representing San Diego International airport and left-click the mouse. Drag the crosshairs south and west to the point representing North Island Naval Air Station and click the mouse there. The distance measurement will appear in the Measure box. If you click on the Info text in the dialog box, you'll see that you are measuring "ellipsoidal" distance, referred to as the "surface" distance, as measured on a sphere. Note that each line of measurement you make with the mouse counts as an individual "segment" and the value listed in the box for "Total" is the sum of all segments.

> **Question 5.11** What is the (ellipsoidal) distance from San Diego International airport to North Island Naval Air Station?

5. You can clear all of the lines of measurement by clicking on **New** in the **Measure** box.

6. Zoom out a bit so that you can see some of the other airports in the region, particularly Montgomery Field and Brown Field Municipal Airport.

> **Question 5.12** What is the total (ellipsoidal) distance in miles from San Diego International airport to Montgomery Field, then from Montgomery Field to Brown Field Municipal Airport, and then finally from Brown Field Municipal Airport back to San Diego International airport?

5.11 Saving Your Work (and Working on It Later)

When you're using QGIS, you can save your work at any time and return to it. When work is saved in QGIS, a "QGS" file is written to disk. Later, you can re-open this file to pick up your work where you left off.

1. Saving to a QGS file is done by selecting the **Project** pull-down menu, then selecting **Save**.

2. Files can be re-opened by choosing **Open** from the **Project** pull-down menu.

3. Exit QGIS by selecting **Exit QGIS** from the **Project** pull-down menu.

Closing Time

This lab was pretty basic, but it served to introduce you to how QGIS operates and how GIS data can be examined and manipulated. You'll be using either QGIS or ArcGIS in the next two labs, so the goal of this lab was to get the fundamentals of the software down. The lab in Chapter 6 takes this GIS data and starts to do spatial analysis with it, while the lab in Chapter 7 will have you starting to make print-quality maps from the data.

5.2 Geospatial Lab Application

GIS Introduction: ArcGIS Pro Version

This lab will introduce you to some of the basic features of GIS. You will be using one of Esri's GIS programs to navigate a GIS environment and begin working with geospatial data. The labs in Chapters 6 and 7 will utilize several more GIS features; the aim of this lab is to familiarize you with the basic functions of the software.

Geospatial Lab Application 5.1: GIS Introduction: QGIS Version asked you to use the free QGIS software; however, this lab provides the same activities but asks you to use ArcGIS Pro.

Important note: This version of the lab uses ArcGIS Pro, but the same lab using ArcMap (from the most recent version of ArcGIS Desktop) is available for you if you'd rather use that. *Geospatial Lab Application 5.3: GIS Introduction: ArcMap Version* is available online at: **www.macmillanlearning.com/catalog/shellitoigt4e**

Objectives

The goals for you to take away from this lab are:

➤ To familiarize yourself with the ArcGIS Pro software environment, including logging in, basic navigation, and tool use

➤ To examine characteristics of geospatial data, such as their coordinate system, datum, and projection information

➤ To familiarize yourself with adding data and manipulating data layer properties, including changing the symbology and appearance of Esri data

➤ To familiarize yourself with data attribute tables in ArcGIS Pro

➤ To join a table of non-spatial data to a layer's attribute table in ArcGIS Pro

➤ To make measurements between objects in ArcGIS Pro

Using Geospatial Technologies

The concepts you'll be working with in this lab are used in a variety of real-world applications, including:

➤ Public works, where GIS is used for mapping infrastructure and things such as sewer lines, water mains, and streetlight locations

➤ Archeology, where GIS is used for mapping boundaries and the spatial characteristics of dig sites, as well as the locations of artifacts found there

Obtaining Software

The current version of ArcGIS is not freely available for use. However, instructors affiliated with schools that have a campus-wide software license may request a 1-year student version of the software online at **www.esri .com/landing-pages/education-promo**.

Important note: Software and online resources can change fast. This lab was designed with the most recently available version of the software at the time of writing. However, if the software or Websites have significantly changed between then and now, an updated version of this lab (using the newest versions) will be available online at **www.macmillanlearning .com/catalog/shellitoigt4e**.

[Bonnie Jo Mount/The Washington Post via Getty Images]

Lab Data

Copy the folder **Chapter5Pro**—it contains a Microsoft Excel file called CAPop which has population information as well as a folder called Cal-Data, which is the file geodatabase (see page 148) you'll be using in this lab. A file geodatabase is a single folder that contains many individual layers—if you look at the contents of the CalData folder in Windows it will contain numerous computer files that look meaningless—but to the GIS software these multiple files make up the GIS data layers we will be using. This lab's data was downloaded from The National Map (see page 11) free of charge and the population information was taken from the U.S. Census Bureau. The GIS layers in the geodatabase have already been projected for you to the U.S. National Atlas Equal Area projection.

Localizing This Lab

The data used in this lab is a variety of features of attributes of the state of California drawn from The National Map and the U.S. Census Bureau. However, you can easily work with the same data from your own state as The National Map makes the same layers available for all 50 states. From The National Map, download the Structures, Boundaries, and Transportation geodatabases for your own state, which contain the layers used for California in this lab. If you are in, for instance, Michigan, you can download county boundaries, the structures layer, the trails, railroads, and airports for Michigan from The National Map. Similarly, the U.S. Census Bureau has population values for each county, which can be placed into an Excel table like the one of California used in this lab.

5.1 An Introduction to ArcGIS Pro

1. Before beginning work with ArcGIS Pro, you'll need to have a license to work with it. Licenses for ArcGIS Pro are managed through your organization's ArcGIS Online setup. In order to get a license, you'll acquire a free Esri Account and have a couple of things done with it. Basically, in order to run ArcGIS Pro you need to log in with your Esri Account then ArcGIS Pro will verify with your organization that you have been assigned a license to run the software.

2. To get your account, point your web browser to: **https://www.arcgis. com/home/signin.html** and select the option for **Create a Public Account**. Follow the steps to set up your Esri account.

3. Next, submit your account information to your instructor or administrator in charge of your organization's ArcGIS Online setup. That individual will work with you to get your account established with your organization's ArcGIS Online account, and then activate your license to use ArcGIS Pro. The good news is you'll only have to do all of this once—after the administrator has assigned you your license, ArcGIS Pro will work just fine until the administrator chooses to take that license away from you. Note however, that the license is tied to your Esri Account. Write down the username and password for it as you'll need it each time you use ArcGIS Pro.

4. Once you have your license properly assigned to you, start **ArcGIS Pro**. The first thing you will be prompted to do is sign in with the username and password you've set up for your Esri Account, so log in using that information. Next, ArcGIS Pro will open and begin by asking you questions about "projects." In ArcGIS Pro, a "project" is the means by which you organize all of the various layers, tables, maps, and so on of your work—because you're just starting, click on **Blank** in the **Create a new project** box.

5. You'll now be prompted to give your Project a name and where it will be located. Call your Project **C5Pro** and for its location use **C:\Chapter5Pro**, then click **OK**. *Important note*: This lab assumes your Chapter5Pro folder is located on the C: drive. If it's somewhere else (for instance on a flash drive or in a different folder) use that drive letter or path instead.

6. ArcGIS Pro will now open. The left-hand column (where you'll see the word Contents) is referred to as the Contents Pane—it's where you'll have a list of available data layers. The blank screen that takes up most of the interface is where GIS data will be displayed. If a tab called Catalog is taking up the center of the screen, you can close it for now.

7. Before you begin adding and viewing data, the first thing to do is examine the data you have available to you. To examine data layers, you can use the Catalog pane, which is designed to allow you to organize and manage GIS data. By default, the Catalog pane will be available on the right-hand side of ArcGIS Pro. If it's not available, go to the **View** tab, within the **Windows** group, and click on the **Catalog** icon, then choose **Catalog Pane**.

8. In the **Catalog Pane**, expand the **Folders** option. You'll see the C5Pro project inside there. We'll now want to connect to the C:\Chapter5Pro folder where this lab's data is located. To do so, right-click on the **Folders** option and choose **Add Folder Connection**.

9. In the **Add Folder Connection** dialog box that opens, navigate to the C: drive and then choose the **Chapter5Pro** folder. (Again, if you've stored this folder somewhere other than C:, such as a flash drive or a different folder, navigate to that—however, this lab will assume that your folder is on your C: drive). Once you've selected the Chapter5Pro folder, click **OK**.

10. Expand the Chapter5Pro folder. You'll see the CAPop.xlsx Microsoft Excel file and the CalData file geodatabase. Expand the file geodatabase to verify that it contains the three feature classes we'll be working with in this lab—CalAirports, CalBounds, and CalTrails.

5.2 Adding Data to ArcGIS Pro and Working with the Contents Pane

1. Back in the main blank screen of ArcGIS Pro, if there are any tabs open (such as a Catalog tab), close them so that you have an empty view with which to work. You'll see that the **Contents Pane** is empty as well, so it's time to add some data to ArcGIS Pro so we can begin working. To begin, the first thing we'll do is insert a map to serve as a backdrop to keep our data in context. Click on the **Insert** tab, then within the **Project** group, click on the **New Map** icon, then select **New Map**. You'll see that the empty space has now been filled with a map of the United States, held in a tab called Map. In ArcGIS Pro, Maps are where you can add your data with which to work. All of the data is held inside of that Map tab, so if you wanted to work with a different location in the same Project, you could insert a second Map tab. However, we'll only be working with one location, and thus, one Map.

2. To add some data to your Map, click on the **Map** tab at the top of the screen, then within the **Layer** group, click on the **Add Data** icon, then select **Data**. In the **Add Data** dialog box, navigate to the **C:** drive, then the **Chapter5Pro** folder, then the CalData geodatabase, and finally select the CalBounds feature class. Click **OK** (or double-click on the name of the feature class).

3. You'll now see the CalBounds feature class in the Contents Pane and its content will be displayed in the Map.

4. In the Contents Pane, you'll see a checkmark in the box next to the CalBounds layer. When the checkmark is displayed, the layer will be shown in the Map, and when the checkmark is removed (by clicking on it) the layer will not be displayed.

5. Now add two more feature classes: **CalAirports** and **CalTrails**. All three of your layers will now be in the Contents Pane.

> **Question 5.1** What objects do each of the following datasets consist of: CalAirports, CalTrails, and CalBounds?

6. You can manipulate the "drawing order" of items in the Contents Pane by grabbing a layer with the mouse and moving it up or down in the Contents Pane. Whatever layer is at the bottom is drawn first, and the layers above it in the Contents Pane are drawn on top of it. Thus, if you move the CalBounds layer to the top of the Contents Pane, the other layers will not be visible in the Map because CalBounds' polygons are being drawn over them.

5.3 Symbology of Features in ArcGIS Pro

1. You'll also notice that the symbology generated for each of the three layers is simple, with each one assigned a random color and appearance. You can significantly modify the appearance of the objects in the Map if you want to customize your maps.

2. Right-click on the **CalBounds** layer and select **Symbology**. You'll see a new pane opened on the right side of the screen, the Symbology Pane. You'll also see at the bottom of the Symbology Pane a separate tab for the Catalog Pane so that you could switch back and forth between them as necessary. ArcGIS Pro allows you to have multiple Panes open but you can choose which one you want to display.

3. In the Symbology Pane, press the large colored rectangle next to the **Symbol** heading—from the new **Format Polygon Symbol** options, you can alter the appearance of the layer. By clicking on **Gallery** you can change the color of the California counties by choosing from one of the many options, and by clicking on **Properties** you can change other things such as the outline color and width of the outline itself.

4. Change the CalBounds layer to something more appealing to you. Then do the same for the CalAirports and CalTrails feature classes. Note that you can alter the points for the airports into shapes like stars, triangles, or actual airport symbols, as well as altering their size and color. Several different styles are available for the lines of the CalTrails layer as well.

5.4 Obtaining Projection Information

1. Chapter 2 discussed the importance of projections, coordinate systems, and datums, and all of the information associated with these concepts can be accessed using ArcGIS Pro.

2. Right-click on the **CalBounds** layer and select **Properties**. Click on the **Source** option.

3. Basic information about the layer is available here—and from scrolling down and expanding the option for Spatial Reference, you can access

information about units, projections, coordinate systems, and datums related to the layer.

4. Click **OK** to exit the CalBounds **Layer Properties** dialog box.

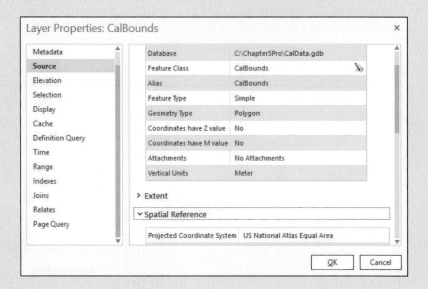

> **Question 5.2** What units of measurement (feet, meters, degrees, etc.) are being used in the CalTrails dataset?

> **Question 5.3** What units of measurement (feet, meters, degrees, etc.) are being used in the CalAirports dataset?

Important note: ArcGIS Pro also allows you to change the projected appearance of the data in the Map from one projected system to another. By altering the properties of the Map, you can change how things appear in the Map. Note that this does not alter the actual projection of the layers themselves, simply how you're seeing them displayed on the Map.

5. In the Contents Pane, right-click on the **Map** and select **Properties**.

6. In the **Map Properties** dialog box, select **Coordinate Systems**. You'll see that currently the Map is set to U.S. National Atlas Equal Area as that's the projected coordinated system used by the three feature classes you've added to it.

7. You can obtain more information about the currently used coordinate system by clicking on **Details**.

Question 5.4: To what "family" of projections does the U.S. National Atlas Equal Area projection belong?

8. You can select other projections for the Map if you wish. Try out some other appearances—for instance, choose the 1984 Web Mercator Auxiliary Sphere projected coordinated system and see how it alters the appearance of California on the Map. When you're done, return to the U.S. National Atlas Equal Area projection for the remainder of the lab.

5.5 Examining Attribute Tables

1. Each of your three layers is represented by points (airports), lines (trails), and polygons (the counties), but each layer also has a corresponding set of attributes that accompanies their spatial features. As noted in the chapter, these non-spatial attributes are stored in an attribute table for each layer. To get started working with this non-spatial data, in the Contents Pane, right-click on the CalBounds layer, then choose **Attribute Table**.

2. A new window with the CalBounds attribute table will open. Each of the rows of the table is a record representing an individual feature (in this case, a single polygon of a California county), while the columns of the table are the fields representing the attributes for each of those records. By perusing the attribute table you can find information about a particular record. However, with 58 records it could take awhile to find some specific information. One way to organize the table is to sort it—scroll across the CalBounds attribute table until you find the field called County_Name. Right-click on the **County_Name** table header and choose the option for **Sort Ascending**, and you'll see that the table will sort itself into alphabetical order by the name of the county.

Right-click on the **County_Name** table header a second time and choose **Sort Descending** and you'll see that the table sorts itself into reverse alphabetical order.

3. With the table sorted, scroll through it to find the record with the County_Name attribute of "San Diego." Each county has a FIPS code assigned to it—this is a numeric value that is a government designation used to identify things such as states, counties, and townships uniquely. This FIPS code information for each California county is located in this attribute table.

> **Question 5.5** What is the county FIPS code for San Diego County?

5.6 Joining a Table to an Attribute Table

1. While there are a lot of records in the CalBounds attribute table, ArcGIS Pro will allow you to add further attributes to them. This can be done by taking another table that contains additional non-spatial information and joining it to the CalBounds attribute table.

2. To begin, the first thing you'll need to do is add a table of non-spatial data with which ArcGIS Pro can work. Return to the **Map** tab, within the **Layer** group, choose the **Add Data** icon, then choose **Data**. Navigate to the C:\Chapter5Pro folder and select the **CAPop.xlxs** file, then select the **Sheet1$** option.

3. A new item called Sheet1$ will be added to the Contents Pane under the heading of Standalone Tables—this is the table containing a new set of attributes. Right-click on Sheet1$ and choose **Open**. You'll see the new non-spatial table open in ArcGIS Pro as a tab alongside the CalBounds attribute table. It contains a field with the county's FIPS code, the name of that county, and the population of that county. In order to join two tables together, they need to contain the same field—in this case, the Sheet1$ table contains a field called CAFIPS that has the FIPS code of the county while your CalBounds layer's attribute table contains a field called County FIPSCode that has the same information. Because these two tables (the spatial data from CalBounds and the non-spatial data from Sheet1$) have a field in common (i.e., they both have a field containing the FIPS code of each county), they can be joined together.

4. To perform the join, right-click on the layer you want to join attributes to (in this case, we want to join information to the **CalBounds** layer), and select **Joins and Relates**, then select **Add Joins**.

5. You'll see a new pane—the Geoprocessing Pane—has opened on the right-hand side of the screen where the Symbology and Catalog Panes

were. The Geoprocessing Pane is used to implement a variety of different ArcGIS Pro tools and functions—in this case, we are using the "Add Join" tool in the Geoprocessing Pane to join our two tables together.

6. In the Add Join dialog box, under **Layer Name or Table View** select **CalBounds** (this is the layer to which we want to join the table).

7. Under **Input Join Field**, choose **County_FIPSCode** (this is the field in the CalBounds layer we will use as the key for the join).

8. Under **Join Table**, choose **Sheet1$** (this is the table with the non-spatial information we want to join).

9. Under **Output Join Field**, choose **CAFIPS** (this is the field in Sheet1$ we want to use for the join).

10. Leave the other options alone and click **Run** in the lower-right hand corner of the Geoprocessing Pane.

11. ArcGIS Pro will run the Add Join tool in a few seconds. When it's done, return to the CalBounds attribute table and you'll see the information from Sheet1$ successfully joined to the attribute table with its fields (such as Pop2010) now fields in CalBounds.

12. Now that Pop2010 is a field in the attribute table, you can sort it from highest to lowest populations (or vice versa) by right-clicking on the field header, and choosing Sort Ascending on Sort Descending (like how you sorted the counties alphabetically by name). Keeping this ability in mind, answer the following questions.

Question 5.6 What is the 2010 population of Los Angeles county?

Question 5.7 What is the 2010 population of the California county with FIPS code 065?

Question 5.8 What county in California has the lowest 2010 population and what is that population value?

Question 5.9 What county in California has the third highest 2010 population and what is that population value?

5.7 Navigating the Map in ArcGIS Pro

1. We'll now just focus on one area of California—San Diego County. In the CalBounds attribute table, scroll through until you find San Diego County, then click once on the header of the record on the far-left of the attribute table. The entire record (the row) will turn a cyan color.

2. Back in the Map, you'll see San Diego County now outlined in a cyan color. The polygon of San Diego County has now been *selected* in ArcGIS Pro.

3. ArcMap provides a number of means for navigating around the data layers in the Map, located in the **Map** tab within the **Navigation** group.

4. The **Explore** icon allows you to "grab" the Map by clicking on it, and to drag the map around the screen for navigation.

5. The globe icon is the **Full Extent** tool, which will zoom the Map to the extent of all the layers. It's good to use if you've zoomed too far in or out in the Map or need to restart.

6. The four pointing arrows are the **Fixed Zoom** tools, which allow you to zoom in (arrows pointing inward) or zoom out (arrows pointing outward) from the center of the Map.

7. The **blue arrows** allow you to step back or forward from the last set of zooms you've made.

8. The magnifying glass is the **Zoom to Selection** tool, which allows to zoom the Map to the extent of any selected features.

9. Note that if your mouse has a center wheel, you can use the wheel to zoom in and zoom out from the Map as well.

10. Because you've already selected San Diego County back in step 2 of this section, click on the **Zoom to Selection** tool and ArcGIS Pro will shift the Map so that San Diego County is in the center and zoomed in. Use the **Explore** icon to shift the Map around so that you have a good view of the entire county.

11. With San Diego County found, return to the CalBounds attribute table and click on the **Clear** button to remove the cyan highlighting, then close the attribute table (as well as the Sheet1$ table) by clicking the "x" in the table's tab.

5.6 Interactively Obtaining Information

1. Even with the Map View centered on San Diego County, there are an awful lot of point symbols there, representing the airports. We're going to want to identify and use only a couple of them. ArcGIS Pro has a searching tool that allows you to locate specific objects from a dataset. To get started, go to the **Map** tab, within the **Inquiry** group, and click on the **Locate** icon. A new Locate Pane will open on the right-hand side of the screen.

2. Locate allows you to search through a layer or search through an external dataset for a particular feature. Because we'll be searching through the CalAirports layer for a specific airport, we first need to tell Locate that's the layer in which we want to search. To begin, in the **Locate Pane**, click on **Settings** – you'll see the two available datasets for you to search through (XY Provider and World). What we want to do is add the CalAirports as a searchable layer—to do so click the "**plus**" button on the Locate Pane then select **Add Layer**.

3. In the Select Layer dialog box, choose **CalAirports** and click **OK**.

4. The Locate Pane will switch to the Configure Layer for Locate option. What you'll be doing here is choosing the fields in the CalAirports layer that you want to be able to search through and how you want to search them. As we will be searching for the name of a specific airport, change the **Name** field to "**Contains**." If you were going to search by something else (such as the FAA Code or the Airport Class) you would have changed their fields instead.

5. Click the **Arrow** button. This will save your changes and return you to the Locate Pane again.

6. In the Locate Pane, make sure that only the CalAirports option has a checkmark in it—remove the checkmarks from all other options. This will ensure that ArcGIS Pro will only search through the CalAirports layer and not other layers.

7. Lastly, click on **Locate** to return to where you can type in what you want to search for. Type: **San Diego International** and hit the **Enter** key on the keyboard. What ArcGIS Pro will do is search through the Name field of the CalAirports layer for text that contains what you just typed.

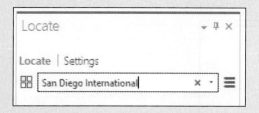

8. However, all that you've done so far is locate an object. To obtain information about it, return to the Map tab, within the Navigate group, and click on the **Explore** icon. Next, click on the point representing San Diego International airport (zooming in as necessary). A new pop-up will appear (you may have to grab the top of the panel and drag it up to expand it so you can see everything that is inside the pop-up), listing all of the attributes for the point you clicked on in the CalAirports layer.

> **Question 5.10** What is the three-letter FAA Airport Classification Code for San Diego International airport?

9. Close the pop-up when you're done answering Question 5.10.

5.9 Labeling Features

Rather than dealing with several points on a map and trying to remember which airport is which, it's easier to simply label each airport so that its name appears in the Map. ArcGIS Pro gives you the ability to do this by creating a label for each record and allowing you to choose the field with which to create the label.

1. Right-click on the **CalAirports** layer in the Contents Pane and select **Label**. You'll see that labels using the Name field of the layer will appear on the map.

2. If you want to change the appearance of the labels (i.e., use a different font to label them or alter where the label appears relative to the feature), right-click on **CalAirports** in the Contents Pane and select **Labeling Properties**.

3. A new Label Class Pane will open on the right-hand side of ArcGIS Pro. In this new Pane, click on **Symbol**, then click on the **icon** with the pen and the letter "A" to access several options for changing the font and appearance of the labels.

5.10 Measurements on a Map

With all the airports now labeled, it's easier to keep track of all of them. With this in mind, your next task is to make a series of measurements between points to see what the Euclidian (straight line) distance is between airports around San Diego.

1. Zoom in tightly so that your Map contains San Diego International and North Island Naval Air Station.

2. From the **Map** tab, within the **Inquiry** group, click on the **Measure** icon.

3. A new item will appear on the Map to help you make measurements. Before getting started, click on **Options**, then select **Distance Units**, then make sure that only **Miles** has a checkmark next to it. This will ensure that when you're making measurements on the Map, you're getting your results back in miles.

4. You'll see that the cursor has turned into a crosshairs with a ruler. Place the crosshairs on the point representing San Diego International airport and left-click the mouse. Drag the crosshairs south and west to the point representing North Island Naval Air Station and click the mouse there. The distance measurement will appear in the new box. Note that each line of measurement you make with the mouse counts as an individual "Segment" and the value listed in the box for "Path" is the sum of all segments.

> **Question 5.11** What is the distance from San Diego International airport to North Island Naval Air Station?

5. You can clear all of the lines of measurement by clicking on the **eraser icon** in the box.

6. Zoom out a bit so that you can see some of the other airports in the region, particularly Montgomery Field and Brown Field Municipal Airport.

> **Question 5.12** What is the total distance in miles from San Diego International airport to Montgomery Field, then from Montgomery Field to Brown Field Municipal Airport, and then finally from Brown Field Municipal Airport back to San Diego International airport?

5.11 Saving Your Work (and Working on It Later)

When you're using ArcGIS Pro, you can save your work at any time and return to it. When work is saved in ArcGIS, the changes are written to your Project (which we called C5Pro back at the start of the lab). Later, you can re-open this Project to pick up your work where you left off.

1. Saving your project is done by selecting the **Project** tab, and then selecting **Save**.

2. Files can be re-opened by selecting the **Project** tab, and then selecting **Open**.

3. Exit ArcGIS Pro by selecting the **Project** tab, and then selecting **Exit**.

Closing Time

This lab was pretty basic, but it served to introduce you to how ArcGIS Pro operates and how GIS data can be examined and manipulated. You'll be using either QGIS or ArcGIS Pro in the next two labs, so the goal of this lab was to get the fundamentals of the software down. The lab in Chapter 6 takes this GIS data and starts to do spatial analysis with it, while the lab in Chapter 7 will have you starting to make print-quality maps from the data.

Using GIS for Spatial Analysis

6

Database Query and Selection, Buffers, Overlay Operations, Geoprocessing Concepts, and Modeling with GIS

Chapter 5 described the basic components of GIS and how it operates; now it's time to start doing some things with that GIS data. Anytime you are examining the spatial characteristics of data or how objects relate to one another across distances, you're performing **spatial analysis**. A very early (pre-computer) example of this is Dr. John Snow's work during an 1854 cholera epidemic in Soho, central London—a poor, overcrowded square mile of narrow streets and alleyways. His analysis was able to identify the infected water from one pump as the source of the outbreak. At some point, Snow mapped the locations of the cholera deaths in relation to the water pumps (see **Figure 6.1**). This type of spatial analysis (relating locations of cholera deaths to the positions of water pumps) was innovative for the mid-nineteenth century, but is commonplace today—with GIS, these types of spatial analysis problems can be addressed easily.

Other common forms of spatial analysis questions include the following:

➤ How many objects of a particular type are within a certain distance of a particular location? Location researchers for a fast food chain might ask, "How many fast food restaurants are within one mile of this interstate exit?" A realtor might ask, "How many rental properties are within two blocks of this house that's for sale?" An environmentalist might ask, "What acreage of wetland will be threatened with removal by this proposed development?"

➤ How do you choose the most suitable locations based on a set of criteria? Many different types of questions related to this topic can be answered, such as those involving where to locate new commercial developments, new real estate options, or the best places for animal habitats.

spatial analysis examining the characteristics or features of spatial data, or how features spatially relate to one another

FIGURE 6.1 ───────○

A version of Dr. John Snow's 1854 map of Soho, London, showing locations of cholera deaths and water pumps. *[Map 1. Published by C. F. Cheffins, Lith, Southhampton Buildings, London, England, 1854, in Snow, John, On the Mode of Communication of Cholera, 2nd ed., John Churchill, New Burlington Street, London, England, 1855]*

This chapter examines how GIS can be used with geospatial data to answer questions such as these in the context of spatial analysis. Keep in mind the two ways that GIS is used to model real-world data from the previous chapter—vector and raster. Due to the nature of some kinds of geospatial data, some types of spatial analysis are best handled with vector data and some with raster data.

How Can Data Be Retrieved from a GIS for Analysis?

One type of analysis is to find which areas or locations meet particular criteria—for instance, to take a map of U.S. congressional districts and identify which districts have a Democratic representative and which have a Republican representative. Alternatively, you might have a map of all residential parcels and want to identify which houses have been sold in the last month. In both of these examples, you would perform a database **query** to select only those records from the attribute table (see Chapter 5) that reflect the qualities of the objects with which you want to work.

query the conditions used to retrieve data from a database

In other cases, before beginning the analysis, you might only want to utilize a subset of your GIS data. For instance, rather than dealing with all congressional districts in the entire United States, you might want to examine only those in Ohio. Or, if you have a dataset of all residential parcels in a county (which is likely thousands of parcels), you might only want to do analysis related to one (or a handful) of the parcels rather than examining all of them. You'd want to go into the dataset and select only those parcels required for your study. Again, in these cases, the query would only select the records you want to deal with.

In GIS, queries are composed in the Structured Query Language (**SQL**) format, like a mathematical function. SQL is a specific format that is used for querying a layer or database to find what attributes meet certain conditions. For example, a layer made up of points representing cities in Ohio also has a number of fields representing its attributes, such as city name, population, and average income. If you just wanted to do analysis with one of these records (the city called "Boardman," let's say), then a query could be built to find all records where the field called CITY_NAME contained the characters of 'Boardman' as: CITY_NAME = 'Boardman'. This should return one record for you to use in your analysis. Another example would be to select all cities in Ohio with a population greater than or equal to 50,000 persons. The attribute table has a field called POP2010 (representing the 2010 city population) that contains this information. Thus, a query of POP2010 >= 50000 could be built, and all records with a population field containing a value of greater than or equal to the number 50,000 would be selected for your use (see **Figure 6.2** for an example of a query in GIS).

SQL Structured Query Language—a formal setup for building queries

FIGURE 6.2

A database query for Ohio cities returning multiple records (selected records and their corresponding cities are in cyan).

relational operator one of the six connectors (=, <>, <, >, >=, or <=) used to build a query

A query will use one of the following **relational operators**:

➤ Equal (=): Use this when you want to find all values that match the query. For instance, querying for CITY_NAME = 'Poland' will locate all records that exactly match the characters in the string that comprise the word 'Poland'. If the record were called 'Poland Township', it would not be located because the character string looks for an exact match.

➤ Not Equal (< >): Use this when you want to find all the records that do not match a particular value. For instance, querying for CITY_NAME < >'Poland' will return all the records that do not have a character string of the word 'Poland' (which would probably be all of the other cities with names in the database).

➤ Greater Than (>) or Greater Than Or Equal To (>=): Use this for selecting values that are more than (or more than or equal to) a particular value.

➤ Less Than (<) or Less Than Or Equal To (<=): Use this for selecting values that are less than (or less than or equal to) a particular value.

compound query a query that contains more than one operator

Boolean operator one of the four connectors (AND, OR, NOT, XOR) used in building a compound query

AND the Boolean operation that corresponds with an intersection operation

intersection the operation wherein the chosen features are those that meet both criteria in the query

OR the Boolean operator that corresponds with a union operation

union the operation wherein the chosen features are all that meet the first criterion as well as all that meet the second criterion in the query

NOT the Boolean operator that corresponds with a negation operation

While a simple query only uses one operator and one field, a **compound query** enables you to make selections using multiple criteria. There are different ways of linking multiple criteria together for creating one of these compound queries, each using a different option of querying (referred to as a **Boolean operator**):

➤ **AND**: Suppose you want to select cities in Ohio that have a population of over 50,000 (a variable called POP2010 in the attribute table) and an average household income (a variable called AVERAGEHI) of more than $30,000. A compound query can combine these two requests by saying: POP2010 >= 50000 AND AVERAGEHI > 30000. Because "AND" is being used as the operator, only records that match both the criteria will be returned by the query. If an attribute only meets one criterion, it won't be selected. An "AND" query is referred to as an **intersection** because it returns what the two items have in common.

➤ **OR**: A different query could be built by saying: POP2010 >= 50000 OR AVERAGEHI > 30000. This would return records of cities with a population of greater than or equal to 50,000, or cities with an average household income of greater than $30,000, or cities that meet both. When an OR operator is being used, records that meet one or both criteria are selected. An "OR" query is referred to as a **union** because it returns all the elements of both operations.

➤ **NOT**: A third query could be built around the concept that you want all of the data related to one criterion, but you want to exclude what relates to the second criterion. For instance, if you want to select cities with a high population, but not those with a higher average household income, you could build a query like: POP2010 >= 50000 NOT AVERAGEHI > 30000 to find those particular records. A "NOT" query is referred to as a **negation** because it returns all the elements of one dataset, but not what they have in common with the other.

➤ **XOR**: A final type of query can be built using the idea that you want all of the data from each layer, except for what they have in common. For instance, if you wanted to find those cities with a high population or those that had high incomes, but not both types of cities, your query would be: POP1990 >= 50000 XOR AVERAGEHI > 30000. XOR works as the opposite of AND because it returns all data, except for the intersection of the datasets. An "XOR" query is referred to as an **exclusive or**, because it acts like a union but leaves out the intersection data.

Compound queries can use multiple operations to select multiple records by adhering to an order of operations, such as: (POP2010 >= 50000 AND AVERAGEHI > 30000) OR CITY_NAME = 'Boardman'. This will select cities that meet both the population and the income requirements, and would select cities named 'Boardman' as well. The result of a query will be a subset of the records that are chosen, and then those records can be used for analysis, rather than the whole data layer. Selected records can also be exported to their own dataset. For instance, the results of the previous query could be extracted to compose a new GIS layer, made up of only those records

negation the operation wherein the chosen features are those that meet all of the first criteria and none of the second criteria (including where the two criteria overlap) in the query

XOR the Boolean operator that corresponds with an exclusive or operation

exclusive or the operation wherein the chosen features are all of those that meet the first criterion as well as all of those that meet the second criterion, except for the features that have both criteria in common in the query

Hands-On Application 6.1

Building SQL Queries in GIS

Durham, North Carolina, has an online GIS mapping utility that allows the user to do a variety of spatial analyses with the city's data, including performing queries. To work with this tool, go to **http://maps2.roktech.net/durhamnc_gomaps**. In the upper left-hand corner there are several tools (such as zooming and panning) to help you get oriented to the GIS data. The tool with the hammer icon allows you to build SQL queries of the data. Click on this icon and a new set of options will appear on the right. Choosing the tab for Query Builder will allow you to construct SQL queries of the GIS. First, select an available layer (for instance, select Stormwater Watersheds). Second, select an attribute to query (for instance, double-click on River). Third, select an operator (for instance, choose the equals "=" symbol). Click on the Get Unique Values button and you'll see the available values appear to finish the SQL statement (for instance, choose CAPE FEAR). Your SQL statement will appear in the box (in this example, it should

read: RIVER = 'CAPE FEAR'). To run the SQL query, click on the Execute button. All records meeting your query will be retrieved (in this case, all stormwater watersheds related to the Cape Fear River). The selected watersheds will be highlighted on the map, while the records (and the attributes that go along with them) will be displayed in a spreadsheet at the bottom of the screen. Investigate some of the other available data layers and the types of queries that can be built with them.

Expansion Questions

➤ How many stormwater watersheds in Durham, North Carolina, are related to the Cape Fear River?

➤ How many stormwater watersheds in Durham, North Carolina, are related to the Neuse River?

➤ Which of the parks in Durham, North Carolina, have boating available? (*Hint:* Use the Parks layer and the boating attribute, then look for records where the boating attribute is equal to "Y.")

(and their corresponding spatial objects) that meet the query's criteria. Once you have your selected records or new feature layers, you can start to work with them in GIS (see *Hands-On Application 6.1: Building SQL Queries in GIS* for an example of doing queries with an online GIS utility).

 How Can You Perform Basic Spatial Analysis in GIS?

Once you have your selected data, it's time to start doing something with it. A simple type of analysis is the construction of a **buffer** around the elements of a data layer. A buffer is an area of proximity set up around one or more objects. For instance, if you want to know how many rental properties are within a half mile of a selected house, you can create a half-mile buffer around the object representing the house and then determine how many rental properties are within the buffer. Buffers can be created for multiple points, lines, or polygons (**Figure 6.3**). Buffers are simple tools to utilize, but they can provide useful information when you're creating areas of analysis for various purposes.

buffer a polygon of spatial proximity built around a feature

GIS can also perform a **dissolve** operation, where boundaries between adjacent polygons that have the same properties are removed, merging the polygons into a single, larger shape. Dissolve is useful when you don't need to examine each polygon, only regions with similar properties. For example, suppose you have a land-use map of a county consisting of polygons

dissolve the ability of the GIS to combine polygons with the same features

FIGURE 6.3

A 50-mile buffer constructed around a major interstate (I-70).

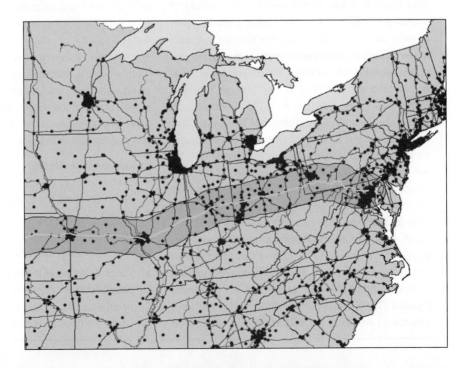

representing each individual parcel and how it's being used (commercial, residential, agricultural, etc.). Rather than having to examine each individual parcel, you could dissolve the boundaries between parcels and create regions of "residential" land use or "commercial" land use for ease of analysis. Similarly, if you have overlapping buffers generated from features, dissolve can be used to combine buffer zones. See **Figure 6.4** for an example of dissolve in action.

Buffers and querying are basic types of spatial analysis, but they only involve a single layer of data (such as creating a buffer around roads or

FIGURE 6.4

Before and after the dissolve operation—Ohio county boundaries are dissolved on the basis of their state's FIPS code.

selecting all records that represent a highway). If you're going to do spatial analysis along the lines of John Snow's hunt for a cholera source, you'll have to be able to do more than just select a subset of the population who died, or just select the locations of wells within the area affected by the disease—you'll have to know something about how these two things interacted with each other over a distance. For instance, you'll want to know something about how many cholera deaths were found within a certain distance of each well. Thus, you won't just want to create a buffer around one layer (the wells)—you'll want to know something about how many features from a different layer (the death locations) are in that buffer. Many types of GIS analysis involve combining the features from multiple layers together to allow examination of several characteristics at once. In GIS, when one layer has some sort of action performed to it and the result is a new layer, this process is referred to as **geoprocessing**. There are numerous types of geoprocessing methods, and they are frequently performed in a series to solve spatial analysis questions.

GIS gives you the ability not just to perform a regular SQL-style query, but also to perform a **spatial query**—to select records or objects from one layer based upon their spatial relationships with other layers (rather than using attributes). With GIS, queries can be based on spatial concepts—to determine things like how many registered sex offenders reside within a certain distance from a school, or which hospitals are within a particular distance of your home. The same kind of query would be done if a researcher wanted to determine which cholera deaths are within a buffer zone around a well. In these cases, the spatial dimensions of the buffer are being used as the selection criteria—all objects within the buffer are what will be selected. See **Figure 6.5**

geoprocessing the term that describes when an action is taken to a dataset that results in a new dataset being created

spatial query selecting records or objects from one layer based upon their spatial relationships with other layers (rather than using attributes)

FIGURE 6.5

Effect of selecting point features (cities) by using a buffer (50 miles around I-70).

Hands-On Application 6.2

Working with Buffers in GIS

Miami-Dade County, Florida, has an online GIS application that looks at multiple GIS layers, including parcels and zoning for the area. On this Website, you can select a location, generate a buffer around it, and select objects (such as parcels, fire stations, and other municipal features) touched by the buffer (and display their parcel records). To use this, go to **http://gisweb.miamidade.gov/emaps**. To begin, we'll start with the Vizcaya Museum and Gardens (a historic landmark in Miami) and use that as the source for analysis with buffers. In the search box, type the address of Vizcaya: 3251 S. Miami Ave, Miami, FL 33129. The viewer will adjust to that address location, so you may have to zoom out a little to see it in relation to its Miami surroundings.

To generate a buffer, select the Map Tools pull-down menu, then choose the option for Buffer. Next you will have to specify a buffer distance (in feet) and then choose which other layer of features

you want to examine in relation to the buffer. For an initial analysis, type 10560 as the Buffer Distance (this is the number of feet equal to two miles) and then choose Municipal Fire Stations as the Target Layer. By clicking on Create Buffer, the 2-mile buffer will be drawn on the screen and all Municipal Fire Station objects that are within the 2-mile buffer will be shown on the screen and also tabulated for you. You can always clear the buffer by selecting the Map Tools pull-down menu and then choosing Clear Selection.

Expansion Questions

➤ How many municipal fire stations are within 2 miles of the Vizcaya Museum and Gardens? How many are within 3 miles (that is, 15840 feet)?

➤ How many public schools are within 2 miles of the Vizcaya Museum and Gardens? How many private schools?

for an example of using a buffer as a selection tool, as well as *Hands-On Application 6.2: Working with Buffers in GIS.*

Other analyses can be done beyond selecting with a buffer or selecting by location. When two or more layers share some of the same spatial boundaries (but have different properties) and are combined together, this is referred to as an **overlay** operation in the GIS. For instance, if one layer contains information about the locations of property boundaries and a second layer contains information about water resources, combining these two layers in an overlay can help determine which resources are located on whose property. An overlay is a frequently used GIS technique for combining multiple layers together for performing spatial analysis (see **Figure 6.6** on page 200).

overlay the combining of two or more layers in the GIS

There are numerous ways that polygon layers can be combined together through an overlay in GIS. Here are some widely used overlay methods (see **Figure 6.7**):

➤ **Intersect:** In this operation, only the features that both layers have in common are retained in a new layer. This type of operation is commonly used when you want to determine an area that meets two criteria—for instance, a location needs to be found within a buffer and also on a plot of land used for agriculture: you would intersect

intersect a type of GIS overlay that retains the features that are common to two layers

FIGURE 6.6

An example of overlaying two layers.

Layer 1

Layer 2

Overlay

the buffer layer and the agricultural layer together to find what areas, spatially, they have in common.

identity a type of GIS overlay that retains all features from the first layer along with the features it has in common with a second layer

➤ **Identity:** In this operation, all of the features of an input layer are retained, and all the features of a second layer that intersect with them are also retained. For example, you may want to examine all of the nearby floodplain and also what portions of your property are on it.

symmetrical difference a type of GIS overlay that retains all features from both layers except for the features that they have in common

➤ **Symmetrical difference:** In this operation, all of the features of both layers are retained except for the areas that they have in common. This operation could show you, for example, all potential nesting areas along with the local plan for development, except where they overlap (so those areas can be taken out of the analysis).

union (overlay) a type of GIS overlay that combines all features from both layers

➤ **Union (overlay):** In this operation, all of the features from both layers are combined together into a new layer. This operation is often used to merge features from two layers (for instance, in order to evaluate all possible options for a development site, you may want to overlay the parcel layer and the water resources layer together).

map algebra combining datasets together using simple mathematical operators

Another way of performing spatial overlay is by using raster data. With raster cells, each cell has a single value, and two grid layers can be overlaid in a variety of ways. The first of these is by using a simple mathematical operator (sometimes referred to as **map algebra**), such as addition or multiplication. To simplify things, we're only going to assign our grids values

FIGURE 6.7

Various types of geoprocessing overlay operations using two sample polygon layers.

of 0 or 1, where a value of 1 means that the desired criteria are met and a value of 0 is used when the desired criteria are not met. Thus, grids could be designed using this system with values of 0 or 1 to reflect whether a grid cell meets or does not meet particular criteria.

Let's start with a simplified example—say you're looking to find a plot of land on which to build a vacation home, and your two key criteria are that it be close to forested terrain for hiking and close to a river or lake for water recreation. What you could do is get a grid of the local land cover and assign all grid cells that have forests on the landscape a value of 1 and all non-forested areas a value of 0. The same holds true for a grid of water resources—any cells near a body of water get a value of 1 and all other cells get a 0. The "forests" grid has values of 1 where there are forests and values of 0 for non-forested regions, while the "water" grid has values of 1 where a body of water is present and values of 0 for areas without bodies of water (see **Figure 6.8** for a simplified way of setting this up).

FIGURE 6.8

A hypothetical landscape consisting of forests, rivers, and non-forested or dry areas. *[(top left) Alex L. Fradkin/Media Bakery (top right) Robert Glusic/Media Bakery (bottom left) Bruce Heinemann/Media Bakery (bottom right) U.S. Geological Survey/photo by Stewart Tomlinson]*

=

1	1
0	1

0	1
0	1

FIGURE 6.9

Raster overlay examples for multiplication and addition. *[Source: (left) Alex L. Fradkin/Getty Images (right) U.S. Geological Survey/photo by Stewart Tomlinson]*

1	1
0	1

×

0	1
0	1

=

0	1
0	1

1	1
0	1

+

0	1
0	1

=

1	2
0	2

Now you want to overlay your two grids and find what areas meet the two criteria for your vacation home, but there are two ways of combining these grids. **Figure 6.9** shows how the grids can be added and multiplied together with varying results. By multiplying grids containing values of 0 and 1 together, only values of 0 or 1 can be generated in the resulting overlay grid. In this output, values of 1 indicate where (spatially) both criteria (forests and rivers) can be found, while values of 0 indicate where both cannot be found. In the multiplication example, a value of 0 could have one of the criteria, or none, but we're only concerned with areas that meet both criteria, or else we're not interested. Thus, the results either match what we're looking for or they don't.

In the addition example, values of 0, 1, or 2 are generated from the overlay. In this case, values of 0 contain no forests or rivers, values of 1 contain one or the other, and values of 2 contain both. By using addition to overlay the grids, a gradient of choices is presented—the ideal location would have a value of 2, but values of 1 represent possible second-choice alternatives, and values of 0 represent unacceptable locations.

This type of overlay analysis is referred to as **site suitability**, as it seeks to determine which locations are the "best" (or most "suitable") for meeting certain criteria. Site suitability analysis can be used for a variety of topics, including identifying ideal animal habitats, prime locations for urban developments, or prime crop locations. When using the multiplication operation, you will be left with sites classified as "suitable" (values of 1) or "non-suitable" (values of 0). However, the addition operation leaves other options that may be viable in case the most suitable sites are unavailable, in essence providing a ranking of suitable sites. For instance, the best place to build the vacation home is in the location with values of 2 because it meets both criteria, but the cells with a value of 1 provide less suitable locations (as they only meet one of the criteria).

site suitability the determination of the "useful" or "non-useful" locations based on a set of criteria

How Can Multiple Types of Spatial Analysis Operations Be Performed in GIS?

There are many criteria that play into the choice of site suitability or into determining what might be the "best" or "worst" locations for something, and often it's not as cut and dried as "the location contains water" (values of 1) or "the location does not contain water" (values of 0). Think about the vacation home example—locations were classified as either "close to water" or "not close to water." When you're choosing a vacation home location, there could be locations right on the water, somewhat removed from it but still close enough for recreation, and some that are much further away from water. Similarly, terrain is so varied that it's difficult to assign it a characteristic of "mountainous" or "not mountainous." Instead, a larger range of values could be assigned—perhaps on a scale of 1 through 5, with 5 being the best terrain for hiking or the closest to the water and 1 being furthest away from the water or the least desirable terrain for hiking.

In order to set up something like this, a distance calculation can be performed. GIS can calculate the distance from one point to all other points on a map, or calculate the distance a location is from all other features. For example, when you're examining a location for building a house in a rural locale, you may want to know the distance that spot is from a landfill or wastewater treatment plant. Alternatively, you may want to know the distance a vacation home is away from a lake or river. A way of conceptualizing distance (as is done in ArcGIS) is to think of it as a continuous surface (see Chapter 5), where each spot on that surface has a value for the distance it is away from a feature. ArcGIS makes these measurements by calculating a distance surface of raster grid cells, with each cell containing a value of the distance from the features in a layer. **Figure 6.10** illustrates this by mapping the location of a set of cities in Ohio and calculating the distance that each cell is away from a city. Next, the distance from the cities is sliced into

FIGURE 6.10

An example of a raster distance surface calculation, where each grid cell contains a value representing the distance from a point (cities in Ohio), and the same grid sliced into 10 ranges.

Distance to Ohio Cities

- 0 - 12,555.70391
- 12,555.70392 - 25,111.40781
- 25,111.40782 - 37,667.11172
- 37,667.11173 - 50,222.81563
- 50,222.81564 - 62,778.51953
- 62,778.51954 - 75,334.22344
- 75,334.22345 - 87,889.92734
- 87,889.92735 - 100,445.6312
- 100,445.6313 - 113,001.3352
- 113,001.3353 - 125,557.0391

Distance Scaled to 10 Ranges

- 1
- 2
- 3
- 4
- 5
- 6
- 7
- 8
- 9
- 10

10 equal ranges (that is, all of the areas closest to a city are given a value of 1, the next closest are given a value of 2, and so forth).

For this example, we'll come up with a new set of grids classified along these lines and throw in a third option as well—distance away from roads (on the assumption that a vacation home would be best located well away from busy roads and the accompanying traffic noise). In all of these, the highest values represent the "best" options (the places furthest away from busy roads), and the lowest values represent the "worst" options (right next to a busy road). **Figure 6.11** shows the new grids and how they are overlaid

FIGURE 6.11

A raster overlay example for combining three criteria together to form a simple suitability index.

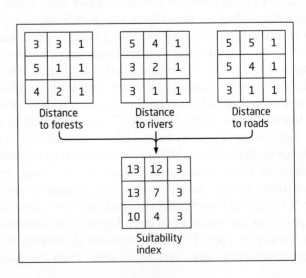

together through the addition operation to create a new output grid. Rather than choices of 0, 1, and 2, the output grid contains a wide range of options, where the highest number reflects the most suitable site based on the criteria and the lowest number reflects the least suitable sites. This new grid represents a very simple **suitability index**, where a range of values is generated, indicating a ranking of the viability of locations (these would be ordinal values, as we discussed back in Chapter 5). In the example in Figure 6.11, the values of 13 represent the most suitable sites, the values of 12 and 10 are possible alternative venues. The value of 7 is likely a mediocre site, and the values of 4 and 3 are probably unacceptable. The advantage of the suitability index is to offer a wider gradient of choices for suitable sites as an alternative to some of the previous options. Note that while this example is using raster data, you can also perform site suitability analysis actions using vector data.

GIS can process numerous large spatial data layers together at once. Thinking about the vacation home example, there are many additional factors that play into choosing the site of a second home—access to recreational amenities, the price of the land, the availability of transportation networks, and how far the site is from large urban areas, for instance. All of these factors (and more) come into play when you're trying to find the optimal location for a vacation home. Similarly, if you work in the retail industry and are assessing which place is the best location for a new store, there are a lot of criteria that will influence your decision—how close the potential sites are to your existing store, how close that area is to similar stores, the population of the area, the average household income for each neighborhood, and the rental price of the property, to name just a few. GIS can handle all of these spatial factors by combining them (see **Figure 6.12** for an example of combining several data layers).

A **GIS model** is a representation that combines these spatial dimensions or characteristics together in an attempt to describe or explain a process or to predict results. For instance, trying to predict where future urban developments will occur based on variables such as a location's proximity to existing urban areas, the price of the land parcels, or the availability of transportation corridors could be performed using GIS modeling, with the end result being a map of the places most likely to convert to an urban land use. There are many different types of GIS models with different features and complexities to them, and it's far beyond the scope of this book to catalog and examine all of them.

GIS models are often used to examine various criteria together to determine where, spatially, some phenomenon is occurring or could potentially occur. These are often used in decision-making processes or to answer questions (like the previous example of where to build a vacation home). Another example of this would be in determining which parcels of land surrounding Civil War battlefields are under the greatest pressure to be converted to urban land use. Battlefield land is often considered a historic

suitability index a system whereby locations are ranked according to how well they fit a set of criteria

GIS model a representation of the factors used for explaining the processes that underlie an event or for predicting results

FIGURE 6.12

An example of combining multiple GIS data layers.

Administrative areas

Site addresses, regulated uses, restrictions

Separated rights and encumbrances

Ownership and tax parcels

Parcel framework

Corners and boundaries

Survey network

Digital orthophotography and hydrography

monument to the men who died in the Civil War, and while the core areas of several sites are under the protection of the U.S. National Park Service, there are many other sites that have been developed or are in danger of residential or commercial development, destroying the "hallowed ground" of history.

GIS can be used to combine a number of factors that influence where new developments are likely to occur (such as proximity to big cities, existing urban areas, or water bodies) in a fashion similar to site suitability in order to try to find the areas under the greatest development pressures. This type of information could be used by preservationists attempting to purchase lands with the intention of keeping them undeveloped.

The factors that contribute to a phenomenon can be constructed as layers within GIS, weighted (to emphasize which layers have a greater effect than others), and combined. The end result will be a ranking of areas to determine their suitability. One example of this process is referred to as multi-criteria evaluation (**MCE**), which takes layers that have been standardized and initially ranked against each other, then weights and combines them. The MCE process produces a map showing various "pressures" at each location (along the lines of the suitability index map) to determine which places have the greatest influence placed on them (based on the combination of all the input layers). This acts like a more complex form of the basic site suitability and can be used in decision-making processes—like determining which areas of battlefield land have the greatest developmental pressures on them to convert to an urban land use. Other problems, such as locating suitable animal habitats, areas of high flood potential, or places for retail sites, can be approached using similar techniques.

Another example of a GIS model is the Land Transformation Model (LTM), a tool used to combine various weighted raster layers in order to determine where areas have changed over to an urban land use, and then to predict where areas are likely to change over to an urban land use in the future. See **Figure 6.13** for an example of the output of the LTM's predictive capability, covering areas in Michigan up to the year 2040. Also see *Hands-On Application 6.3: The Land Transformation Model* for further examination of the LTM as a GIS model.

MCE multi-criteria evaluation—the use of several factors (weighted and combined) to determine the suitability of a site for a particular purpose

FIGURE 6.13

The use of the Land Transformation Model to examine and predict areas that will change to an urban land use in the future.

Hands-On Application 6.3

The Land Transformation Model

The Land Transformation Model is headquartered at Purdue University. For further background on LTM as an example of a GIS model, go to **http://ltm.agriculture.purdue.edu/default_ltm.htm**. The actual model itself (and sample datasets) can be downloaded, as well as a selection of graphics and papers related to the model, if you're feeling adventurous.

Expansion Questions

➤ For what types of applications has the Land Transformation Model been used?

➤ What kinds of GIS layers are used in this model for forecasting urban development?

Thinking Critically with Geospatial Technology 6.1

What Are Potential Societal or Policy Impacts of GIS Models?

The output of GIS models is often used to aid in decision-making processes, and could have a significant impact in areas of concern to all of us. For instance, trying to determine which sites would make the most suitable locations for a new casino development would greatly affect the social, economic, and land use conditions of an area. If the end result of a model is to determine what areas of a residential area are under the greatest potential threat of flooding, it could have a strong impact on an area's real estate market. What kinds of impacts can these types of models have on local, state, or federal policy?

There is also this question to consider: What if the model's predictions are incorrect? If a model predicts that one area will be the optimal location for a new shopping center, and it turns out not to be, how could this kind of error affect this area and future decision making? Keep in mind that you could have a great model or design, but have inaccurate data for one (or more) of the layers. If that's the case, how can having this kind of poor-quality data in a model affect the output (and what kinds of impacts could the results of models using inaccurate data have on policy decisions)?

Chapter Wrap-Up

Spatial analysis is really at the core of working with geospatial data and GIS. While there are many more types of spatial analysis that can be performed with GIS, this chapter simply introduces some basic concepts of how it can be performed (and some applications of it). Check out *Spatial Analysis Apps* for some mobile apps related to this chapter's concepts, as well as *Spatial Analysis in Social Media* for some blogs and YouTube videos concerning spatial analysis. In the next chapter, we're going to look at taking GIS data (for instance, the results of an analysis or a model) and using it to make a map.

This chapter's labs will return you to working with GIS software (either QGIS or ArcGIS), to start in on more functions, such as querying and buffers.

Important note: The references for this chapter are part of the online companion for this book and can be found at **www.macmillanlearning .com/catalog/shellitoigt4e**.

Spatial Analysis Apps

Here's a sampling of available representative GIS and spatial analysis apps for your smartphone or tablet. Note that some apps are for Android, some are for Apple iOS, and some may be available for both.

➤ **Crime Maps:** An app that contains GIS maps and location information for numerous types of crimes across the United States

➤ **mySoil:** An app from the British Geological Survey that incorporates soils data from across Britain and Europe

Spatial Analysis in Social Media

Here's a sampling of some representative YouTube videos and blogs related to this chapter's concepts.

You Tube On YouTube, watch the following videos:

➤ **GIS Spatial Analyst Tutorial Using John Snow's Cholera Data** (a video showing ArcGIS spatial analysis techniques using Snow's nineteenth-century data): **www.youtube.com/ watch?v=isVD8u6WrG4**

➤ **Ten Questions for Esri—Spatial Analysis** (a video answering introductory and advanced questions concerning GIS applications of spatial analysis): **https://www.youtube.com/ watch?v=nY6QNZwZYp8**

For further up-to-date info, read up on this blog:

➤ **ArcGIS Blog—Analysis & Geoprocessing Category** (blog posts concerning spatial analysis and geoprocessing topics): **http://blogs. esri.com/esri/arcgis/category/subject-analysis-and-geoprocessing**

Key Terms

GIS Spatial Analysis: QGIS Version

This chapter's labs build on the basic concepts of GIS from Chapter 5 by applying several analytical concepts to the GIS data that you used in that chapter's labs. This lab also expands on Chapter 5 by introducing several more GIS tools and features and explaining how they are used.

Similar to the three Chapter 5 Geospatial Lab Applications, three versions of this lab are also provided. The first version (*Geospatial Lab Application 6.1: GIS Spatial Analysis: QGIS Version*) uses the free QGIS package. The second version (*Geospatial Lab Application 6.2: GIS Spatial Analysis: ArcGIS Pro Version*) provides the same activities for use with ArcGIS Pro. There is also a third version (*Geospatial Lab Application 6.3: GIS Spatial Analysis: ArcMap Version*) available online. Although ArcGIS Pro contains more functions than QGIS, these two programs share several useful spatial analysis tools.

Objectives

The goals for you to aim for in this lab are:

➤ To build simple and compound database queries

➤ To extract the results of a query into separate GIS layers

➤ To count the number of features within a feature

➤ To overlay two features for spatial analysis

➤ To create and use buffers around objects for spatial analysis

Using Geospatial Technologies

The concepts you'll be working with in this lab are used in a variety of real-world applications, including:

➤ Law enforcement, where officers use spatial analysis techniques for examining "hot spots" of crimes as well as examining where different types of crimes are occurring in relation to other locations

➤ Emergency management services, which use spatial analysis for a variety of applications, such as researching what populated areas are covered (or not covered) by tornado sirens

[© ZUMA Press, Inc./Alamy]

Obtaining Software

The version of QGIS used in this lab is 2.18, and is available for free download at **http://qgis.org/downloads**.

Important note: Software and online resources can change fast. This lab was designed with the most recently available version of the software at the time of writing. However, if the software or Websites have significantly changed between then and now, an updated version of this lab (using the newest versions) will be available online at **www.macmillanlearning.com/catalog/shellitoigt4e**.

Lab Data

Copy the folder **Chapter6QGIS**—it contains a folder called CalSpatial which is the file geodatabase (see page 148) you'll be using in this lab, in the same way you used a file geodatabase in Chapter 5. This lab's data were downloaded from The National Map (see page 10) at no charge and the population information was taken from the U.S. Census Bureau. The GIS layers in the geodatabase have already been projected for you to the U.S. National Atlas Equal Area projection.

Localizing This Lab

The data used in this lab are a variety of features of attributes of the state of California drawn from The National Map and the U.S. Census Bureau. However, you can easily work with the same data from your own state as The National Map makes the same layers available for all 50 states. From The National Map, download the Structures and Boundaries geodatabases for your own state, which contain the layers used for California in this lab. If you are in, for instance, Arkansas, you can download county boundaries and the structures layer for Arkansas from The National Map. Similarly, the U.S. Census Bureau has population values for each county, which can be joined to the county boundaries layer.

6.1 Some Initial Pre-Analysis Steps in QGIS

If needed, refer back to *Geospatial Lab Application 5.1: GIS Introduction: QGIS Version* for specifics on how to do the following.

1. Start **QGIS**.
2. From the **C:\Chapter6QGIS** folder, add the following feature classes: **CalBounds** and **CalStructs**.
3. Change the color schemes and symbology of each layer to make them distinctive from each other when you'll be doing analysis.
4. Position the layers in the Map Legend so that all layers are visible (that is, the CalBounds layer is at the bottom of the Map Legend and the CalStructs layer arranged on top of it).

6.2 Building Queries in QGIS

To begin GIS analysis, we'll start by building a query, which is a standard feature in QGIS. We'll start off with a simple query (that is, a query that uses only one expression).

1. To build a query in QGIS, select the layer you want to query (in this case, CalBounds), right-click on its name in the Map Legend, and choose **Properties**. In the Layer Properties dialog box, select the **General** tab, and at the bottom of the dialog box, click on the **Query Builder** button.

2. For your first simple query, we're going to find out how many counties in California have a population of more than half a million people. To build this query, first select the Field to use. In this case, double-click **Pop2010**.

3. Next, select an **Operator** to use. In this case, select the > operator.

4. Lastly, select the appropriate **Values**. For this query, we won't be using any values from the Pop2010 field in the attribute table, so you can just type: **500000**.

5. Your final query should read: "Pop2010" > 500000. Click **OK** to run the query. After it runs, the query builder will close and you'll return to the **Layer Properties** dialog box. Close that dialog box as well.

6. Open the CalBounds' layer's attribute table by right-clicking on its name in the Map Legend and selecting **Open Attribute Table**. The number of filtered records (that is, those that met the results of your query) will be shown at the top of the attribute table dialog box—you'll also see that only those counties that met the results of your query are displayed in the Map View.

> **Question 6.1** How many counties in California had a 2010 population of more than 500,000 persons?

7. Return to the Query Builder for the CalBounds layer and click the **Clear** button at the bottom to remove your previous query and also clear the filtered results and reset things to all 58 counties.

8. Create a fresh new query, this time querying for counties with a population of less than 50,000. Your final query should read: "Pop2010" < 50000. Both the results in the attribute table and of the county polygons appearing in the Map View will be filtered to show only the records/objects that meet the results of the query.

> **Question 6.2** How many counties in California had a 2010 population of less than 50,000 persons?

9. Simple queries are very straightforward, but often you'll need to use more than one query expression, therefore you'll need to build a compound query. To get started with a compound query, return to the CalBounds layer's Query Builder and Clear the previous query.

10. Build a query that reads: **"Pop2010" > 50000** but don't click OK just yet. This is just one part of the compound query.

11. Next click on the **AND** button. This will add a Boolean connector to your query.

12. Last, after the AND, build the other half of the compound query that reads: **"Pop2010" < 100000**.

13. When your compound query is ready, click **OK**.

> **Question 6.3** How many counties in California had a 2010 population of more than 50,000 persons but also less than 100,000 persons?

14. Return to the Query Builder for the CalBounds layer and Clear the current query. We'll now build one more compound query from the CalBounds layer that will bring together three query statements instead of just two. To build the first part of the query, click on **County_Name** in the **Fields** box. Then click on the = button, and then in the Values box, click on **All**. This will show all of the entries in the County_Name column of the attribute table and will allow you to choose directly from the list. From the list of Values, click on **Los Angeles**. At this point, your query should read: "County_Name" = 'Los Angeles'.

15. Because we're building a compound query, we'll need a Boolean connector, so click on **OR**.

16. Next, build the second part of the compound query so that it reads: "County_Name" = 'San Diego'.

17. Next, click on **OR** to add another Boolean connector.

18. Last, build the third part of the compound query so that it reads: "County_Name" = 'Orange'.

19. At this point, your compound query should read: "County_Name" = 'Los Angeles' OR "County_Name" = 'San Diego' OR "County_Name" = 'Orange'. When the query is set, click **OK**.

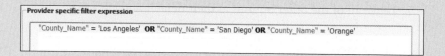

20. You'll see that the attribute table will be filtered to only the three records corresponding to the counties you queried for.

> **Question 6.4** What is the combined total 2010 population of these three counties?

> **Question 6.5** Why did we use "OR" as the Boolean connector between the queries each time instead of "AND?" How would the results have been different if we had used "AND" instead of "OR" each time in this compound query?

21. Return to the Query Builder and Clear the filtered results, then close any open dialog boxes.

6.3 Exporting Query Results to Their Own Layer in QGIS

1. When you build a query you can see the results in the Map View and the attribute table, but those results remain part of the layer you're querying. If you wanted to create a map of just those results or use those results in future analysis, you'd have to move them into their own layer. That's what we'll do next—build a query, then export the results of that query to its own layer. To begin, right-click on the **CalStructs** layer, select **Properties**, then click on **Query Builder**.

2. Using the techniques from the previous section, build a query that finds all of the values in the field called FCode that are equal to 74026. This is the code assigned to buildings that are classified as Fire Stations/EMS Stations.

3. When the query is ready, click **OK** in the Query Builder to run the query. You'll see the filtered results shown in the Map View (that is, the only points being shown are those that represent Fire Stations/EMS Stations).

4. To take these points and put them into their own layer, right-click on **CalStructs** and choose **Save As…**.

5. In the **Save vector layer as…** box that opens, under Format, select **ESRI Shapefile**. This means the results will be exported into Shapefile format.

6. Next to File name, click the **Browse** button and navigate to the **C:\Chapter6QGIS** folder, and name the file **Firestations**.

7. Leave all of the other options at their default and click **OK**.

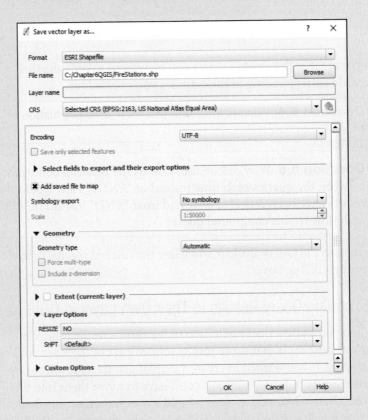

8. You'll see a new layer added to the Layers Panel called Firestations that only contains the points you queried for and then exported.

> **Question 6.6** How many points are in the Firestations layer?

9. Return to the Query Builder and Clear the filter. Build a new query that finds all of the values in the field called FCode that are equal to 80012. This is the code assigned to buildings that are classified as Hospitals/Medical Centers.

10. After running the query, export the results to their own ESRI Shapefile, save the file in your **C:\Chapter6QGIS** folder, and call the file **Hospitals**. You'll then have the new layer added to the Layers Panel.

> **Question 6.7** How many points are in the Hospitals layer?

11. Clear the results of the query from the query builder and close any open tables or windows. You can also turn off the CalStructs layer from the Layers Panel.

6.4 Spatial Analysis: Counting Points in Polygons in QGIS

1. Our first type of spatial analysis is to figure out how many hospitals and fire stations are in each county. Doing this visually would be a long and imprecise process (just imagine zooming in to each county, manually counting up the hospital points in each, having to differentiate points that are close together, and closely zoom in on county borders to make sure a point is in one county and not another). Spatial analysis techniques give us a much better, much more accurate means of doing this. What we will be doing is joining the FireStations points layer to the CalBounds polygon layer and creating a third layer in the process. To begin, from the **Vector** pull-down menu, select **Analysis Tools**, then select **Count points in polygon**.

2. For the Polygons option, select **CalSpatial CalBounds** (this is the polygon boundaries of the counties).

3. For the Points option, select **FireStations** (these are the points representing Fire Station locations).

4. For the Count field name type **FIREPOINTS** (this will be what the new field is called that contains the value of how many Fire Stations are in each county).

5. Leave the other options alone and click **Run**.

Count points in polygon	? ✕
Parameters Log Run as batch process...	**Count points in polygon**
Polygons	This algorithm takes a points layer and a polygon layer and counts the number of points from the first one in each polygons of the second one.
CalSpatial CalBounds [EPSG:2163] ▾ ... 🔊	
Points	A new polygons layer is generated, with the exact same content as the input polygons layer, but containing an additional field with the points count corresponding to each polygon.
FireStations [EPSG:2163] ▾ ... 🔊	
Count field name	
FIREPOINTS	
Count	
[Create temporary layer] ...	
✖ Open output file after running algorithm	
0%	
	Run Close

6. A new polygon layer called COUNT is now added to the Layers Panel. Open its attribute table and scroll across to the far right end of it. It will be the same attribute table as CalBounds but with a new field called Firepoints that will be a count of how many fire station points are within each county polygon.

> **Question 6.8** How many fire stations are in Ventura County? How many fire stations are in Sutter County?

7. Redo the "Count points in polygon" operation, but this time we want to count how many hospitals are in each county. For the Polygons option, select **CalSpatial CalBounds**, for the Points option, select **Hospitals**, and for the Count field name type **HOSPITALPOINTS** (this will be what the new field is called that contains the value of how many Hospitals are in each county).

> **Question 6.9** How many hospitals are in Ventura County? How many fire stations are in Sutter County?

6.5 Spatial Analysis: Overlaying Features in QGIS

1. Our next type of spatial analysis is to figure out which county each of the hospitals and fire stations is in. Like counting the points, doing this process visually would take a long time and be very imprecise (just imagine zooming in to each point, manually clicking on the point and its county, recording that information, and then repeating that process hundreds of times). Again, spatial analysis techniques give us a much better, much more accurate means of doing this. What we're going to do is overlay two layers to get this information (more specifically, we're going to perform a point in polygon overlay). To begin, from the **Vector** pull-down menu, select **Geoprocessing Tools**, then select **Intersection**.

2. For the Input layer option, select **FireStations** (this is the layer that will be doing the intersecting).

3. For the Intersect layer option, select **CalSpatial CalBounds** (this is the layer with which FireStations will be intersecting).

4. Place a checkmark in the box next to **Open output file after running algorithm.**

5. Leave the other options alone and click **Run.**

6. A new point layer called Intersection will be added to the Layers Panel. This new layer is the result of the point in polygon overlay we just performed. Each point will have the attributes of the polygon that it intersected. Open up the new layer's attribute table and you'll see that each record represents a fire station point—then scroll across to the far right of the table and you'll see the attributes of the counties that intersect with that point.

> **Question 6.10** In what county is the Yurok Volunteer Fire Department located?

7. Close the Intersection attribute table and redo the Intersection analysis, this time using Hospitals as the Input layer and again using CalSpatial CalBounds as the Intersect layer. This will intersect the hospital points with the county polygons and create a new point layer of hospitals that have the attributes of the counties with which they intersect.

> **Question 6.11** In what county is Ukiah Valley Medical Center located?

8. Close any open attribute tables when you're done.

6.6 Spatial Analysis: Working with Buffers in QGIS

1. QGIS allows you to create buffers around objects for examination of spatial proximity and gives you the ability to use the spatial dimensions of the buffer as a means of selecting other objects. For instance, in this section, we're going to figure out how many fire stations are within 1 mile of a hospital in California. To get started and create a buffer,

from the **Vector** pull-down menu, select **Geoprocessing Tools**, then select **Fixed Distance Buffer**.

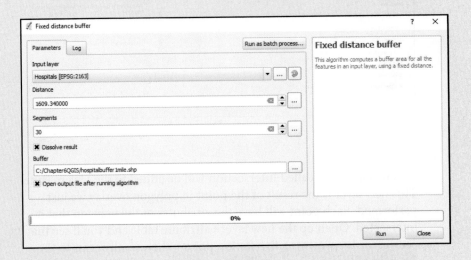

2. We'll start with a 1-mile buffer around the hospitals of California, so for Input layer choose **Hospitals**.

3. For Distance type in **1609.34**. This value is the number of meters equal to 1 mile. Because your Hospitals data are being measured in meters, QGIS will use that for the buffer's units of measurement.

4. Set the Segments option to **30**. This is a measure of the "smoothness" of the buffer being created by QGIS.

5. Place a checkmark in the box next to **Dissolve buffer results**.

6. Under the option for Buffer, click **the button with the three dots**, then choose the option for **Save to file** and navigate to your **C:\Chapter6QGIS** folder. Name the Output shapefile (the buffer that will be created) **Hospitalbuffer1mile.shp** and Save it.

7. Place a checkmark in the box next to **Open output file after running algorithm**.

8. Back in the **Fixed distance buffer dialog** box, click **Run** to create the buffer.

9. After the process runs, a new layer called Buffer will be added to the Layers Panel. Zoom in on some of the hospitals in the Map View to see the circular 1-mile buffers that have been created.

10. Next, QGIS allows you to select features based on their proximity to other features. For instance, you can find out how many fire stations are within the 1-mile buffer you just created and create a

new selection of these fire stations. To do this, select the **Vector** pull-down menu, then choose **Research Tools**, and then choose **Select by location**.

11. For **Layer to select from**, choose **FireStations** (this is the layer from which you want to select features).

12. For **Additional layer (intersection layer)** choose **Buffer** (this is the layer whose shape you want to use to do the selecting).

13. Under **Geometric predicate**, choose **within** (this is the kind of selection you want to do, that is find the hospitals that are within the buffer).

14. For **Precision**, change it to **1.000000**.

15. For **Modify current selection by**, choose **creating new selection**.

16. When all the settings are ready, click **Run**.

17. When the process is done, zoom in closely on some of your buffered areas. You'll see that any fire stations that are within the 1-mile buffer around a hospital are now shown in yellow, indicating that they have been *selected* by QGIS. Open the FireStations attribute table and from the pull-down menu at the bottom of the attribute table, choose **Show Selected Features**. Only those records that correspond with fire stations within the 1-mile buffer around hospitals will be shown in the attribute table.

Question 6.12 How many fire stations are within a 1-mile radius of the hospitals?

18. Next, you'll want to clear those selected features so that no fire stations are selected by clicking on the **Deselect All** button in the FireStations attribute table.

19. Close any open attribute tables or dialog boxes before proceeding.
20. Also, at this point, if you're satisfied with your answers and progress, you can delete both Intersection layers and the Buffer layer from the Layers Panel (to delete a layer, right-click on it and choose **Remove**).

6.7 Spatial Analysis: Putting It All Together in QGIS

1. For this last section you'll be applying the analysis techniques of this lab to answering some questions concerning the state capitol of California. First, build a query of the CalStructs layer where the FCode field is equal to **83006**. That is the classification code for the state capitol building. There will only be one point of the entire layer that matches this query.
2. Next, export that single point into its own layer, as an ESRI shapefile, and call that layer **SC**.

3. Next, we'll want to know in what county that point representing the state capitol building is. Use the intersection tool to do this, by intersecting the SC layer and the CalBounds layer.

> **Question 6.13** In which county is the California state capitol building?

4. Next, we want to know something about the emergency facilities within 5 miles of the state capitol building. Create a 5-mile buffer (note that 5 miles is equal to 8046.72 meters) again using 30 segments around the state capitol.

5. Next, select all of the hospitals within this 5-mile buffer.

> **Question 6.14** How many hospitals are within 5 miles of the California state capitol building?

6. Finally, select all of the fire stations within this 5-mile buffer.

> **Question 6.15** How many fire stations are within 5 miles of the California state capitol building?

Closing Time

You can exit QGIS by selecting **Exit QGIS** from the **Project** pull-down menu. There's no need to save any data in this lab. Now that you have the basics of working with GIS data down and know how to perform some initial spatial analysis, the Geospatial Lab Application in Chapter 7 will describe how to take geospatial data and create a print-quality map, using either QGIS or ArcGIS.

6.2 Geospatial Lab Application

GIS Spatial Analysis: ArcGIS Pro Version

This lab builds on the basic concepts of GIS from Chapter 5 by applying several analytical concepts to the GIS data you used in *Geospatial Lab Application 5.2: GIS Introduction: ArcGIS Pro Version*. This lab also expands on Chapter 5 by introducing several more GIS tools and features and explaining how they are used.

The previous lab (*Geospatial Lab Application 6.1: GIS Spatial Analysis: QGIS Version*) uses the free QGIS software package; however, this lab provides the same activities for use with ArcGIS Pro. Although ArcGIS Pro contains more functions than QGIS, the two programs share several useful spatial analysis tools.

Important note: This version of the lab uses ArcGIS Pro, but the same lab using ArcMap (from the most recent version of ArcGIS Desktop) is available for you if you'd rather use that. *Geospatial Lab Application 6.3: GIS Introduction: ArcMap Version* is available online at: **www.macmillanlearning.com/catalog/shellitoigt4e**

Objectives

The goals for you to aim for in this lab are:

➤ To build simple and compound database queries

➤ To extract the results of a query into separate GIS layers

➤ To count the number of features within a feature

➤ To overlay two features for spatial analysis

➤ To create and use buffers around objects for spatial analysis

Using Geospatial Technologies

The concepts you'll be working with in this lab are used in a variety of real-world applications, including:

➤ Law enforcement, where officers use spatial analysis techniques for examining "hot spots" of crimes as well as examining where different types of crimes are occurring in relation to other locations

➤ Emergency management services, which use spatial analysis for a variety of applications, such as researching what populated areas are covered (or not covered) by tornado sirens

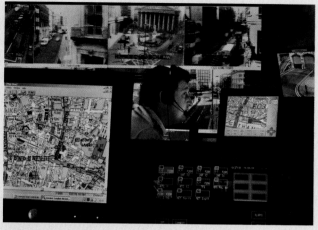

[Barry Lewis/Getty Images]

Obtaining Software

The current version of ArcGIS Desktop (and thus, ArcGIS Pro) is not freely available for use. However, instructors affiliated with schools that have a campus-wide software license may request a 1-year student version of the software online at **http://www.esri.com/landing-pages/education-promo**.

Important note: Software and online resources can change fast. This lab was designed with the most recently available version of the software at the time of writing. However, if the software or Websites have changed significantly between then and now, an updated version of this lab (using the newest versions) will be available online at **www.macmillanlearning.com/catalog/shellitoigt4e**

Lab Data

Copy the folder **Chapter6Pro**—it contains a folder called CalSpatial which is the file geodatabase (see page 148) you'll be using in this lab, like how you used a file geodatabase in Chapter 5. This lab's data were downloaded from The National Map (see page 10) free of charge and the population information was taken from the U.S. Census Bureau. The GIS layers in the geodatabase have already been projected for you to the U.S. National Atlas Equal Area projection.

Localizing This Lab

The data used in this lab are a variety of features of attributes of the state of California drawn from The National Map and the U.S. Census Bureau. However, you can easily work with the same data from your own state as The National Map makes the same layers available for all 50 states. From

The National Map, download the Structures and Boundaries geodatabases for your own state, which contain the layers used for California in this lab. If you are in, for instance, Arkansas, you can download county boundaries and the structures layer for Arkansas from The National Map. Similarly, the U.S. Census Bureau has population values for each county, which can be joined to the county boundaries layer.

6.1 Some Initial Pre-Analysis Steps

If needed, refer back to *Geospatial Lab Application 5.2: GIS Introduction: ArcGIS Pro Version* for specifics on how to do the following.

1. Start **ArcGIS Pro**.
2. When prompted, log in with your Esri Account username and password.
3. When ArcGIS Pro opens, start with a new blank Project and call it **C6Pro** and place it in your **C:\Chapter6Pro** folder.
4. From the **Insert** tab, within the **Project** folder, click on the **New Map** icon and choose **New Map**.
5. From the **C:\Chapter6Pro** folder, add the following feature classes from the CalSpatial geodatabase to the Contents Pane: **CalBounds** and **CalStructs**.
6. Change the color schemes and symbology of each layer to make them distinctive from each other when you'll be doing analysis.
7. Position the layers in the Contents Pane so that all layers are visible (that is, the CalBounds layer is at the bottom of the Contents Pane and the CalStructs layer is on top of it).

6.2 Building Queries in ArcGIS Pro

To begin GIS analysis, we'll start by building a query, which is a standard feature in ArcGIS Pro. We'll start off with a simple query (that is, a query that uses only one expression).

1. To build a query in ArcGIS Pro, select the **Map** tab, within the **Selection** group, and click on the **Select By Attributes** icon.
2. The Geoprocessing Pane will open on the right-hand side of ArcGIS Pro with the Select Layer By Attribute tool.

3. First we'll choose which layer to query. Under **Layer Name or Table View** choose **CalBounds**.

4. Under **Selection type**, choose **New selection**.

5. Next, we will build the SQL query—to do so, click on **Add Clause** to use the interactive query building features of ArcGIS Pro. A new pop-up window will open in which to build the query.

6. From the first pull-down menu, you'll choose the Field you want to use. In this case choose **Pop2010**.

7. From the second pull-down menu, you'll choose the Operator you want to use. In this case choose **is Greater Than**.

8. From the last pull-down menu, you would choose the Value you want to use. For this query, we won't be using any values from the Pop2010 field in the attribute table, so you can just type: **500000**.

9. Your final query should read: Pop2010 is Greater Than 500000. Click **Add** to add this expression to the tool. Back in the Select Layer By Attribute dialog box click **Run** to execute the SQL query.

10. Open the CalBounds' layer's attribute table by right-clicking on its name in the Contents Pane and selecting **Attribute Table**. The number of selected records (that is, those that met the results of your query) will be shown in the attribute table's window and you'll also see that counties that met the results of your query are selected in the Map as well. Both selected records in the attribute table and features in the Map will be shown in cyan.

> **Question 6.1** How many counties in California had a 2010 population of more than 500,000 persons?

11. In the CalBounds attribute table press the **Clear** button at the top of the table to clear the results of the selection and restore the table to not having any records/features selected.

12. Return to the Geoprocessing Pane and the Select Layer By Attributes tool to build a fresh new query. Because we're starting with a brand-new query, you'll want to clear the existing expression by clicking on the **Clear** icon (the eraser) in the tool.

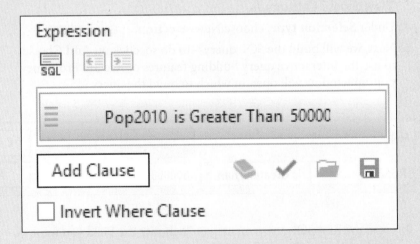

13. Build a new query but this time query for counties with a population of less than 50,000. Your final query should read: Pop2010 is Less Than

50000. **Run** this query and you'll see the results selected in both the CalBounds attribute table and the features in the Map.

> **Question 6.2** How many counties in California had a 2010 population of less than 50,000 persons?

14. Simple queries are very straightforward, but often you'll need to use more than one query expression, therefore you'll need to build a compound query. To get started with a compound query, return to Select Layer By Attributes tool and Clear the previous query.

15. Build a query that reads: **Pop2010 is Greater Than 50000** and click **Add**, but don't click Run just yet. This is just one part of the compound query.

16. Next click on **Add Clause** to add the second part of the compound query.

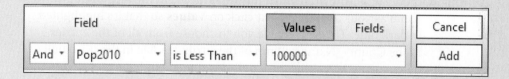

17. In the new dialog box, from the first pull-down menu choose **And**. This is the Boolean connector that will join the first part of the query with the second part of the query.

18. From the other pull-down boxes complete the query so that it reads: **Pop2010 is Less Than 100000**. Then click **Add**.

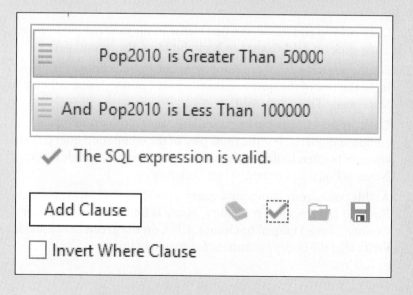

19. Back in the Select Layer By Attribute tool, click on the **green checkmark** and ArcGIS Pro will verify that everything is okay with your compound query.

20. Click **Run** to execute your compound query. Again, you'll see the selected records and polygons in cyan.

> **Question 6.3** How many counties in California had a 2010 population of more than 50,000 persons but also less than 100,000 persons?

21. Return to the Select Layer By Attribute tool and Clear the current query. We'll now build one more compound query from the CalBounds layer that will bring together three query statements instead of just two. Click **Add Clause** and to build the first part of the query, begin with choosing the field called **County_Name** from the first pull-down menu and **is Equal to** from the second pull-down menu as your operator. Next click on **Values** so that it's highlighted in dark blue. This will enable you to choose from all of the entries in the County_Name column of the attribute table in the third pull-down menu. From this third pull-down menu of Values, click on **Los Angeles**. At this point, your query should read: County_Name is Equal to Los Angeles.

22. Click **Add** to add this as the first clause in the new compound query. Then click on **Add Clause** to begin building the second part.

23. For the second part of the compound query, first choose **Or**, then build the rest of the query so that it reads County_Name is Equal to San Diego. Then click **Add**.

24. Click on **Add Clause** to begin building the third part of this compound query. For the third part of the compound query, first choose **Or**, then build the rest of the query so that it reads County_Name is Equal to Orange. Then click **Add**.

25. At this point, your compound query should read: County_Name is Equal to Los Angeles Or County_Name is Equal to San Diego Or County_Name is Equal to Orange. Click on the **green checkmark** to verify that the query is constructed correctly.

26. When the query is set, click **Run**. Again, you'll see the three selected records and polygons in cyan.

> **Question 6.4** What is the combined total 2010 population of these three counties?

> **Question 6.5** Why did we use "OR" as the Boolean connector between the queries each time instead of "AND?" How would the results have been different if we had used "AND" instead of "OR" each time in this compound query?

27. Press the **Clear** button in the CalBounds attribute table to clear all of the selected records and features, then close the CalBounds attribute table.

28. In the Select Layer By Attributes tool, clear the query as well.

6.3 Exporting Query Results to Their Own Layer in ArcGIS Pro

1. When you build a query you can see the selected results in the Map and the attribute table, but those results remain part of the layer you're querying. If you wanted to create a map of just those results or use those results in future analysis, you'd have to move them into their own layer. That's what we'll do next—build a query, then export the results of that query to its own layer. To begin, return to the Select Layer By

Attribute tool, select **CalStructs** as the layer you want to build a query for, and then click **Add Clause** to start a new query.

2. Using the techniques from the previous section, build a query of the **CalStructs** layer so that it finds all of the values in the field called FCode that are equal to 74026. This is the code assigned to buildings that are classified as Fire Stations/EMS Stations.

3. When the query is ready, click **Run** to execute the query. Open the CalStructs attribute table and you'll see the selected records in the table and also their corresponding selected points shown on the map (that is, the cyan points being shown are those that represent Fire Stations/EMS Stations).

4. To take these points and put them into their own layer, right-click on **CalStructs** in the Contents Pane and choose **Data**, then choose **Export Features**.

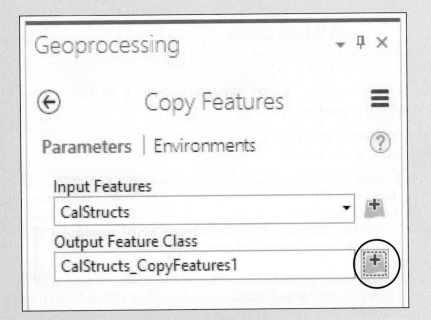

5. The Geoprocessing Pane will open again, this time with the Copy Features tool. Under Input Features, choose **CalStructs**—this is the layer from which you want to take features.

6. Under Output Feature Class, the default name for the exported selected features is CalStructs_CopyFeatures1, which is less than helpful. To change this, press the + **button** next to this option.

7. In the **Output Feature Class** dialog box, navigate to your **C:\Chapter6Pro** folder, then choose the **CalSpatial.gdb** geodatabase as the location where you want to place the exported features.

8. Give this new exported feature class the name FireStations.

9. When you have the new exported layer named correctly and being exported to the right location in the right geodatabase, click **Save**.

10. Back in the Copy Features tool, verify that your Input Features option is **CalStructs** and your new Output Feature Class is called **FireStations**, then click **Run**.

11. You'll see a new layer added to the Contents Pane called Firestations, which only contains the points you selected via the query for and then exported.

> **Question 6.6** How many points are in the FireStations layer?

12. Clear the selected results in the CalStructs attribute table.

13. Return to the Select Layer By Attribute tool and clear the existing query. Build a new query that finds all of the values in the field called FCode that are equal to 80012. This is the code assigned to buildings that are classified as Hospitals/Medical Centers.

14. After running the query, export the selected results to their own feature class and save it in the same place as the FireStations feature class—in your **C:\Chapter6Pro** folder, then in the **CalSpatial.gdb** geodatabase—and call the new output feature class **Hospitals**. You'll then have the new feature class with only the points representing hospitals available to you in the Contents Pane.

> **Question 6.7** How many points are in the Hospitals layer?

15. Clear the results from the query in the Select Layer By Attributes tool.

16. Clear the selected results from the CalStructs attribute table, then you can close that attribute table as well. Turn off the CalStructs layer in the Contents Pane but leave the Hospitals and FireStations layers turned on.

6.4 Spatial Analysis: Counting Points in Polygons in ArcGIS Pro

1. Our first type of spatial analysis is to figure out how many hospitals and fire stations are in each county. Doing this visually would be a long and imprecise process (just imagine zooming in to each county, manually counting up the hospital points in each, having to differentiate points that are close together, and closely zoom in on county borders to make sure a point is in one county and not another). Spatial analysis techniques give us a much better, much more accurate means of doing this. What we will be doing is joining the FireStations points layer to the CalBounds polygon layer and creating a third layer in the process. To begin, from the **Analysis** tab, then within the **Tools** group you'll see a number of analytical tools. To see more (including the tool we want), click on the **down-arrow** of the window for more options.

2. From the expanded list of options that appears, click on the **Summarize Within** icon.

3. The Geoprocessing Pane will open again with the Summarize Within tool.

4. For **Input Polygons**, select **CalBounds** (this is the polygons layer that we will be counting within).

5. For **Input Summary Features**, select **FireStations** (this is the point layer that we want to count).

6. For **Output Feature Class**, press the plus button and then navigate to your **C:\Chapter6Pro** folder, within the **CalSpatial.gdb** geodatabase, and name the new output feature class **SumFireStations**.

7. Put a checkmark in the box next to **Keep all input polygons**.

8. Leave the other options the way they are and then click **Run** to start the process.

9. A new polygon feature class called SumFireStations will be added to the Contents Pane. Open its attribute table and scroll across to its far right-hand side. You'll see that each record is a different California county and a field called Count of Points has been added, which is the total of how many fire station points are within each county polygon.

> **Question 6.8** How many fire stations are in Ventura County? How many fire stations are in Sutter County?

10. Redo the Summarize Within operation, but this time we want to count how many hospitals are in each county. For the **Input Polygons** option, again select **CalBounds**, for the **Input Summary Features** option, select **Hospitals**, and for the **Output Feature Class**, press the **+ button** and then navigate to your **C:\Chapter6Pro** folder, within the **CalSpatial.gdb** geodatabase, and name the new output feature class **SumHospitals**.

> **Question 6.9** How many hospitals are in Ventura County? How many fire stations are in Sutter County?

11. Close any open attribute tables, then turn off both the SumHospitals and SumFireStations layers.

6.5 Spatial Analysis: Overlaying Features in ArcGIS Pro

1. Our next type of spatial analysis is to figure out in which county each of the hospitals and fire stations is located. Like counting the points, doing this process visually would take a long time and be very imprecise (just imagine zooming in to each point, manually clicking on the point and its county, recording that information, and then repeating that process hundreds of times). Again, spatial analysis techniques give us a much better, much more accurate means of doing this. What we're going to do is overlay two layers to get this information (more specifically, we're going to perform a point in

polygon overlay). To begin, from the **Analysis** tab, within the **Tools** group, click on the **Intersect** icon.

2. From the pull-down menus under Input Features, choose **FireStations** for one input feature, then choose **CalBounds** for a second input feature. These are the two layers you're going to overlay using the Intersect tool.

3. For Output Feature Class, press the + **button** and then navigate to your **C:\Chapter6Pro** folder, within the **CalSpatial.gdb** geodatabase, and name the new output feature class **FireIntersect**.

4. Click **Run** to run the Intersect tool.

5. A new point layer called FireIntersect will be added to the Contents Pane. This new layer is the result of the point in polygon overlay we just performed. Each point will have the attributes of the polygon that it intersected. Open up the new layer's attribute table and you'll see that each record represents a fire station point—then scroll across to the far right of the table and you'll see the attributes of the counties that intersect with that point.

> **Question 6.10** In what county is the Yurok Volunteer Fire Department located?

6. Close the FireIntersect attribute table and redo the Intersect analysis, this time using **Hospitals** and **CalBounds** as the Input Features, and create a new Output Feature Class called **HospIntersect**. This will intersect the hospital points with the county polygons and create a new point layer of hospitals that have the attributes of the counties with which they intersect.

> **Question 6.11** In what county is Ukiah Valley Medical Center located?

7. Close any open attribute tables when you're done.

6.6 Spatial Analysis: Working with Select By Location in ArcGIS Pro

1. ArcGIS Pro allows you to select features based on their proximity to other features. For instance, in this section, we're going to figure out how many fire stations are within 1 mile of a hospital in California. To get started, choose the **Map** tab, within the **Selection** group, then click on the **Select By Location** icon.

2. The Geoprocessing Pane will open again, this time with the Select Layer By Location tool. In Select Layer By Location, you'll specify one or more layers as the Input Feature layer (the layer whose spatial dimensions you'll be using to perform the selection), and a second layer as the Selecting Features layer (the layer from which you'll be selecting features). There are several different ways these two layers can spatially intersect with each other—however, in this case, you'll be finding which of the features from the fire stations layer (the Selecting Features layer) are within a 1-mile distance of the hospitals (the Input Feature layer).

3. For the Input Feature Layer, choose **FireStations**.

4. For the Relationship, choose **Within a distance**.

5. For the Selecting Features, choose **Hospitals**.

6. For the Search Distance, choose **1** and then choose **Miles**.

7. For Selection type, choose **New selection**.

8. When all the settings are ready, click **Run**.

> **Question 6.12** How many fire stations are within a 1-mile radius of the hospitals?

9. Next, you'll want to clear those selected features so that no fire stations are selected. To do so, choose the **Map** tab, within the **Selection** group, and click on the **Clear** icon.

10. Close any open attribute tables or dialog boxes before proceeding.

11. Also, at this point, if you're satisfied with your answers and progress, you can delete the following layers from the Contents Pane: FireIntersect, HospIntersect, SumFireStations, and SumHospitals (to delete a layer, right-click on it and choose **Remove**).

6.7 Spatial Analysis: Putting It All Together in ArcGIS Pro

1. For this last section you'll be applying the analysis techniques of this lab to answering some questions concerning the state capitol of California. First, build a query of the CalStructs layer where the FCode field is equal to **83006**. That is the classification code for the state capitol building. There will only be one point of the entire layer that matches this query.

2. Next, export that single point into its own layer, as a feature class within the **CalSpatial.gdb** geodatabase, and call that layer **SC**.

3. Next, we'll want to know in what county that point representing the state capitol building is. Use the Intersect tool to do this, by intersecting the SC layer and the CalBounds layer.

> **Question 6.13** In which county is the California state capitol building located?

4. Next, we want to know something about the emergency facilities within 5 miles of the state capitol building. Use Select Layer By Location to select all of the hospitals within 5 miles of the state capitol building.

> **Question 6.14**   How many hospitals are within 5 miles of the California state capitol building?

5. Finally, select all of the fire stations within 5 miles of the state capitol building.

> **Question 6.15**   How many fire stations are within 5 miles of the California state capitol building?

Closing Time

Exit ArcGIS Pro by selecting the **Project** tab, and then selecting **Exit**. There's no need to save any data in this lab. Now that you have the basics of working with GIS data down and know how to perform some initial spatial analysis, the Geospatial Lab Application in Chapter 7 will describe how to take geospatial data and create a print-quality map, using either QGIS or ArcGIS.

Using GIS to Make a Map

Scale, Map Elements, Map Layouts, Type, Thematic Maps, Data Classification Methods, Color Choices, and Digital Map Distribution Formats

One thing that makes spatial data unique is that it can be mapped. Once you've used GIS to create the boundaries of a Civil War battlefield and analyzed its proximity to new housing and commercial developments, the next step is to make a map of the results of your work. However, making a good map with GIS involves more than taking the final product of the analysis, throwing on a title and legend, clicking on PRINT, and then calling it a day. There are numerous design elements and considerations that should be taken into account when you're laying out a map. In fact, there's a whole art and science of mapmaking (called **cartography**), involving things such as color selection, the positioning of items on the map, and what kind of message the final map product is intended to convey to the reader.

A **map** is a representation of spatial data, which is designed to convey information to its reader. Making digital maps with GIS offers a wide variety of options for creating an end product, but there are a few considerations to take into account before you start the mapping process. For example, say there's a park nearby—a municipal wooded area that's used by the local populace for walking trails, biking, picnicking, or walking dogs. The only map of note is an old hand-drawn one you can pick up at a kiosk at the park entrance, and the local government wants to update it. As you're the local geospatial technology expert, they come to you to do the job. Before you break out your GIS tools, there are some basic questions to ask that are going to define the map you design.

First, what's the purpose of the map, and who will be using it? If the park map you've been asked to design is intended for the park's visitors to use, then some of the important features the map should contain are locations of park entrances, parking, comfort stations, and designated picnic areas. It should be at an appropriate scale, the trails should be accurately mapped, and trailhead locations and points of interest throughout the park

cartography the art and science of creating and designing maps

map a representation of geographic data

Thinking Critically with Geospatial Technology 7.1

Why Is Map Design Important?

Why is a map's appearance so important? After all, you could argue that a poorly put together map of New Zealand is still (at the end of the day) a map of New Zealand. However, a well-designed map is going to be more useful than a poorly designed one—but why? Why are things such as the choice of colors, lettering, and placement of items on the map critical to the usefulness of a map? Consider what kind of effect a poorly designed map might have, and what sort of impact bad design could make on the usefulness of, say, a map of a local county fair, a real estate map of a new subdivision, a promotional map showing the location of a small business, or a map of a proposed new urban development readied for promotional purposes. How can the design of a map influence the way the map's information is perceived by its readers?

visual hierarchy how features are displayed on a map to emphasize their level of prominence

should be highlighted. If the map is to be used for zoning purposes, such things as exact boundaries, road systems, and parcel information will likely be some of the most important factors. This is part of the **visual hierarchy** of the map, which determines those elements of the map that are most important (and thus, most prominently displayed) and what information is less important (and therefore not so visually conspicuous).

Second, is the information on the map being effectively conveyed to the map reader? For instance, if you're designing a trail map, should all trails be named or marked, or will that cause too much congestion and clutter? Will the colors and symbols be easily understood by novice map readers who just want to find their way around the park? The trail map will likely not need an inset map of the surrounding area to put the park into a larger context, nor should the map reader be left guessing the meanings of the symbols used on the map.

Third, is the map well designed and laid out properly as a representation of the park? A good map should be well balanced with regard to the placement of the various map elements, for best ease of use by the reader. This chapter will examine several of these cartographic design and data display functions, and by the time we reach the labs, you'll be designing a professional-looking map of your own using GIS.

How Does the Scale of the Data Affect the Map (and Vice Versa)?

geographic scale the real-world size or extent of an area

A basic map item should be information about the scale of the map—and there are a couple of different ways of thinking about scale. First, there's the **geographic scale** of something—things that take up a large area on the ground (or have large boundaries) would be considered large in terms of geographic scale. Study of a global phenomenon occurs on a much larger geographic scale than study of something that exists at a city level, which would be a much smaller geographic scale. Something different is **map scale**, a value representing that x number of units of measurement on the map equals y number of units in the real world. This relationship between the real world and the map can be expressed as a representative fraction (**RF**). An example of an RF would be a map scale of 1:24000—a measure of 1 unit on the map would be equal to 24,000 units in the real world. For instance, measuring 1 inch on the map would be the same as 24,000 inches in the real world, or 1 foot on the map is equal to 24,000 feet in the real world (and so on).

Maps are considered **large-scale maps** or **small-scale maps** depending on that representative fraction. Large-scale maps show a smaller geographic area and have a larger RF value. For instance, a 1:4000-scale map is considered a large-scale map—due to the larger scale, it shows a smaller area. The largest-scale map you can make is 1:1—where 1 inch on the map is equal to 1 inch of measurement in the real world (that is, the map will be the same size as the ground you're actually mapping—a map of a classroom will be the same size as the classroom itself). Conversely, a small-scale map has a smaller RF value (such as 1:250,000) and shows a much larger geographic area.

For instance, on a very small-scale map (such as one that shows the entire United States), cities will be represented by points, and likely only major cities will be shown. On a slightly larger-scale map (one that shows all of the state of New York), more cities are likely to be shown as points, along with other major features (additional roads can be shown as lines, for example). On a larger-scale map (one that shows only Manhattan), the map scale allows for more detail to be shown—points will now show the locations of important features and many more roads will be shown with lines. On an even larger-scale map (one that shows only a section of lower Manhattan) buildings may now be shown as polygon shapes (to show the outline or footprint of the buildings) instead of points, and additional smaller roads may also be shown with lines.

The choice of scale will influence how much information the map will be able to convey and what symbols and features can be used in creating the map in GIS. **Figure 7.1** shows a comparison between how a feature (in this case, Salt Lake City International Airport) is represented on large-scale and small-scale maps. The actual sizes of the maps vary greatly, but you can see that more detail and definition of features is available on the larger-scale map than on the smaller-scale one (also see *Hands-On Application 7.1: Powers of 10—A Demonstration of Scale* for a cool example of visualizing different scales).

The same holds true for mapping of data—for instance, the smaller-scale map of the whole state of New York could not possibly show point locations of all of the buildings in Manhattan. However, as the map scale grows larger, different types of information can be conveyed. For instance, in Figure 7.1,

map scale a metric used to determine the relationship between measurements made on a map and their real-world equivalents

RF representative fraction—a value indicating how many units of measurement in the real world are equivalent to how many of the same units of measurement on a map

large-scale map a map with a higher value for its representative fraction; such maps will usually show a smaller geographic area

small-scale map a map with a lower value for its representative fraction; such maps will usually show a larger geographic area

Salt Lake City International Airport—a comparison of its representation on a 1:24,000 scale map (left) and a 1:100,000 scale map (right). *[Source: Utah State Geographic Information Database (SGID)]*

cartographic generalization the simplification of representing items on a map

the large-scale map can convey much more detail concerning the dimensions of the airport runways, while the smaller-scale map has to represent the airport as a set of simplified lines. This kind of **cartographic generalization** (or the simplification of representing items on a map) is going to affect the kind of GIS data you're able to derive from it. For instance, if you digitize the lines of the airport runways, you'll end up with two very different datasets (one more detailed, one very generalized).

This relationship between the scale of a map and data that can be derived from a map can be a critical issue when you're dealing with geospatial data. For instance, say you're using GIS for mapping a university campus. At this small geographic scale, you're going to require detailed information that fits your scale of analysis. If the hydrologic and transportation data you're working with is derived from 1:250,000-scale maps, it's likely going to be way too coarse to use. Data generated from smaller-scale maps is probably going to be incompatible with the small geographic scale with which you're working. For instance, digitized features on a small-scale map (like a 1:250,000 scale) are going to be much more generalized than data derived from larger-scale maps (like a 1:24,000 scale), or from sources such as aerial photos taken at a larger scale (for example, 1:6000). (See Chapter 9 for more information on using aerial photos for analysis.)

Hands-On Application 7.1

Powers of 10—A Demonstration of Scale

Though it's not a map, an excellent demonstration of scale and how new items appear as the scale changes is available at **http://micro.magnet.fsu.edu/primer/java/scienceopticsu/powersof10**.

This Website (which requires Java to run properly) shows a view of Earth starting from 10 million light-years away. Then the scale changes to a view from 1 million light-years away, and then 100,000 light-years away (a factor of 10 each time). The scale continues changing until it reaches Earth—and then continues all the way down to sub-atomic particles. Let it play through, and then use the manual controls to step through the different scales for a better view of the process.

Expansion Questions

➤ Examine the video at the following scales: 1000 km, 100 km, 10 km, 1 km, and 100 m. The imagery at each of these scales could be used as a basemap for digitizing GIS data—what kinds of features and phenomena would be most appropriate to map at each scale?

➤ Using the same five scales as the previous question, what kinds of features and phenomena would be the least appropriate to map at each scale?

What Are Some Design Elements Included in Maps?

There are several elements that should show up on a good map. For instance, someone reading a map should be able to figure out what scale the map is fairly quickly. A map element might simply be text of the RF (such as 1:24,000, 1:100,000, or whatever the map scale might be). A **scale bar** is a graphical representation of equivalent distances shown on a map (**Figure 7.2**). The scale bar provides a means of measuring the map scale

scale bar a graphical device used on a map to represent map scale

FIGURE 7.2

Examples of various scale bars. *[Used by permission. Copyright © 2015 Esri. All rights reserved.]*

Chapter 7 USING GIS TO MAKE A MAP

FIGURE 7.3

Examples of various north arrows. *[Used by permission. Copyright © 2015 Esri. All rights reserved.]*

itself, except that a measurement of *x* distance on the scale bar is equal to *y* units of distance in the real world.

A second map element is a **north arrow**, a graphical device used to orient the direction of the map. However the map is oriented, the north arrow is used to point in the direction that is due north. A north arrow may sometimes be drawn as a compass rose, which shows all cardinal directions on the map. North arrows can be as simple as an arrow with the letter "N" attached, or as complex as a work of art. See **Figure 7.3** for examples of north arrows used in map design.

Another item is the map's **legend**—a guide to what the various colors and symbols on the map represent. A good legend should be a key to the symbology of the map (see **Figure 7.4** for an example of a map legend). Because the legend is the part of the map where you can explain things to the map's reader, you may want to call it something other than "legend." Instead, consider calling the legend something more useful like "park trail guide" or "county population change."

The choice of **type** used on the map is also important, as it's used for such things as the title, the date of the map's creation, the name of the map's creator, the origin of the data sources used to make the map, and the

north arrow a graphical device used to show the orientation of the map

legend a graphical device used to explain what the various map symbols and colors represent

type the lettering used on a map

FIGURE 7.4

Examples of various legend items. *[Used by permission. Copyright © 2015 Esri. All rights reserved.]*

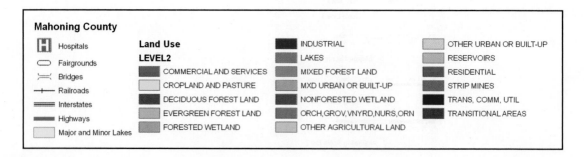

Mahoning County			
Hospitals	**Land Use**	INDUSTRIAL	OTHER URBAN OR BUILT-UP
Fairgrounds	**LEVEL2**	LAKES	RESERVOIRS
Bridges	COMMERCIAL AND SERVICES	MIXED FOREST LAND	RESIDENTIAL
Railroads	CROPLAND AND PASTURE	MXD URBAN OR BUILT-UP	STRIP MINES
Interstates	DECIDUOUS FOREST LAND	NONFORESTED WETLAND	TRANS, COMM, UTIL
Highways	EVERGREEN FOREST LAND	ORCH,GROV,VNYRD,NURS,ORN	TRANSITIONAL AREAS
Major and Minor Lakes	FORESTED WETLAND	OTHER AGRICULTURAL LAND	

Map Source: Esri Data   Map Source: Esri Data

Map Source: Esri Data   **Map Source: Esri DATA**

Map Source: Esri Data   **Map Source: Esri Data**

Map Source: Esri Data   Map Source: Esri DATA

Map Source: Esri Data   Map Source: Esri Data

FIGURE 7.5

Examples of various fonts available for type in map layouts.

lettering attached to the map's features. GIS programs will usually allow you to **label** map features (adding things like the names of rivers, roads, or cities to the map) using an automatic placement tool or allowing you to select, move, and place map labels interactively.

label text placed on a map to identify features

When you're selecting the type to use for particular map elements, a variety of different **fonts** (or lettering styles) are available. GIS programs (like word-processing programs such as Microsoft Word) will often have a large number of fonts to select from—everything from common fonts (like Arial or Times New Roman) to much flashier fonts (such as Mistral or Papyrus). See **Figure 7.5** for some examples of different versions of type fonts available from Esri. Often, a map will contain only two different fonts, carefully selected to fit well with each other without clashing. Using too many fonts, or crafting a map's type out of several of the more elaborate fonts, is likely to make a map difficult to read and reduce its usefulness.

fonts various styles of lettering used on maps

In GIS, a map is put together by assembling all of the elements together in a **layout**. A good way to think of a layout is as a digital version of a blank piece of paper on which you'll arrange the various map elements together in a cartographic design to create a version of what will ultimately be your map. Sometimes, the software (such as ArcGIS) will include several **map templates**, which provide pre-created designs for your use. Using a template will take your GIS data and place things like the title, legend, and north arrow at pre-determined locations and sizes on the layout. Templates are useful for creating a quick printable map layout, but GIS will also allow you to design a map the way you want it with regard to size, type, and placement of elements.

layout the assemblage and placement of various map elements used in constructing a map

map templates pre-made arrangements of items in a map layout

When you're designing a layout, it's important to not simply slap a bunch of elements on the map and decide that the map's ready. Balancing the placement of items is important, partly so as not to overload the map with too much information, but also to provide a useful, readable map. Good map balance provides a means of filling in the "empty" spaces on the map, so that it doesn't have all its information crammed into the top or bottom and leave other parts of the map blank. Items should be of uniform size (instead of, say, making the north arrow gigantic just to fill some empty space) and placed in proportion to one another (see **Figure 7.6** for an example of trying to balance out the placement, size, and proportion of items in a map layout with a template).

FIGURE 7.6

Example of balancing the placement of items on a map layout.

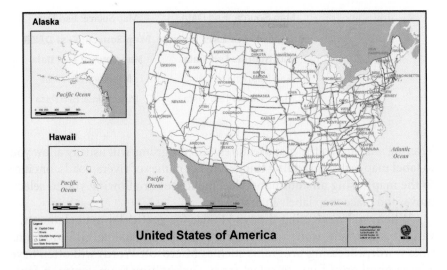

How Is Data Displayed on a GIS Map?

reference map a map that serves to show the location of features, rather than thematic information

There are several different types of maps that can be made using GIS tools. The purpose of a **reference map** (like you'd see in an atlas or a road map) is to give location information and to highlight different features. A map of park trails, a Manhattan restaurant guide map, a map of the casinos on the Las Vegas Strip, or a zoning map of your neighborhood are all examples of reference maps. Topographic maps (that also show landforms) are another example of reference maps (we'll deal with them in Chapter 13).

thematic map a map that displays a particular theme or feature

Another main type of map is the **thematic map**, which is geared toward conveying one particular theme to the reader. Thematic maps can be used to show things such as the state-by-state increase in U.S. population over a particular period, or presidential election results county by county (see *Hands-On Application 7.2: Presidential Election Thematic Maps* for examples of types of thematic maps online). In GIS, a layer's attributes are used for displaying information on the map. For instance, to create a map of the 2016 U.S. presidential election results by county, each polygon representing that county will have an attribute designating whether Donald Trump or Hillary Clinton had a higher number of votes in that county (colored red for counties won by Trump and blue for counties won by Clinton). This attribute will then be the data being displayed on the map (see **Figure 7.7**). Other thematic maps may use **graduated symbols** for display. In this case, points (or other symbols) of differing sizes are plotted to represent the thematic factors.

graduated symbols the use of different-sized symbology to convey thematic information on a map

Of course, many thematic maps don't rely on a simple two-choice attribute like the election map (in which each county will be marked Trump or Clinton—or else contain no data). Attributes such as the percentage of colleges and universities per state that are using this textbook in a class

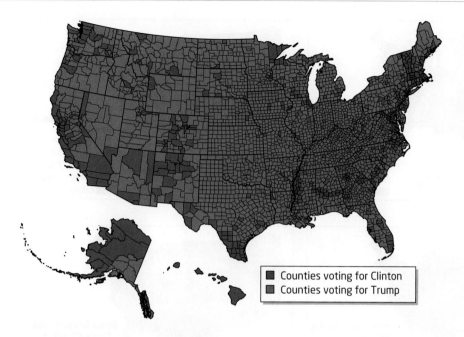

FIGURE 7.7

A thematic map showing the results of the 2016 U.S. presidential election by county (the vote in red counties going for Donald Trump and the vote in blue counties going for Hillary Clinton).

■ Counties voting for Clinton
■ Counties voting for Trump

Hands-On Application 7.2

Presidential Election Thematic Maps

The results of a U.S. presidential election by state can be presented as a thematic map. A state would either be colored red if its Electoral College favored the Republican candidate or blue if they favored the Democratic candidate. Presidential election thematic maps can be seen online at: **www.270towin. com**. This thematic map currently shows the results of the 2016 presidential election between Donald Trump and Hillary Clinton, but you can use this map to make predictions of the 2020 election. Click on a state to mark its status as red, blue, or yellow (undecided). From the Select a starting view pull-down menu, you can examine thematic maps various scenarios, including the actual 2012 state by state results. Check out the various options and create your own for a possible 2020 election result.

Expansion Questions

➤ How did the 2012 and 2016 presidential elections change from state to state (in which

states had their electoral votes go blue and which go red)?

➤ Under the Electoral College system, the candidate with the greatest number of electoral votes is the winner, not the candidate who wins the greatest number of states. Turn on the option to display the number of electoral votes per state and re-examine the presidential election thematic maps over the years. Which elections saw the winning candidate take a smaller number of states but a larger number of electoral votes?

➤ From a mapping context, would it be more informative to map presidential election results as red states and blue states, or should the states be mapped according to their electoral vote count?

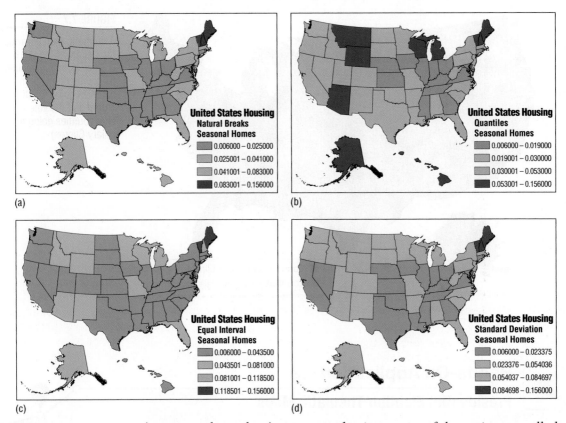

Four choropleth map examples created using the same data (the year 2000 percentage of the total number of houses that are considered seasonal or vacation homes), but employing different data classification methods, as follows: (a) natural breaks, (b) quantiles, (c) equal interval, and (d) standard deviation.
[Source: Esri]

choropleth map a type of thematic map in which data is displayed according to one of several different classifications

data classification various methods used for grouping together (and displaying) values on a choropleth map

(or some other values) are mapped using a type of thematic map called a **choropleth map**. In order to display this information at its best on a map, the data will have to be classified or divided into a few categories. GIS gives several options for **data classification** of multiple values—each method classifies data differently, which can result in some very different results being displayed on the maps. (See **Figure 7.8** for four different maps of the percentage of the total number of houses per state that are considered seasonal homes or vacation homes by Census, each created using a different data classification technique.) Note that each map has the data broken into the same number of classes (four) for comparison purposes. Also keep in mind that choropleth maps already have the boundaries of their polygons set in place (elements such as the states' boundaries being already defined) and the attributes being mapped do not define these boundaries. For instance, the states' boundaries are already delineated and the choropleth map is showing a classification of a value of seasonal home percentages utilizing all of these pre-determined boundaries.

The first of these data classification methods is called **natural breaks** (also referred to as the Jenks optimization). As the name implies, this method takes all of the values being mapped and looks at how they're grouped together. The spaces in between the data values are used to make different classes (or ranges of data) that are displayed. This is shown in Figure 7.8a—states with the lowest percentages of seasonal home values (such as Nebraska, Oklahoma, and Texas) end up in one class, and states with the highest percentages of seasonal homes (such as Maine, Vermont, and New Hampshire) end up together in another class.

The map in Figure 7.8b shows the results of using the **quantile** method of data classification. This method takes the total number of data values to be mapped and splits them up into a number of classes. It tries to distribute values so that each range has a similar number of values in it. For instance, with 50 states being mapped (plus the District of Columbia as a 51st area), each of the four ranges will have about 13 counties' worth of data being shown in each class. Because the breakpoints between the ranges are based on the total number of items being mapped (that is, the number of states ending up in each range), rather than the actual data values being mapped, the quantile method causes a relatively even distribution of values on the map.

The third method (shown in the map in Figure 7.8c) uses **equal intervals** for data classification. It works like it sounds—it creates a number of equally sized ranges and then splits the data values among these ranges. The size of each range is based on the total span of values to be mapped. For instance, in the seasonal home map, the data is divided into four classes, and the range of values goes from the state with the lowest seasonal home percentage (0.6% of the total housing stock in Illinois) to the state with the highest seasonal home percentage (15.6% in Maine). Equal interval takes the complete span of data values (there is 15% separating the lowest and highest values) and divides it by the number of classes (in this case, four), and that value (in this case, 3.75%) is used to compute the breaking point between classes. Therefore the first class represents states that have a seasonal home value of 3.75% more than at the lowest end (for instance, the first class would have values between 0.6% and 4.35%). Note that this method simply classifies data based on the range of all values (including the highest and lowest) but does not take into account clusters of data or how the data is distributed. As such, only a few states end up in the upper class because their percentages of seasonal homes were greater than three-fourths of the total span of values.

The final method uses the concept of **standard deviation** (see Figure 7.8d). A standard deviation is the average distance that a single data value is away from the mean (the average) of all data values. The breakpoints for each range are based on these statistical values. For instance, the GIS would calculate the average of all United States seasonal home values (3.9%) and the standard deviation for them (3.1%). Thus, when it comes to the percentage of the total housing stock that is seasonal, each state's percentage is an average of 3.1% away from the average state's percentage. These values for the mean and standard deviation of the values are used to set up the breakpoints. For instance, the breakpoint of the first range is of all states whose seasonal home values are less than half a standard deviation value lower than the mean—those states with a seasonal home percentage of less than the

natural breaks a data classification method that selects class break levels by searching for spaces in the data values

quantile a data classification method that attempts to place an equal number of data values in each class

equal interval a data classification method that selects class break levels by taking the total span of values (from highest to lowest) and dividing by the number of desired classes

standard deviation a data classification method that computes class break values by using the mean of the data values and the average distance a value is away from the mean

mean minus 0.5 times the standard deviation (1.55%), or 2.34%. The fourth range consists of those counties with a value greater than 1.5 times the standard deviation away from the mean. The other ranges are similarly defined by the mean and standard deviation values of the housing data.

As Figure 7.8 shows, the same data can produce choropleth maps that look very different and carry different messages, depending on which method is used to classify the data. For instance, the map in Figure 7.8d (standard deviation) shows that most states have roughly an average (or below average) percentage of seasonal homes, while the other maps show various distinctions between the states that are classified as having a higher or lower percentage of seasonal homes. Thus, the same data can result in different maps, depending on the chosen method of classification. When you're selecting a method, having information about the nature of the data itself (that is, if it is evenly distributed, skewed toward low or high numbers, or all very similar values) will help you to determine the best kind of mapping.

You should keep in mind that while different data classification methods can affect the end result shown on the choropleth map, the type of data values being mapped can also greatly affect the outcome of the choropleth map. An issue involved with choropleth mapping is displaying the values of data that can be counted (for instance, the total population values per state or the total number of housing sales per county) when the sizes of the areas being mapped are very different. If you're mapping the number of vacation homes in each state, a very large state such as California is probably going to have a larger number of homes (12,214,549 homes in total, with 239,062 of them classified as seasonal, according to the 2000 Census) than a smaller state such as Rhode Island (439,837 homes in total with 13,002 of them classified as seasonal). Thus, if you're making a choropleth map of the number of vacation homes in each state, California will show many more vacation homes than Rhode Island, just because it has a lot more houses.

A better measure to map is the percentage of the total number of houses that are vacation homes—in this way, the big difference in housing counts between California and Rhode Island won't be a factor in the map. Instead, you'll be mapping a phenomenon that can be comparably measured between the two states. When you map the percentages of total housing considered vacation homes, California's seasonal homes only make up about 2% of the total, while Rhode Island's seasonal homes make up about 3%. To make a choropleth map of count data, the data must first be **normalized**—that's to say, it must have all count values brought onto the same level. For instance, dividing the number of seasonal homes by the total number of houses would be a way of normalizing the data. See *Hands-On Application 7.3: The Census Data Mapper* for online tools used for creating choropleth maps.

normalized altering count data values so that they are at the same level of representation (such as using them as percentages instead of regular count values)

Hands-On Application 7.3

The Census Data Mapper

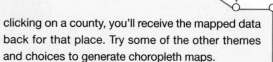

The U.S. Census Bureau has an online utility for mapping demographic data from the U.S. Census as choropleth maps. To get started with it, go to **http://datamapper.geo.census.gov/map .html**—this is the Website for the Census Data Mapper. When the Website completely loads, you'll go through a quick six-step process to create a choropleth map. First, select the theme you want to map, such as Age and Sex, Population and Race, or Family and Housing. Choose the Population and Race option for your first map. Next, you'll choose the data table (or the attribute) associated with that theme that you'll make a map of—choose Percent Black or African American. Third, choose an appropriate color palette for displaying the data. Fourth, select the number of classes you want to display—for instance, for this first map, choose 5. Fifth, choose the classification method you want to use—for instance, choose Quantile. Lastly, click on Make Your Map to generate the choropleth map of your choices. The choropleth map is also interactive; you can zoom in and out of areas and by

clicking on a county, you'll receive the mapped data back for that place. Try some of the other themes and choices to generate choropleth maps.

Expansion Questions

➤ In Cook County, Illinois (which contains the city of Chicago), what is the percentage of the population that is Black or African American? What is the percentage of the population listed as White, as well as the percentage of the population listed as Latino or Hispanic?

➤ In Wayne County, Michigan (which contains the city of Detroit), what is the percentage of the population listed as male, as female, and as 65 years and over?

➤ What is the percentage of occupied housing units in Allegheny County, Pennsylvania (where Pittsburgh is located), Laramie County, Wyoming (where Cheyenne is located), and Clark County, Nevada (where Las Vegas is located)?

What Kinds of Colors Are Best to Use with GIS Maps?

The choice of colors for a map is also important—first to create something pleasing for the map reader, but also in terms of graphic design and usability in a digital format. For instance, the choropleth maps in Figure 7.8 use a range of values of dark green for low percentages and dark blue for high percentages, but there are a lot of ways those classifications could have been colored. There are different color schemes for use in GIS maps—one of these is **RGB**, a setup used for design on computer screens. RGB utilizes the colors red, green, and blue as the primary colors, mixing them together as needed to create other hues. Another is **CMYK,** used in graphic design, which uses the colors of cyan, magenta, yellow, and black (the "K" of the acronym) for creating hues. Chapter 10 will deal further with generating different colors from an initial set of primary colors in the context of examining digital imagery.

RGB a color scheme based on using three primary colors: red, green, and blue

CMYK a color scheme based on using the colors cyan, magenta, yellow, and black

FIGURE 7.9

Examples of various color ramps.

Single-Hue Color Ramps

Multiple-Hue Color Ramps

With a reference map, different colors are used to represent the various objects and symbols. On a choropleth map (such as those shown in Figure 7.8), a range of colors is needed to show the lowest classes of data through the highest classes. In this case, a **color ramp** is used to show a set of colors to represent the classes of values. A color ramp is a selected set of colors that will show the different color changes of a color scheme—for instance, using a single hue of a color in a color ramp might run the range of lightest to darkest blue or light green to very dark green. Other color ramps (like those shown in Figure 7.8) incorporate multiple hues ranging from green through blue. Several different examples of color ramps are shown in **Figure 7.9**.

Part of why color choice becomes important with a map is the wide range of visualization media that are available for viewing it. Colors of map items may appear one way on a computer monitor, another way when they're projected onto a screen with an LCD projector, and appear differently yet again when they're printed on a laser printer. Just because you like the look of the shades of blue and green on the GIS map on your monitor, you shouldn't assume that the same colors will translate the same way to a printed page or to a photocopy of the printout. Your colors might easily

color ramp a particular range of colors applied to the thematic data on a map

 Hands-On Application 7.4

ColorBrewer Online

To investigate further the impact of color choices on a map, check out ColorBrewer 2.0, which is available online at **http://colorbrewer2.org**. ColorBrewer is a great tool for color choice and color ramp selection for a variety of different map data and formats. ColorBrewer 2.0 sets up a predetermined choropleth map while allowing you to examine different color selections, see how different color choices on choropleth maps can be distinguished from one another, or examine the best choices for print, photocopy, and color-blind-safe schemes. Try out several of the options to determine (based on the pre-set choropleth maps) what the best color scheme would be for your map. You can use ColorBrewer to determine the optimum setup of colors and then apply those settings to your own maps.

Expansion Questions

➤ What color schemes are suggested as being photocopy-able? Why would these schemes be

more useful than other schemes when you're producing a map that will be photocopied?

➤ If you were producing a map with five data classes and needed it to be print-friendly only, which classes does ColorBrewer recommend? Of these sequential or diverging schemes, which would you choose for your map, and why?

have a washed-out appearance when the map is projected onto a big screen. Check out *Hands-On Application 7.4: ColorBrewer Online* to investigate the effect that different color choices will have on a final map.

How Can GIS Maps Be Exported and Distributed?

Once a map has been designed and formatted the way you want it, it's time to share and distribute the results. Rather than just print a copy of the map, there are several digital formats to which maps can be quickly and simply converted for ease of distribution. A simple way is to export the map as a graphical raster file—this saves a "snapshot" of the map as a digital graphic that can be viewed like a picture, either as a file or an image placed on a Website. There are many different formats for map export—two common ones are **JPEG** (Joint Photographic Experts Group) and **TIFF** (Tagged Image File Format). Images saved in JPEG format can experience some data loss due to the file compression involved. Consequently, JPEG images usually have smaller file sizes for viewing or downloading. Images saved in TIFF format have a much larger file size but are a good choice for clearer graphics.

The clarity of an image is a function of what **DPI** (dots per inch) setting is used when the map is exported to a graphic. The lower the value of DPI (such as a value of 72), the less clarity the resultant image will have. Very low values of DPI will result in the exported image being very blocky or pixelated. Maps exported with higher values of DPI (such as 300) will be very crisp and clear. However, the higher the DPI value, the larger the file size will be, which becomes important when you're distributing map data online—the larger the file is, the more time will be needed to transfer or display the images. For professional-print-quality maps, the TIFF file format is used at a higher DPI value (such as 300).

JPEG the Joint Photographic Experts Group image or graphic file format

TIFF the Tagged Image File Format used for graphics or images

DPI dots per inch—a measure of how coarse (lower values) or sharp (higher values) an image or map resolution will be when exported to a graphical format

GeoPDF a format that allows maps to be exported to a PDF format, yet contain multiple layers of geographic information

A **GeoPDF** is another option for exporting and distributing maps. A GeoPDF allows the user to export a GIS map in the commonly used PDF file format, which can be opened by the Adobe Reader software using a special free plug-in. A GeoPDF differs from a regular PDF in that it allows the user to interact with map layers and get information about the coordinates of locations shown in the PDF. By using the GeoPDF format, you can click on a location and get the latitude and longitude of that point, as well as make measurements of lengths or areas on the map in real-world units. Also, a GeoPDF can contain multiple layers of data that the user can turn on and off (such as annotations for road names or a separate layer of the roads themselves) as in GIS (see Chapter 13 for more uses of GeoPDFs with other maps).

Chapter Wrap-Up

Once GIS data is created, compiled, or analyzed, the results are usually best communicated using a map, and the ability to create a map of data is a standard feature in GIS software packages such as ArcGIS or QGIS. However, for producing a useful, readable, well-balanced map, there are several choices that go into design, layout, and the presentation of data, colors, and symbols.

This chapter presented an overview of several cartographic and map-design concepts, and this chapter's *Geospatial Lab Applications* will have you take the data you've been analyzing in Chapters 5 and 6 to create a professional-quality map. In the next chapter, we're going to examine some specific types of maps—road network maps—and see how GIS and geospatial technologies are used to design street maps that can be used to locate addresses and compute shortest paths (and directions) between locations. Before that, check out *Cartography Apps* for some downloadable apps, as well as *Cartography in Social Media* for some mapping blogs and related Twitter accounts.

Important note: The references for this chapter are part of the online companion for this book and can be found at **www.macmillanlearning.com/catalog/shellitoigt4e.**

Cartography Apps

Here's a sampling of available representative cartographic apps for your smartphone or tablet. Note that some apps are for Android, some are for Apple iOS, and some may be available for both.

➤ **Atlas Lite:** A free app for displaying pre-generated maps

> **Atlas of the World:** An app for displaying pre-made maps of world population by country and also by state within the United States

> **Cartovista:** An app for displaying pre-made datasets as choropleth maps

> **Census Quickfacts Data Browser:** An app for examining population maps of Census data of the United States by state or county

Cartography in Social Media

Here's a sampling of some representative Facebook and Twitter accounts, along with some blogs related to this chapter's concepts.

On Facebook, check out the following:

> **The History of Cartography Project:** www.facebook.com/HistoryofCartographyProject

Become a follower on Twitter of:

> **British Cartographic Society:** @bcsweb

For further up-to-date info, read up on these blogs:

> **ArcGIS Blog—Mapping Category** (blog posts concerning mapping and cartography topics): **http://blogs.esri.com/esri/arcgis/category/mapping**

> **Strange Maps** (a blog dedicated to all types of weird and cool maps): http://bigthink.com/articles?blog=strange-maps

Key Terms

cartographic generalization (p. 246)
cartography (p. 243)
choropleth map (p. 252)
CMYK (p. 255)
color ramp (p. 256)
data classification (p. 252)
DPI (p. 257)
equal intervals (p. 253)

fonts (p. 249)
geographic scale (p. 244)
GeoPDF (p. 258)
graduated symbols (p. 250)
JPEG (p. 257)
label (p. 249)
large-scale maps (p. 245)
layout (p. 249)
legend (p. 248)

GIS Layouts: QGIS Version

This lab will introduce you to the concept of taking GIS data and creating a print-quality map from it. This map should contain the following:

➤ The population per square kilometer for all counties in California, set up in an appropriate color scheme

➤ The data displayed in a projection with units of measurement other than decimal degrees (the default units used by the lab data are meters—be sure that the scale bar reflects this information)

➤ An appropriate legend (make sure your legend items have regular names, and that the legend is not called "legend")

➤ An appropriate title (make sure that your map title doesn't include the word "map" in it)

➤ A north arrow

➤ A scale bar

➤ Text information: your name, the date, and the sources of the data

➤ Appropriate borders, colors, and design layout (your map should be well designed, and not look like map elements were thrown together randomly)

Important note: At the end of this lab, there is a checklist of items to aid you in making sure the map you make is complete and of the best quality possible.

Similar to the three Chapter 6 *Geospatial Lab Applications*, three versions of this lab are also provided. The first version (*Geospatial Lab Application 7.1: GIS Layouts: QGIS Version*) uses the free QGIS package. The second version (*Geospatial Lab Application 7.2: GIS Layouts: ArcGIS Pro Version*) provides the same activities for use with ArcGIS Pro. There is also a third version (*Geospatial Lab Application 7.3: GIS Layouts: ArcMap Version*) available online. Although ArcGIS Pro contains more functions than QGIS, these two programs share several useful cartographic tools.

Objectives

The goals for you to take away from this lab are:

➤ To familiarize yourself with the map composer functions of QGIS

➤ To arrange and print professional-quality maps from geographic data using the various layout elements

Using Geospatial Technologies

The concepts you'll be working with in this lab are used in a variety of real-world applications, including:

➤ Urban planning, where GIS is used to compose maps showing data about housing and population distribution, transportation routes, land parcels, zoning information, and other matters

➤ Law enforcement, where agents use GIS to create maps for distribution among police officers and the community showing the major locations of particular types of crimes, analysis of crime hot spots, and locations of 911 calls

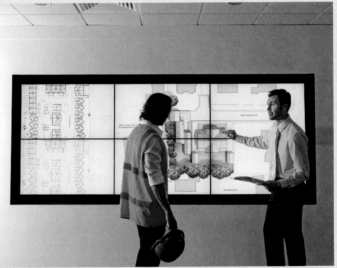

[Monty Rakusen/© Cultura Creative (RF)/Alamy]

Obtaining Software

The version of QGIS used in this lab is 2.18, and is available for free download at **http://qgis.org/downloads**.

Important note: Software and online resources can change fast. This lab was designed with the most recently available version of the software at the time of writing. However, if the software or Websites have changed significantly between then and now, an updated version of this lab (using the newest versions) will be available online at **www.macmillanlearning.com/catalog/shellitoigt4e**.

Lab Data

Copy the folder **Chapter7QGIS**—it contains a folder called "CalMap" which is the file geodatabase you'll be using in this lab, like how you used file geodatabases in Chapters 5 and 6. This lab's data was downloaded from The National Map (see page 10) free of charge, and the population information was taken from the U.S. Census Bureau. The GIS layers in the geodatabase have already been projected for you to the U.S. National Atlas Equal Area projection.

Localizing This Lab

The data used in this lab is the boundaries and attributes of California drawn from The National Map and the U.S. Census Bureau. However, you can easily work with the same data from your own state as The National Map makes the same layers available for all 50 states. From The National Map, download the Boundaries geodatabases for your own state, which contain the layers used for California in this lab. If you are in, for instance, Florida, you can download the county boundaries layer for Florida from The National Map. Similarly, the U.S. Census Bureau has population values for each county, which can be joined to the county boundaries layer. A value for population per square kilometer for each county was then computed by dividing the census population value by the National Map's value for area in square kilometers.

7.1 Initial Pre-Mapping Tasks in QGIS

If needed, refer back to *Geospatial Lab Application 5.1: GIS Introduction: QGIS Version* for specifics on how to do the following.

1. Start **QGIS**.

2. Add the **CalBounds** feature class from the CalMap geodatabase in the **C:\Chapter7QGIS** folder. Leave the CalBounds layer's symbology alone for now—you'll change it in the next step.

3. Pan and zoom the Map View so that all of California is shown filling up the Map View.

   Important note: The data used in this lab application has already been projected to the U.S. National Atlas Equal Area projection for you to use. However, there are many more projections to choose from if you desire— you can change the projection of a data layer by right-clicking on it, selecting **Save As**, and then choosing a projected coordinate system in which to change the layer.

4. The next step is to set the properties of the project you're working with so that QGIS will be able to render some of the map elements you'll be working with (such as the scale bar) properly. From the **Project** pull-down menu, choose **Project Properties**.

5. Select the **CRS** tab.

6. Place a checkmark in the **Enable 'on the fly' CRS transformation (OTF)** box.

7. In the box next to Filter, type **U.S. National Atlas Equal Area**. This will search through all available projections and find the one you'll need.

8. In the **Coordinate reference systems of the world box**, click on the **US National Atlas Equal Area** option, then click **Apply**. This has set the CRS of the project environment to the chosen projection. Click **OK** to close the dialog box.

7.2 Setting Graduated Symbology in QGIS

1. QGIS gives you the ability to change a feature's symbology from a single symbol to multiple symbols or colors, and allows for a variety of different data classification methods.

2. Right-click on the **CalBounds** in the Map Legend and select **Properties**. Click on the **Style** tab. To display the states as graduated symbols, use the following settings:

a. From the pull-down menu to the right of the **Style** tab, select **Graduated**.

b. The **Column** to use (in this lab) is the **PopDens** (each county's population per square kilometer from the year 2010).

c. Use **5** for the number of **Classes**.

d. Use **Quantile (Equal Count)** for the **Mode**.

e. For the color ramp options, use the pull-down menu to select an appropriate choice.

3. When you have things arranged how you want them, click **Apply** to make the changes. Take a look at the map and make any further color changes you think are needed.

4. Click **OK** to close the dialog box.

5. Now, the symbology of CalBounds has changed in the Map View and the values that make up each one of the breaks can be seen in the Layers Panel. If the breaks and classes are not already displayed, you can show them by pressing the **plus** button, which you'll find to the left of the CalBounds layer in the Layers Panel.

7.3 The Composer in QGIS

To begin laying out the print-quality version of the map, you'll need to begin working in the Composer. This mode of QGIS works like a blank canvas, allowing you to construct a map using various elements.

1. To begin, select **New Print Composer** from the **Project** pull-down menu.

2. Before the Composer opens, you'll be prompted to give your composition a title. Type in a descriptive title, such as "California Population Density 2010" or something similar. Click **OK** after you've put in your title.

3. A new window, the Composer, will open. In the Composer, the screen represents the printed page of an 8½ × 11 piece of paper, so be careful when you're working near the edges of the page and keep all elements of the map within that border.

4. The composer has its own toolbar across the top with a new set of tools—locate and examine these navigation tools.

5. Starting at the left and moving right, the tools and their uses are as follows:

   a. The blue disk is used to save a composition.

   b. The white paper will create a new composition.

   c. The white paper over the blue paper is used to create a duplicate composition.

   d. The white paper with the wrench will open the Composer Manager.

   e. The yellow folder will allow you to add a template to the composition.

f. The blue disk with the green bar will allow you to save a template.

g. The printer icon is used when you're printing (see later in the lab).

h. The next three icons will allow you export your composition to either (1) an image, (2) SVG format, or (3) PDF format.

i. The two curved arrows allow you to either revert to the last change you made or restore the last change you made (which is useful when you need to "back up" a step in your map design).

j. The magnifying glass with three arrows will zoom to the full extent.

k. The magnifying glass with the 1:1 text will zoom the map to 100%.

l. The plus and minus magnifying glasses are used to zoom in and out of the layout.

m. The twin curved blue lines icon is used to refresh the view.

n. The square and circle icons allow you to gather up several items and treat them as a single group (group items) or to turn a group of items back into individual items (ungroup).

o. The lock and unlock icons allow you to fix items in place in the composition (lock) or remove this fix so they can be moved (unlock).

p. The last two icons with the blue and yellow boxes are used for raising or aligning map elements.

6. There is also a vertical toolbar down the left-hand side of the screen with additional tools:

   a. The hand is used for panning and adjusting the content of a particular window—for instance, for moving the position of what's showing on the map without changing the map itself.

   b. The magnifying glass is used to zoom in on elements on the map.

   c. The cursor pointing to the yellow box is used to move map elements to different places in the layout.

   d. The icon with the blue arrows on a piece of paper is used to move the content of items.

   e. The icon with the hammer and the three blue points joined together is used for editing nodes.

   f. The other tools are used for the addition of map elements to the layout, which we'll use later in the lab (and you'll be referred back to these icons). You can add a new map, an image, a new label, a new legend, and a new scale bar, as well as ellipses, nodes, and arrow shapes for annotating the map, the attribute table of a layer, and an HTML frame for displaying Web content.

7.4 Choosing Landscape or Portrait Mode in QGIS

1. Before going any further with your map, the first choice you'll have to make is whether the map should be oriented in Portrait or Landscape mode. Portrait is aligning the layout vertically (so that it's longer than it is wide), while Landscape aligns the layout horizontally (so that it's

wider than it is long). If you were mapping the United States, you'd likely choose Landscape to place the entire country to fill the page. However, with mapping California, you'll likely use Portrait mode as it better fits the dimension of the state. To select whether you want to use Portrait or Landscape, click on the **Composition** tab on the right-hand side of the composer.

2. Under **Page size**, next to **Orientation**, choose **Portrait**. You'll see the composition shift to the vertically oriented Portrait mode. By selecting Landscape or Portrait from this option, you can change the orientation of the printed map.

7.5 Map Elements in QGIS

1. Again, think of the layout (or composition) as a blank sheet of paper that you'll use to construct your map. There are numerous map elements that can be added, including the map canvas, scale bars, north arrows, and a legend. Each element has properties (such as creating borders, filling colors, or fixing size and position) that can be accessed by selecting the **Item Properties** tab on the right side of the composer.

2. Map elements can be resized and moved by selecting them with the mouse and resizing as you would any object.

3. Map elements can be deleted by selecting them with the cursor and pressing the **Delete** key on your keyboard.

7.6 Placing a Map Canvas into the Composition in QGIS

1. The first element to add to the layout is the default map canvas. This will be an element that shows all of the visible layers in the Map Legend.

2. Click on the **Add new map** icon from the vertical toolbar. This will allow you to draw a box on the layout that will display the current Map View. Draw a rectangular box on the layout to create the canvas.

3. Once it's drawn, you can click and drag the canvas around the map or resize it as you see fit (using the four blue tabs at its corners), and you can treat the entire canvas as if it's a single map element. When you're adjusting the size, keep in mind that you'll be adding several more map elements (such as a legend or a title), so you'll need to adjust the size, spacing, and balance of the elements accordingly.

4. You can't manipulate individual layers (for instance, you can't click and drag a particular county to somewhere else), but using the **Move item content** icon on the vertical toolbar will allow you to drag the contents of the canvas around and alter what's being displayed inside the canvas.

5. To make further adjustments to the canvas itself (such as creating a border or altering the color of the border), select the **Item Properties** tab on the right side of the screen and scroll down to see what choices are available (such as expanding the **Frame** option to change the appearance of the map's border or the **Background** option to change the fill color of the map itself). Investigate the other choices under Item Properties (such as the options under **Main Properties** for adjusting the scale of the layers being displayed on the canvas) to set up the canvas the way you'd like it to appear on the final printed map.

7.7 Placing a Scale Bar into the Composition in QGIS

1. To add a scale bar to the map, select the **Add new scalebar** icon from the toolbar, then click on the composition where you want the scale bar to be added.

2. Once the scale bar appears, click on it so that the four blue boxes appear at the corners. In the **Item Properties** box on the right-hand side of the screen, you'll see several options for altering the scale bar. Use the options to change the following properties of the scale bar until it looks appropriate:

a. The number of segments on the left or the right side of the 0 value on the scale bar

b. The size of each segment

c. The number of map units per bar units (to even out the breakpoints on the scale bar)

d. The physical dimensions of the scale bar itself (height, line, width, etc.)

e. The font and color used

7.8 Placing a North Arrow into the Composition in QGIS

1. To add a north arrow to the map, select the **Add Image** icon from the toolbar, then click on the composition where you want the north arrow to be added and draw a box the size of how large you want to north arrow symbol to be.

2. An empty box and several different graphics options will appear in the Item Properties pane. The Add Image option allows you to add a graphic of your own to the map or to select from several pre-made graphics, including several different north arrows. Expand the options

for **Search directories** and scroll part-way down and you'll see a number of options for north arrows.

3. Select an appropriate north arrow and it will appear in the empty box on the composition. Use the cursor to position or resize the north arrow.

7.9 Placing a Legend into the Composition in QGIS

Each layer you use in QGIS (such as CalBounds) has a set of layer properties. Anything changed in the layer properties will be reflected in changes to the composition. For instance, if you change the symbology of a layer in its properties, its appearance will be transferred over to a legend in the composition. Similarly, whatever name is given to a layer in the Map Legend will carry over to a legend in the composition.

1. Return to the regular QGIS window. Currently, the CalBounds layer is named "CalMap–CalBounds–Multipolygon." However, you can change this name to something more descriptive before adding a legend to the composition. In the QGIS Map Legend, right-click on the name you want to change (the **CalBounds** layer) and select **Rename**.

2. In the Map Legend itself, you can type a new name for the layer (like "Population Per Square Km"). Click the **Enter** key when you've typed in a new name.

3. Return to the composer. To add a legend to the map, select the **Add new legend** icon from the toolbar, then click on the composition where you want the legend to be added.

4. A default legend will be added to the map, consisting of all the layers in the Map Legend with whatever names are assigned to them. Use the cursor to move and resize the legend as needed.

5. To make changes to the default legend, click on the **legend** itself, and there will be several options available in the Item Properties. You can change the name of the legend (don't just call your legend "Legend"), its font, symbol width, and appearance, as well as spacing between items.

7.10 Adding Text to the Composition in QGIS

1. To add text to the map (such as a title, or any other text you may want to add), select the **Add new label** icon from the toolbar, then click on the composition where you want the text to be added and draw a box of the size you want the text to be.

2. Note that when text is added, a small box filled with the word "QGIS" will be added to the composition. The lettering in the box, as well as its font, size, and color, can be changed in the Item Properties.

3. Type the text you want to appear in the **Item Properties** box under the Label heading. To alter the font, color, and other properties of the text, click the **Font...** and **Font color** options. More options will appear— note that you can alter each text box separately, create larger fonts for titles and smaller fonts for type, and so on. You can also resize each text box and move it to a new position.

7.11 Other Map Elements in QGIS

1. Though they're not used in this lab, other map elements can be added, including:

   a. A shape: This allows you to add the outline of a shape (an ellipse, rectangle, or triangle) to the composition for additional graphics or annotation. Use the **Add ellipse** icon to do this.

   b. An arrow: This allows you to add a graphic of an arrow (not a north arrow) so that you can point out or highlight areas to add further annotation to your map. Use the **Add arrow** icon to do this.

c. An attribute table: This allows you to add a graphic of a layer's attribute table to the map to enable the presentation of additional information. Use the **Add attribute table** icon to do this.

d. An HTML frame: This allows you to add a frame that will contain a Web page or other Web content. In the Item Properties you can specify the URL for what you want to display inside the frame. Use the **Add HTML frame** icon to do this.

7.12 Printing the Composition in QGIS

1. When you have constructed the map the way you want, choose the **Composer** pull-down menu and choose **Page Setup**. This will allow you to specify if you want your map printed in Landscape or Portrait format or add margins for printing.

2. When the composition is ready, select the **Print** icon from the toolbar.

3. In the Print dialog box, click **Print** and QGIS should print to your computer's default printer.

4. If you want to convert the composition to a digital form instead of printing it, click on the **Composer** pull-down menu and choose one of the available options: **Export as Image** (this will allow to convert the composition to a format such as JPG or PNG), **Export as PDF**, or **Export as SVG**.

Closing Time

Close the composer by selecting the **Composer** pull-down menu and then selecting **Quit**. Exit QGIS by selecting the **Project** pull-down menu and selecting **Exit QGIS**. The geospatial lab application in Chapter 8 will return to using Google Earth Pro for analysis. You'll be using Google Earth Pro to take a series of addresses and plot them on a map, and then do some analysis of the results. You'll also use Google Maps and Google Earth to analyze the shortest paths between these plotted points.

Final Layout Checklist

_____ The state of California's population per square kilometer, classified, and displayed with an appropriate color scheme

_____ Appropriate map legend, with all items listed with normal names (the legend should not be called "legend" and should be appropriate size, font, number of columns, and so on)

_____ Scale bar (with appropriate divisions and numerical breaks)

_____ The proper units displayed on the scale bar

_____ Appropriate title (don't use the words "map" or "title" in your title) and appropriate size and font

_____ A north arrow of appropriate size and position

_____ Type (your name, the date, the source of the data) in an appropriate size and font

_____ Overall map design (appropriate borders, color schemes, balance of items and placement, and so on)

GIS Layouts: ArcGIS Pro Version

This lab will introduce you to the concepts of taking GIS data and creating a print-quality map from it. This map should contain the following:

➤ The population per square kilometer for all counties in California, set up in an appropriate color scheme

➤ An appropriate legend (make sure your legend items have regular names and that the legend is not called "legend")

➤ An appropriate title (make sure that your map doesn't include the word "map" in it)

➤ A north arrow

➤ A scale bar

➤ Text information: your name, the date, and the source of the data

➤ Appropriate borders, colors, and design layout (your map should be well designed instead of map elements thrown together randomly)

Important note: At the end of this lab, there is a checklist of items to aid you in making sure the map you make is complete and of the best quality possible.

Geospatial Lab Application 7.1: GIS Layouts: QGIS Version used the free QGIS software package; however, this lab provides the same activities for use with ArcGIS Pro.

Important note: This version of the lab uses ArcGIS Pro, but the same lab using ArcMap (from the most recent version of ArcGIS Desktop) is available for you if you'd rather use that. *Geospatial Lab Application 7.3: GIS Layouts: ArcMap Version* is available online at **www.macmillanlearning .com/catalog/shellitoigt4e**.

Objectives

The goals for you to take away from this lab are:

➤ To familiarize yourself with the layout functions of ArcMap

➤ To arrange and print professional-quality maps from geographic data using the various layout elements

Using Geospatial Technologies

The concepts you'll be working with in this lab are used in a variety of real-world applications, including:

➤ Urban planning, where GIS is used to compose maps showing data about housing and population distributions, transportation routes, land parcels, and other matters

➤ Law enforcement, where agents use GIS to create maps for distribution among police officers and the community showing the major locations of particular types of crimes, analysis of crime hot spots, and locations of 911 calls

[© Monty Rakusen/© Cultura Creative (RF)/Alamy]

Obtaining Software

The current version of ArcGIS Desktop (and thus, ArcGIS Pro) is not freely available for use. However, instructors affiliated with schools that have a campus-wide software license may request a 1-year student version of the software online at **http://www.esri.com/landing-pages/education-promo**.

Important note: Software and online resources can change fast. This lab was designed with the most recently available version of the software at the time of writing. However, if the software or Websites have changed significantly between then and now, an updated version of this lab (using the newest versions) will be available online at **www.macmillanlearning.com/catalog/shellitoigt4e**.

Lab Data

Copy the folder **Chapter7Pro**—it contains a folder called "CalMap" which is the file geodatabase you'll be using in this lab, like how you used file geodatabases in Chapters 5 and 6. This lab's data was downloaded from The

National Map (see page 10) free of charge, and the population information was taken from the U.S. Census Bureau. The GIS layers in the geodatabase have already been projected for you to the U.S. National Atlas Equal Area projection.

Localizing This Lab

The data used in this lab is the boundaries and attributes of California drawn from The National Map and the U.S. Census Bureau. However, you can easily work with the same data from your own state as The National Map makes the same layers available for all 50 states. From The National Map, download the Boundaries geodatabases for your own state, which contain the layers used for California in this lab. If you are in, for instance, Florida, you can download the county boundaries layer for Florida from The National Map. Similarly, the U.S. Census Bureau has population values for each county, which can be joined to the county boundaries layer. A value for population per square kilometer for each county was then computed by dividing the census population value by the National Map's value for area in square kilometers.

7.1 Initial Pre-Mapping Tasks

If needed, refer back to *Geospatial Lab Application 5.2: GIS Introduction: ArcGIS Pro Version* for specifics on how to do the following.

1. Start **ArcGIS Pro**.
2. When prompted, log in with your Esri account username and password.
3. When ArcGIS Pro opens, start with a new blank Project and call it **C7Pro** and place it in your **C:\Chapter7Pro** folder.
4. If the **Project** tab is open, close it so that you have an empty Map screen with which to work.
5. From the **Insert** tab, within the **Project** folder, click on the **New Map** icon and choose **New Map**.
6. From the **C:\Chapter7Pro** folder, add the CalBounds feature class from the CalSpatial geodatabase to the Contents Pane.

Important note: The data used in this lab application has already been projected to the U.S. National Atlas Equal Area projection for you to use. However, there are many more projections to choose from if you desire, such as the Lambert Conformal Conic projection (see *Geospatial Lab Application 5.2: ArcGIS Pro Version* for more about changing the projected appearance of layers in ArcGIS Pro). Keep in mind that your final map will be of California if you select a regional or world projection. You should use a Projected Coordinate System for this lab.

7. Leave the CalBounds symbology alone for now—you'll change it in the next step.

7.2 Setting Graduated Symbology in ArcGIS Pro

ArcGIS Pro gives you the ability to change an object's symbology from a single symbol to multiple symbols or colors and allows for a variety of different data classification methods.

1. Right-click on the **CalBounds** shapefile in the Contents Pane and select **Symbology**. The Symbology Pane will open on the right-hand side of ArcGIS Pro.

a. Under the **Symbology** options, select **Graduated Colors**.

b. The **Field** to use in this lab is **PopDens** (the counties' population per square kilometer from the year 2010).

c. For **Normalization**, leave it at **<None>**.

d. For **Method**, choose **Natural Breaks**.

e. Use **5** for the number of **Classes**.

f. For the color scheme, select an appropriate choice from the available options.

2. Try some of the various color schemes. ArcGIS Pro will update the CalBounds layer in the Map as you make changes.

7.3 Creating an Initial Layout in ArcGIS Pro

To begin laying out the print-quality version of the map, you'll need to work with a Layout instead of the Map. This mode of ArcGIS Pro works like a blank canvas, allowing you to construct a map using various elements. A Layout in ArcGIS Pro will be a separate tab so that you can switch back and forth between the Map and the Layout as needed.

1. To begin, select the **Insert** tab, within the **Project** group, and click on **New Layout**.

2. ArcGIS Pro will give you a large number of choices about what your layout will be—everything from orientation to size. Before getting any further with your map, the first choice you'll have to make is whether the map should be oriented in Portrait or Landscape mode. Portrait aligns the layout vertically (so that it's longer than it is wide), while Landscape aligns the layout horizontally (so that it's wider than it is long). If you were mapping the United States, you'd likely choose Landscape to place the entire country to fill the page. However, with mapping California, you'll likely use Portrait mode as it better fits the dimension of the state. For this Lab, choose the option for **ANSI— Portrait–Letter 8.5" × 11"**.

3. A new **Layout** tab will open next to your Map tab. The Layout tab works like a blank canvas and allows you to construct a map using various elements and you can switch back and forth from the Map to the Layout by clicking on the appropriate tab.

4. *Important note:* If you find that later on during map design you want to change the orientation from portrait to landscape (or vice versa) you don't have to start from scratch with a new blank layout. Instead, choose the **Layout** tab, within the **Page Setup** Group, and click on the **Orientation** icon. There you can choose Portrait or Landscape orientation for your layout.

7.4 Inserting a Map Frame into the Layout in ArcGIS Pro

1. The first element to add to the layout is the default map frame. This will be an element that shows all of the visible layers in the Contents Pane of the **Map** tab.

2. From the **Insert** tab, within the **Map Frames** group, click on the **Map Frame** icon and then select the **Map** option for the Map with the scale values after it. This is a reference to the open Map tab that contains your CalBounds layer—by choosing this option, you're letting ArcGIS Pro know that you want to add the visible layers from that Map to this Layout's Map Frame.

3. Back in the **Layout** tab, you'll see that the Map Frame has been added and filled with the visible layers of the Map (i.e., the CalBounds layer and the Topographic Map background).

7.5 Working with Layouts and Map Elements in ArcGIS Pro

1. Choosing the **Layout** tab in the main Ribbon at the top of the screen will give you a new set of tools for working with your Layout.

2. In the **Navigate** group, the tools and their uses are as follows:

   a. The Navigate tool is used to pan around the Layout and move the Layout as a whole.

   b. The magnifying glass icon is used to zoom the view to the extent of any selected features.

   c. The icon with the four triangles pointing outward is used to zoom the map to its full extent.

   d. The 1:1 icon is used to zoom the layout to 100%.

   e. The white pages with the two black arrows are used for zooming in to the extent of the entire width of the page.

   f. The white pages with the blue arrows are used for switching between previous extents that you may have zoomed to.

3. Again, think of the layout as a blank sheet of paper that you'll use to construct your map and there are numerous map elements that can be added, including scale bars, north arrows, and a legend. For instance, the Map Frame itself that you added is considered a map element on your Layout. Each element has properties (such as creating borders, filling colors, or fixing size and position) that can be accessed individually after you have selected the element. In the **Elements** group, there are tools and buttons that let you interact with the map elements as follows:

   a. The Select Elements tool will allow you to click on an element to work with it or to select multiple elements at once.

   b. The Select All button allows you to choose all elements on the Layout.

   c. The Clear Selection button allows you to de-select all selected elements.

   d. The Delete button removes all selected elements.

4. Once map elements have been selected, they can be resized and moved as you would any object. They can also be deleted by selecting them with the cursor and pressing the **Delete** key on the keyboard.

7.6 Working with the Map Frame in the Layout in ArcGIS Pro

1. The first element that is added to the layout is the Map Frame, which contains all of the visible layers in the Contents Pane. You can't manipulate parts of individual layers (for instance, you can't click and drag a particular county to somewhere else), but you can treat the entire Map Frame as if it's a map element. From the **Layout** tab, within the **Selection** group click on the **Select** button. Then click and drag the **Map Frame** around the Layout or resize it as you see fit (using the eight white boxes at its corners and midpoints). When you're placing and resizing the Map Frame, keep in mind that you'll have several other map elements on the final layout, therefore be mindful of the position and size of the frame, as well as the available blank white space.

2. Resizing the Map Frame may also alter the scale of the map or how much is being displayed in the Map Frame, therefore you will want to adjust and fix the scale of the map interactively. In the lower left-hand corner of the Layout you'll see the pull-down menu for the scale box. You can select one of the pre-set map scales or type in your own value to adjust the scale of the map. Try some of the different values and see how they affect the map's appearance, and when you're done, type **1:6000000** for the map scale to use.

3. Even with changing the scale though you've resized the Map Frame, you may want to reposition the content of the Map Frame, or change what's visible in the Map Frame. To adjust what is displayed within the Map Frame, on the Ribbon, select the **Layout** tab within the **Map** group, then click on **Activate**.

4. With the Map Frame *activated*, you'll have access to the regular Map Navigation tools for panning the content of the Map Frame around (see *Geospatial Lab Application 5.2: ArcGIS Pro Version* for more information on how to use the Navigation tools). Adjust the

appearance of the California counties within the Map Frame as necessary. When you have everything ready the way you want it, and want to return to working with the content of the Layout, on the Ribbon select the **Layout** tab within the **Map** group, and click on the **Close Activation** button.

5. You can also access the Map Frame's properties to change (or remove) borders around the outline of the Map Frame itself. Right-click on the **Map Frame** element and select **Properties**.

6. A new pane, the Format Map Frame Pane, will open on the right-hand side of ArcGIS Pro. From the blue **Map Frame** pull-down menu, the available options will allow you change the **Background** appearance of the Map Frame (such as changing the color of the Map Frame's background), alter the **Border** around the Map Frame itself (the color or thickness of the border, or to make the border invisible), or add a **Shadow** to the Map Frame.

7.7 Placing a Scale Bar into the Layout in ArcGIS Pro

1. To add a scale bar to the map, choose the **Insert** tab within the **Map Surrounds** group, and click on the **Scale Bar** button.

2. A number of options for scale bars will appear.

3. Additional scale bars are available by clicking the **More...** option.

4. Select an appropriate scale bar by clicking on it. It will be added to the Layout as a map element that you can select and move around as previously described. You'll see that in the Layout's Contents Pane, the scale bar will be placed in the Drawing Order above the Map Frame. Just like with adding layers to the Map's Contents Pane, elements will be placed on top of or underneath other map elements depending on where they are ranked in the Layout's Contents Pane.

5. To change the appearance of the scale bar right-click on the scale bar element and select **Properties**.

6. A new pane, the Format Scale Bar Pane, will open on the right-hand side of ArcGIS Pro. From the blue **Scale Bar** pull-down menu, the available options will allow you to change everything from the appearance of the scale bar itself to the appearance and placement of the numbers shown on it.

7.8 Placing a North Arrow into the Layout in ArcGIS Pro

1. To add a north arrow to the map, choose the **Insert** tab within the **Map Surrounds** group, and click on the **North Arrow** button.

2. A number of options for north arrows will appear.

3. Additional north arrows are available by clicking the **More...** option.

4. Select an appropriate north arrow by clicking on it. It will be added to the Layout as a map element that you can select and move around as previously described.

5. To change the appearance of the north arrow, right-click on the north arrow element and select **Properties**.

6. A new pane, the Format North Arrow Pane, will open on the right-hand side of ArcGIS Pro. From the blue **North Arrow** pull-down menu, the available options will allow you to change everything from the color of north arrow to the border around it.

7.9 Placing a Legend into the Layout in ArcGIS Pro

1. To add a legend to the map, return to the **Layout** tab, then on the ribbon choose the **Insert** tab within the **Map Surrounds** group, and click on the **Legend** button. The legend will be added to the Layout as a map element that you can select and move around as previously described.

2. When you inserted the legend, a new pane, the Format Legend Pane, opened on the right-hand side of ArcGIS Pro. From the blue **Legend** pull-down menu in this pane, options will be available to change the appearance of all the different things within the legend, including its border or adding a shadow.

3. In the Format Legend Pane, you can change the name of the legend in the **Title** box. For this map, give your legend a better name than "Legend"—type in **Population Per Square Km**. Also, to display this new title, put a checkmark in the **Show** box.

4. By clicking on the **Show Properties…** button, you will bring up the Format Legend Item Pane that will allow you to choose exactly what is shown in the legend and how things are arranged in columns in the legend.

5. For instance, your legend currently should read Population Per Square Km with a heading under it that reads CalBounds (the name of the feature class), then another heading under it that reads PopDens (i.e., the name of the field from the layer's attribute table). Having these extra headings that say CalBounds and PopDens on your map probably wouldn't mean much to the casual map user, and besides, the descriptive title you gave the legend will convey to the reader what the map is. To remove these items from the legend, take out the checkmarks next to **Layer Name** and **Headings**. Your legend should now just simply read Population Per Square Km with the appropriate values beneath it.

7.10 Adding Text and a Title to the Layout in ArcGIS Pro

1. To add text to the Map (for instance, for the name or the date), on the Ribbon choose the **Insert** tab within the **Text** group, and click on the **New Text** button.

2. Then click on the Layout where you want to place the text. A small box that reads "Text" will be added to the Layout.

3. When you inserted the text, a new pane, the Format Text Pane, opened on the right-hand side of ArcGIS Pro. First, click on the blue word **Text**. Then, expand the options next to **General** and put a checkmark in the box next to **Visible** (this will ensure that your text is visible on the Layout). Next, in the box next to **Text**, type the text you want to add to the Layout. You'll see that in the Layout, the word "Text" has changed to whatever you typed.

4. To change the size and font of the typed text, click on **Text Symbol**, then expand the options for **Appearance**. New options for the font to use, the font style, the size of the font, the color of the text, and so on, will be available to you.

5. Under the other headings—Position, Rotation, Halo, Shadow, and Callout—are more options to change the appearance of your text.

6. After you have the text set up the way you want it, you'll need to add a new set of text for the title of the map. To add more text, return to the Ribbon and choose the **Insert** tab within the **Text** group, and click on the **Add Text** button. Click on the Layout and a new text box with the word "Text" will be added. Use the same techniques you used to modify your previous text to set up a title for the map with appropriate size and font. Make sure that your title is descriptive of what your layout is showing and that it doesn't use the words "map" or "title."

7.11 Other Map Elements in ArcGIS Pro

1. Though they're not used in this lab, other map elements can be added from the **Insert** tab:

a. In the **Map Frames** group, a **Grid** can be added to your layout to showcase things such as latitude/longitude lines on the map.

b. In the **Text** group, **Symbol** allows you to add specific symbols such as road signs or callouts to the layout.

c. Also in the **Text** group, **Dynamic Text** allows you to insert a variety of different types of text including the map's representative fraction, the current date, the coordinates of the map frame, the coordinate system of the layers, and so on.

d. In the **Graphics** group, the various options (**Rectangle, Line**, and **Point**) allow you to draw things onto the map that are not GIS layers (such as drawing a circle around a point of interest).

e. Also in the **Graphics** group, **Picture** allows you to add an image to your map, including downloaded pictures, images from your camera or smartphone, or similar graphics.

7.12 Printing or Exporting a Layout in ArcGIS Pro

1. When the layout is the way you want it, choose the **Share** tab in the Ribbon.

2. To print the layout, within the **Print** group click on the **Layout** button. Printer options will appear.

3. To export the layout to another digital format (like a JPEG or a PDF), within the **Export** group click on the **Layout** button. You'll be able to specify where you want to save the exported version of the layout, what file name you want to give it, and from the **Save As Type** pull-down menu what you want to save the file as (a graphic format such as PNG, JPG, or TIFF, or else a PDF).

Closing Time

Exit ArcGIS Pro by selecting the **Project** tab, and then selecting **Exit**. There's no need to save any data in this lab. The geospatial lab application in Chapter 8 will return to using Google Earth Pro for analysis. You'll be using Google Earth Pro to take a series of addresses and plot them on a map, and then do some analysis of the results. You'll also use Google Maps and Google Earth to analyze the shortest paths between these plotted points.

Final Layout Checklist

_____   All the counties in California displaying their population per square kilometer, classified, and displayed with an appropriate color scheme

_____   Appropriate map legend, with all items listed with normal names (the legend should not be called "legend" and should be appropriate size, font, number of columns, and so on)

_____   Scale bar (with appropriate divisions and numerical breaks)

_____   Appropriate title (don't use the words "map" or "title" in your title) and appropriate size and font

_____   A north arrow of appropriate size and position

_____   Type (your name, the date, the source of the data) in an appropriate size and font

_____   Overall map design (appropriate borders, color schemes, balance of items and placement, and so on)

Getting There Quicker with Geospatial Technology

8

Satellite Navigation Systems, Road Maps in a Digital World, Creating a Street Network, Geocoding, Shortest Paths, and Street Networks Online

Satellite navigation systems (the kind of devices made by companies such as Garmin, Magellan, and TomTom) are really revolutionary technologies. One small device mounted on the dashboard will find your precise location, plot it on a map, determine where the nearest gas stations are, then compute the quickest route to get you there—all with turn-by-turn directions that announce the names of streets and the distance to the next turn (see **Figure 8.1**). Many smartphone apps can do the same things, effectively turning your phone or mobile device into a satellite navigation system. The position determination is straightforward—the device has a GPS receiver in it that finds your location on Earth's surface, using the methods we discussed in Chapter 4. In fact, most people who own a device with a navigation system refer to it simply as a "GPS," as in "Punch our destination into the GPS" or "What does the GPS say is the shortest route?" However, it's doing a disservice to these devices to call them simply a "GPS"—they do so much more than a regular GPS receiver does.

These devices rely on a GIS-style system at their core—hence their ability to handle spatial data in the form of road-network maps, use those maps for routing, determine the shortest path between two (or more) points, and match an address to a spatial location. This same type of data is used to route emergency vehicles to the site of a 911 emergency phone call or to manage a fleet of delivery vehicles. You'll find the same sort of system at the heart of online mapping applications such as MapQuest or Google Maps, and in the location and mapping apps of a smartphone. This type of technology is changing fast and getting better at what it does.

A March 2009 article in *USA Today* described an incident where a vehicle navigation device instructed the driver to turn off a road and follow a snowmobile trail toward a destination—this ended with the car stuck in

satellite navigation system a device used to plot the user's position on a map using GPS technology

FIGURE 8.1

A Garmin GPS satellite navigation system. *[Edward J Bock III/Dreamstime.com]*

the snow and the state police being called for emergency help. Similarly, a March 2009 article of the *Daily Mail* relates a story of a vehicle navigation system that directed a driver along a footpath, which ended in a sheer drop of 100 feet off the edge of a cliff (luckily, the driver stopped in time). When errors like these occur, it's usually not because the GPS receiver is finding the incorrect position from the satellite information. The problem is much more likely to be with the base network data itself. These types of systems are only as accurate as the base network data they have available to them. This chapter delves into how these types of geospatial technology applications function, how they're used, what makes them tick, and why they may sometimes lead you astray.

Thinking Critically with Geospatial Technology 8.1

What Happens When the Maps Are Incorrect?

How many times has this happened to you when you're using some sort of mapping service—either the directions take you to the wrong place, or they indicate that you should turn where you can't, or they just can't find where you want to go (or the road you want to go on)? When these things happen, the GPS location is probably correct, but the maps being used for reference are likely either outdated or have errors. How much does the usefulness of this aspect of geospatial technology rely on the base data? For example, when Apple released its Maps component of iOS 6 in 2012, several location inaccuracies were reported, including the mislabeling of a supermarket as a hospital. If the basemaps are incomplete or not updated, then how useful is a mapping system to the user?

It should also be noted that many vehicle navigation systems will recompute a new route on the fly if you miss a turn or start to take another route to your destination. However, if the system has taken you on an improper route, how useful will the system be in getting you back to where you want to go? A key purpose of mapping systems is to navigate through unfamiliar territory—but if they don't fulfill that function properly, what can travelers rely on? Are there any sorts of liability issues that could result from inaccurate mapping and misleading instructions?

How Do You Model a Network for Geospatial Technology?

Back in Chapter 5, we discussed how real-world items are modeled, or represented, using GIS. Any type of **network** is going to involve connections between locations, whether those locations are streams, power lines, or roads. Thus (to use Esri terminology), in GIS a network in its most basic form is represented by a series of **junctions** (point locations or nodes) that are connected to each other by a series of **edges** (lines or links). For instance, in a road network, junctions might be the starting and ending points of a road, or they might be points of intersection with other roads, while the edge is the line that represents the road itself. When you're designing a road network, keep in mind that there may be many types of edges and junctions to represent. For example, a city's road network will have edges that represent streets, highways, railroads, light-rail systems, subway lines, or footpaths, while junctions may represent not only the beginnings and endings of streets, but also highway entrances and exits, freeway overpasses and underpasses, subway stops, or rail terminals.

When you're dealing with all these different types of edges and junctions, the **connectivity** of the network in GIS is essential when you come to model it. With proper connectivity, all junctions and edges should properly connect to one another, while things that should not connect do not. For example, if a freeway crosses over a road via an overpass, the network connectivity should not show a valid junction where the street is allowed access to the freeway. If your vehicle navigation system leads you to this point, then instructs you to do the impossible and "turn right onto the highway," it's because the network data is set up to make it think you can. In the same way, a railroad line may intersect with a street, but the network should not have a connection showing that the street has direct practical access to the railroad line as they cross. If this kind of connection is built into the data, you could conceivably be routed to turn onto the railroad line and continue on it toward your destination. It sounds silly to think of driving your car on the railroad tracks, but if the network data is incorrect, this line might simply represent the next road to take to get to your destination.

Thinking along these lines, other features of a road network must also be included in the model. For instance, some streets may be one way, some junctions may not allow left-hand turns, or U-turns may not be permitted. These types of features need to be modeled properly for the system to be an accurate, realistic model of the road network. You can see the "one way" street sign when you're driving, but if that feature hasn't been properly set up in the network, the system has no way of knowing it shouldn't route cars in both directions along the street. As noted, an overpass or underpass should not show up as a viable option for a turn—if a device or GIS instructs you

network a series of junctions and edges connected together for modeling concepts such as streets

junction a term used for the nodes (or places where edges come together) in a network

edge a term used for the links of a network

connectivity the linkages between edges and junctions of a network to make a right-hand turn onto the freeway that you're currently driving under, then something's gone wrong with some aspect of the technology.

When you're modeling a network, you should consider each edge as a separate entity, though not necessarily each individual street. A long city street may be modeled in the GIS as several **line segments**, with each segment representing a section of the street. A major urban street may be made up of more than 100 line segments—each segment being a different section of the street (with each segment being delineated by roads that intersect it). For example, **Figure 8.2** shows a geospatial road network in Virginia Beach, Virginia. The road cutting through the center of the city (highlighted in blue) is Virginia Beach Boulevard, a major multi-lane city street, with numerous intersecting roads and streetlights. Although we think of Virginia Beach Boulevard as one big, long street, the system models it as 272 line segments, with each line segment representing a portion of its length.

Breaking a road up into individual line segments allows the modeling of different attributes for each segment. Attributes such as the road's name, the address ranges for each side of the road, the suffix of the road's name (Drive, Avenue, Boulevard, etc.), the type of road it is (residential street, interstate, highway, etc.), and the speed limit are examples of values that can

line segment a single edge of a network that corresponds to one portion of a street (for instance, the edge between two junctions)

FIGURE 8.2

Virginia Beach Boulevard (shown in blue) in a GIS network. Although it is only one road, it consists of 272 line segments as it crosses the city. *[U.S. Census/Esri]*

be assigned to individual segments. Thus, an attribute table of this layer in a GIS would consist of 272 records, each with multiple attributes.

Several different types of geospatial road network files are available. A **street centerline** file models each city road as a line and contains lines for the different types of roads. Street centerline files are produced as GIS line layers that represent roads. While this type of data is available from numerous sources, the U.S. Census Bureau regularly issues this type of road network data in a format usable by geospatial technology software as **TIGER/Line** files. TIGER stands for Topologically Integrated Geographic Encoding Referencing, and the files delineate different boundaries throughout the United States (such as block groups or congressional districts) in addition to containing road line data. (See *Hands-On Application 8.1: The U.S. Census TIGERweb* for more information.)

Each record in a street centerline file (such as a TIGER/Line file) represents a small segment of a road, and thus each segment (record) can have multiple attributes (fields) assigned to it. **Figure 8.3** shows a TIGER/Line file of Virginia Beach and a portion of the attribute table of those selected segments that make up Virginia Beach Boulevard. Note how many attributes there are (information gets encoded into each road segment—the entire Virginia Beach TIGER/Line file is made up of over 19,000 segments). These attributes describe the characteristics of each segment and are also useful in referencing locations (such as address ranges) with each segment. These kinds of attributes include things such as the name of the road segment,

street centerline a file containing line segments representing roads

TIGER/Line a file produced by the U.S. Census Bureau that contains (among other items) the line segments that correspond with roads all over the United States

Hands-On Application 8.1

The U.S. Census TIGERweb

TIGER files are made freely available from the Census Bureau via the Web. Go to **http://tigerweb. geo.census.gov/tigerweb** to see the U.S. Census Bureau's online viewer for TIGER data. By zooming in on an area, you can select the TIGER layers you wish to display from the choices on the left-hand side of the screen (such as roads, school districts, hydrography, and various types of boundaries).

Note that here you can view data, but if you wish to download the actual TIGER files themselves, go to **https://www.census.gov/geo/maps-data/ data/tiger-line.html**. This is the part of the U.S. Census Bureau's Website for downloading TIGER/ Line files. There's also full documentation of TIGER files available in PDF format on the Website. Files

for U.S. counties can be downloaded in shapefile format (see Chapter 5 for more info about shapefiles) to be used in GIS products, such as ArcGIS or QGIS. You can also download several other types of TIGER files aside from road-network data, including census block information, hydrography, landmarks, and American Indian reference data— check to see what types of datasets are available for download.

Expansion Questions

➤ What kinds of transportation data are contained within TIGER/Line files aside from roads?

➤ What kinds of TIGER data are available for your county?

FIGURE 8.3

The attributes of the selected records making up Virginia Beach Boulevard from the Virginia Beach TIGER/Line file in ArcGIS. *[U.S. Census Bureau/Esri]*

TABLE 8.1	Commonly used road segment attributes and how they are named in TIGER 2000 and the TIGER 2016 Address Range Feature County layers.

Attribute	TIGER 2000	TIGER 2016
Prefix direction of road (N. Smith St.)	FEDIRP	n/a
Name of road (N. **Smith** St.)	FENAME	FULLNAME
Type of road (N. Smith **St.**)	FETYPE	n/a
Suffix direction of road (Canal Street **E.**)	FEDIRS	n/a
Start of address ranges on left side of road	FRADDL	LFROMHN
End of address ranges on left side of road	TOADDL	LTOHN
Start of address ranges on right side of road	FRADDR	RFROMHN
End of address ranges on left side of road	TOADDR	RTOHN
Zip code on left side of road	ZIPL	ZIPL
Zip code on right side of road	ZIPR	ZIPR
Code to determine the kind of road (residential, highway, etc.)	CFCC	ROAD_MTFCC

the suffix designation of the road segment (such as a lane, avenue, drive, etc.). Different street centerline sources will use different names to represent these commonly used attributes. See **Table 8.1** for some examples on how these types of attributes are named by different sources.

These kinds of attributes define the characteristics of each road segment. Similar attributes would be found in road data, such as other street centerline files. If these attributes are incorrect, then the base network map will be incorrect. If the vehicle navigation system gives you incorrect street names or calls a road "east" when it's really west, it's likely that there are incorrect values in the base network data's attributes.

Attributes such as those from the TIGER/Line files, The National Map, or similar street-network data created by others concerning specific address ranges, zip-code information, and detailed data for the names of roads can be used as a basemap source for other applications, such as pinpointing specific addresses on a road. It's this source data that allows for a match of a typed street address to a map of the actual location.

How Is Address Matching Performed?

address matching another term for geocoding

geocoding the process of using the text of an address to plot a point at that location on a map

Whenever you use a program such as MapQuest or Google Maps to find a map of a location, you're typing in something (like "1600 Pennsylvania Avenue, Washington, D.C.") and somehow the Website translates this string of characters into a map of a spatial location (like the White House). The process of taking a bunch of numbers and letters and finding the corresponding location that matches up with them is called **address matching** or **geocoding** (**Figure 8.4**). Although the process seems instantaneous, there are several steps involved in geocoding that are happening "behind the scenes" when you use an address-matching system (like those in GIS).

FIGURE 8.4

A map generated by querying MapQuest for "1600 Pennsylvania Avenue, Washington, D.C."

reference database the base network data used as a source for geocoding

parsing breaking an address up into its component parts

address standardization setting up the components of an address in a regular format

First, you need to have some sort of **reference database** in place—this is a road network to which the addresses will be matched. The National Map transportation data, a TIGER/Line file, or similar type of street centerline file (many are produced commercially) is needed here. What's essential is that the line segments contain attributes along the lines of those found in a TIGER/Line file—for example, street direction, name of the street, address ranges on the left and right sides of the street, street suffix, and zip codes on the left and right sides of the street. This information will be used as the source to match addresses to as well as a source for the final plotted map.

Next, the address information is subjected to **parsing**, or breaking it up into its component pieces. For instance, the global headquarters of Google is located at 1600 Amphitheatre Parkway, Mountain View, CA 94043. When this address is parsed, the address number is split off from the rest of the address, along with the name of the street, the street type, the city, the state, and the zip code. The parsed version of the address would be something like: 1600 | Amphitheatre | Parkway | Mountain View | CA | 94043.

Next, the geocoding process has to have a way of handling different items in addresses that mean the same thing. For instance, if you're using Google Maps to plot the location of Google's headquarters, you might type in either "Mountain View, CA" or "Mountain View, California." Similarly, the White House is located at 1600 Pennsylvania Avenue NW in Washington D.C. However, when you're trying to find this on Google Maps, you might type in "Pennsylvania Avenue," "Pennsylvania Ave," or "Pennsylvania Av." Even though all of these things mean the same thing to us, the GIS also has to understand that they mean the same thing as well. In some cases, **address standardization** is performed to set up data in a consistent format. In this case, the geocoding process needs to standardize addresses to match a location properly using its appropriate attributes in the reference database—for instance, it might substitute the word "AVE" in place of "Avenue," "Ave," or "Av" so that all those representations of the address are the same.

Other geocoding systems may generate a list of multiple representations of the address, so that the GIS will treat "Avenue," "Ave," "AVE," or "Av" as the same thing. Sometimes the system will have an alternate list of road names stored in its reference database if a road is known locally by one name but is part of a larger road system—for instance, in Ohio, State Route 224 runs the length of Mahoning County, but the section of it that passes through the town of Canfield is known as "Main St." The geocoding system would use an alternate list of addresses to know that these two names mean the same place along the road.

Table 8.2 shows a number of locations in the Washington, D.C., area with their addresses, as well as these addresses parsed and standardized. For instance, in the National Gallery of Art's address, the street name is "Constitution." When the address matches, the system refers to line segments with a name attribute of "Constitution" and those segments with a street-type

TABLE 8.2 **Addresses that have been parsed into their component parts and standardized.**

Location	Address	Prefix	Number	Street name	Street type	Suffix
White House	1600 Pennsylvania Avenue NW		1600	Pennsylvania	AVE	NW
National Gallery of Art	401 Constitution Avenue NW		401	Constitution	AVE	NW
U.S. Capitol	1 1st Street NE		1	1st	ST	NE
Office of the Federal Register	800 North Capitol Street NW	N	800	Capitol	ST	NW
Washington National Cathedral	3101 Wisconsin Avenue NW		3101	Wisconsin	AVE	NW
United States Holocaust Memorial Museum	100 Raoul Wallenberg Place SW		100	Raoul Wallenberg	PL	SW

attribute of "AVE" (rather than ST, BLVD, LN, or any other) and a suffix direction attribute of NW (instead of some other direction). The street number, 401, is used to determine which road segments match an address range (on the left or right side of the street, depending on whether the number is odd or even).

After the address has been parsed and standardized, the matching takes place. The geocoding system will find the line segments in the reference database that are the best match to the component pieces of the address and rank them in ArcGIS. For instance, in trying to address match the National Gallery of Art, a line segment with attributes of Name = "Constitution," Type = "AVE," Suffix Direction = "NW," and an address range on the left side of "401–451" would likely be the best (or top-ranked) match. A point corresponding with this line segment is placed at the approximate location along the line segment to match the street number. For instance, our address of 401 would have a point placed near the start of the segment, while an address of 425 would be placed close to the middle. The method used to plot a point at its approximate distance along the segment is called **linear interpolation**.

Keep in mind that the plotting is an approximation of where a specific point should be. For instance, if a road segment for "Smith Street" has an address range of 302 through 318, an address of 308 will be placed near the middle. However, if the actual real-world location of house number 308 is closer to the end of the street, then the placement of the plotted point will not necessarily match up with the actual location of the house. Where streets contain only a handful of houses that correspond with the address range in the reference file, plotted locations may be estimated incorrectly. See **Figure 8.5** for an example of plotting a geocoded point on a road network in GIS.

linear interpolation a method used in geocoding to place an address location among a range of addresses along a segment

FIGURE 8.5

Plotting an address point on a line segment in GIS.

In GIS, a satellite navigation system, or a smartphone equipped with geospatial technology, whenever you specify an address, the system will match the address and fix it as a destination point to be found. The process of geocoding multiple addresses at once is referred to as **batch geocoding.** For instance, in batch geocoding, you can have a list of the addresses of all the coffee shops in Seattle, and the GIS will match every address on the list, so you won't have to input the addresses one at a time (for an example, see *Hands-On Application 8.2: Geocoding Using Online Resources*). If no match can be found, or if the ranking is so poor as to be below a certain threshold for a match, sometimes a point will not be matched at all for that address (or it may be matched incorrectly, or placed at something like the center of a zip code). You will sometimes be prompted to recheck the address, or to try to match the address interactively by going through the process manually.

batch geocoding matching a group of addresses together at once

Finally, when an address is plotted, the system may have the capability to calculate the *x* and *y* coordinates of that point and return those values to the user. For example, the street address of the Empire State Building in New York City is 350 5th Ave., New York, NY 10018. From address matching, using the free online **gpsvisualizer.com** utility, the GIS coordinates that fit that address are computed to be latitude 40.74827 and longitude 273.98531 (**Figure 8.6**). Thus, geocoded points have spatial reference attached to them for further use as a geospatial dataset.

Geocoding is a powerful process, but it's not infallible. There are several potential sources of error that cause an address to be plotted in an incorrect location. Because the addresses you're entering will be parsed to match up with the segments in the reference database, the result can sometimes

Hands-On Application 8.2

Geocoding Using Online Resources

Many online mapping resources (see *Hands-On Application 8.4: Online Mapping and Routing Applications and Shortest Paths* for details on what they are and how they're used) will geocode a single address, or a pair of addresses (so that you can calculate directions between the two). You can use GIS software to geocode multiple addresses, or you can use another resource, such as the Batch-Geo Website. Open your Web browser and go to **www.batchgeo.com**—this is a free online service that allows you to geocode an address (or multiple addresses in batches), and then view the plotted point or points on a map. Set up a batch of three or more addresses (such as home addresses for a group of family members, a group of friends, or several workplaces) and run the batch geocode utility. The Website will give you examples of how the addresses need to be formatted. From there, you can view your results on the Google Map that the Website generates.

Expansion Questions

➤ Assemble a list of five addresses familiar to you (such as your home address, your work address, or a local park or attraction) and batch geocode them using the Website. Zoom in closely on the resulting Google Map that is produced for you and examine the geocoded points—how closely do the points match up with where the address should be?

➤ If any points are geocoded incorrectly, how are the locations improperly matched?

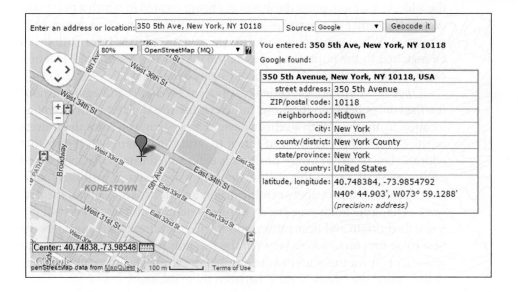

be an error in matching. For instance, the address of the White House in Table 8.2 is listed as "1600 Pennsylvania Avenue NW." The line segments that match up with this street should have the name listed with "Avenue" (or "Ave.") as a suffix. If you input something different, for instance "1600 Pennsylvania Street" or "1600 Penn Avenue," the location could potentially receive a lower ranking, end up plotted somewhere else, or otherwise fail to get properly matched. With more complete information (such as "1600

FIGURE 8.6

The geocoded result for the Empire State Building in New York City, with corresponding latitude and longitude calculation.

Pennsylvania Avenue NW, Washington, D.C., 20500"), the system will be better able to identify an accurate match.

It should also be kept in mind that the geocoding system can only properly identify locations if the line segments are in the reference database. If you're searching for an address established after the reference database network file was put together, the system will not be able to match the address properly. If the point is plotted on the correct street but at an incorrect location on the street, it's likely because of a problem with address range data in the reference database and how it reflects the real world (for instance, a house at 50 Smith Street is not necessarily halfway down a street segment that begins with 2 and ends with 100). The geocoding will usually only be as accurate as the base data to which it's being matched. If the reference database doesn't contain all line segments (for instance, if it's missing subdivisions, streets, or new freeway bypasses that haven't yet been mapped and added), or if its attributes contain inaccurate address ranges or other incorrect information, or if any of its attributes are missing, the geocoding process will probably be unable to match the addresses or may plot the addresses in the wrong location.

New methods for geocoding addresses to get a more accurate match have been developed. Rather than using a line segment and interpolating the address location, point databases are being created in which a point represents the center of a parcel of land (for instance, a house or a commercial property). When geocoding with this point data, address information can be matched to the point representing the parcel, and the address location can be found for the road immediately next to the parcel.

Once locations are geocoded, the system (or GIS) can begin to examine the routes between locations to determine the shortest path from one location to another. With a vehicle navigation system, you enter the address of the destination you want to travel to, and the system will match that address. The device's current position is determined using GPS and plotted on the network map—this will be the origin. The system will then compute the shortest route between the origin and destination across the network. The same holds true for an online system to find directions—it has a matched origin and destination, and it will compute what it considers the best route for you to follow between the two points. With so many different ways to get from the origin to the destination, the system now needs a way to determine the "shortest path" between these locations.

 ## How Are Shortest Paths Found?

When you leave your home to go to work, you probably have several different ways you can go. Some of them are pretty direct and some of them are more roundabout, but you have plenty of options available. If you want to take the "shortest" path from home to work, you'll probably focus on some of the more direct routes and eliminate some of the longer or more circuitous routes. However, a **shortest path** can mean different things. For instance, driving through city streets may be the shortest physical driving distance (in terms of mileage), but if those streets have lots of stoplights and traffic congestion, then this "shortest path" could easily take longer in time spent driving than a longer distance on a highway that doesn't have these impediments. If you want to minimize the time spent driving, the highway route will probably get you to work faster, but you'd actually be traveling a longer distance. Major city streets can take a long time to navigate at rush hour, but you'll probably sail through them much more quickly if you make the same journey late at night.

All of these things have to be considered when you're figuring out the shortest path (or "best route") to take when traveling from an origin to a destination. People have their own decision-making criteria and stick to their own rules when it comes to determining what they judge to be the shortest path from one place to another—things such as "always keep to the main roads," or "use highways whenever you can," or "never make a left turn." This is why vehicle navigation systems often offer multiple options, such as "shortest distance" or "shortest driving time," and sometimes apply conditions such as "avoid highways" to compute the best route between points.

Within a vehicle navigation system (or GIS), each line segment has a **transit cost** (or impedance) assigned to it. The transit cost reflects how many units (of things like distance and travel time) it takes to go the length of a particular edge. The transit cost may reflect the actual distance in miles from one junction to another along the edge. The transit cost could also be the equivalent time it takes to drive that particular segment. Whatever transit cost is used, that value will be utilized in the shortest path computation. Other impedance attributes can be modeled as well—segments could have different transit costs under certain conditions (such as heavy traffic or construction). These types of impedance factors can help in making a network more realistic for use.

The shortest path is then calculated using an **algorithm**, a step-by-step mathematical procedure designed to solve a particular kind of problem—in this case, to determine the overall lowest transit cost to move along the network from a starting point to a destination. There are various types of shortest path algorithms, including **Dijkstra's Algorithm**, which will compute the path of lowest cost to travel from a starting point to any destination in the network. For instance, say you have three different destinations you plan to travel to (work, the pizza shop, and the grocery store). Dijkstra's Algorithm will evaluate the overall transit cost from your home to each destination, and find the shortest path from your home to work, from your home to the pizza shop, and from your home to the grocery store.

shortest path the route that corresponds to the lowest cumulative transit cost between stops in a network

transit cost a value that represents how many units (of time or distance, for example) are used in moving along a network edge

algorithm a set of steps used in a process designed to solve a particular type of problem (for example, the steps used in computing a shortest path)

Dijkstra's Algorithm an algorithm used in calculating the shortest path between an origin node and other destination nodes in a network

Whatever type of algorithm is used, the system will compute the shortest path, given the constraints of the network (such as transit cost or directionality of things like one-way streets). The system can then generate directions for you by translating the selected path into the various turns and names of streets that you'll take as you follow the path through to your destination. See **Figure 8.7** for an example of using an online geospatial technology utility to compute the shortest path or best route between two locations. Also check out *Hands-On Application 8.3: Online Mapping and Routing Applications and Shortest Paths* for some other Web tools for online directions and routing.

Of course, sometimes you have more than one location to visit—say you have three places you want to stop at (shoe store, music store, and bookstore) and you want to drive the overall shortest route to hit all three places. When you're using geospatial technology to compute a shortest route between several of these **stops**, there are two types of scenario from which to choose:

1. In the first scenario you need to find the shortest path when you're going to be visiting stops in a pre-defined order: This means you have to stop at the shoe store first, the music store second, and the bookstore last. In this case, you'd want to find the shortest paths from your house to the shoe store, from the shoe store to the music store, and finally from the music store to the bookstore.

2. The second scenario involves finding the shortest path when you can arrange the order of visiting the stops: For instance, if the bookstore and the shoe store are near each other, it makes sense to rearrange your travels so that you visit them one after the other and then drive to the music store.

stops destinations to visit on a network

FIGURE 8.7

The multi-colored line marks the shortest path from the White House to Georgetown University in Washington, D.C. (computed from MapQuest).

Hands-On Application 8.3

Online Mapping and Routing Applications and Shortest Paths

There are many online applications available for creating a map of an address, and then generating a shortest path and directions from that address to another one. Examine the functionality of some of the following online mapping sites:

1. Bing Maps: **www.bing.com/maps**
2. Google Maps: **www.google.com/maps**
3. HERE: **https://wego.here.com**
4. MapQuest: **www.mapquest.com**

Try inputting the same pair of locations (origin and destination) into each Web service and compare the shortest paths and routes they generate (some of them may be similar, some may be different, and some will likely give you more than one option for the "shortest path"). In addition, the services will also allow you to alter the route interactively by moving and repositioning junctions of the path, so that you can tailor the route more to your liking.

Expansion Questions

➤ Set up a shortest route between two locations familiar to you (such as home and work, home and school, or two well-known destinations in your local area), then position the cursor at junctions along the path. You can drag and drop the nodes to new locations and thus change the route to your liking. By altering the route, how specifically did it change in terms of time and distance?

➤ Each of the five mapping sites will give you information back about the source of the maps being generated (usually in small type at the bottom of the map). What is the copyrighted source of the maps for each of the four mapping sites?

Geospatial technology applications can help determine some solutions to both of these scenarios. To visit stops in a pre-determined order, the system will find the shortest route (of all possible routes) from the origin to the first stop, then choose the shortest route to travel from the first stop to the second stop, and so on. An example of this in vehicle navigation systems is being able to set up a "Via Point" (another stop) between your origin and final destination—the device will then calculate the route from the origin (or current location) to the Via Point, then from the Via Point to the final destination.

With the ability to rearrange the order of stops, a system will evaluate options by changing the order of visiting stops to produce the overall shortest route. See **Figure 8.8** for an example of how the "shortest path" changes between visiting stops in order as opposed to being able to rearrange the order of visiting the stops. Additional constraints can be placed on the problem—for instance, you may have to return home (the starting point) after visiting all of the stops. In that scenario, when you rearrange the stops, the starting point also becomes the last stop, which may affect how the reordering is done. Conversely, you may want to have your starting point and ending point different from one another, and these types of parameters will have to be placed on the problem. Keep in mind that if you can arrange the order of stops, the program will likely give a decent solution to the problem; however, the only way to truly identify which configuration is the best is for the program to work through every possible arrangement—something that

FIGURE 8.8

Two different shortest paths through Youngstown, Ohio, both involving multiple stops at public libraries. The stops on the first path are made in a pre-determined order (from #1 through #5 on the map). In the second map, the library visits are rearranged to find the shortest route (without a fixed starting or ending point).

Hands-On Application 8.4

Finding the Best Route for Multiple Stops

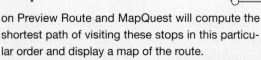

A Web application such as Google Maps or Bing Maps allows you to compute the shortest path between two stops. However, when you have multiple stops, a different tool can be used to find the shortest path among them. For an example of this, go to **www.mapquest.com/routeplanner**. This is the Website for the MapQuest Route Planner, which will allow you to begin with two stops, add others, and then allow MapQuest to reorder your stops to find the shortest route among all of them.

We'll find the shortest route between a set of addresses in Washington, D.C., from Table 8.2. Use the address for the White House (1600 Pennsylvania Avenue NW, Washington, D.C.) as the first address and the Office of the Federal Register (800 North Capitol Street NW, Washington, D.C.) as the second address. Then click on the option for Add another stop under the two addresses to add a third address, that of the U.S. Capitol (1 1st Street NE, Washington, D.C.) and then add a fourth address, that of the National Gallery of Art (401 Constitution Avenue NW, Washington, D.C.). Click

on Preview Route and MapQuest will compute the shortest path of visiting these stops in this particular order and display a map of the route.

However, MapQuest can also rearrange the stops—slide the lever that says "Allow us to re-order your route" and you'll see that the route has changed and the stops have been rearranged for an overall shorter path.

Expansion Questions

➤ How did the shortest paths change between the stops in order and the rearranged order of stops?

➤ Clear the stops and add them again, but this time use the other two addresses from Table 8.2 as new fourth and fifth stops in between the White House (first stop) and the Office of the Federal Register (now the sixth stop). How does the shortest path change the stops in order? How does it change after allowing MapQuest to rearrange the stops?

would be impossible with a larger number of stops. See *Hands-On Application 8.4: Finding the Best Route for Multiple Stops* for an online tool that works with these types of conditions.

 ## How Are Networks Used in Geospatial Technology?

All of the things discussed in this chapter are used for a variety of applications with geospatial technology. Through GIS, these types of network data can be created and utilized in various ways. The vehicle navigation systems integrate many of these concepts—using GPS to pinpoint the device's location on a map, then using network base data, geocoding, and shortest paths to navigate through locales worldwide. Other options on these systems involve utilizing real-time data broadcast into the device to determine areas of congestion, heavy traffic, construction, or road accidents. This data can then be used in the shortest-path process to route you around these impediments. These same maps and technological resources have been integrated into smartphones and mobile devices to put GPS locations, mapping, geocoding, shortest paths, and real-time routing information capabilities into the palm of your hand (**Figure 8.9**). As we have noted, however,

FIGURE 8.9

A smartphone running a mapping application.
[iPhone/Alamy]

FIGURE 8.10

The U.S. Capitol Building as seen from Google Maps Street View.

Street View a component of Google Maps and Google Earth that allows the viewer to see 360-degree imagery around an area on a road

these devices are only as accurate as the basemaps they're utilizing, so regular updates are usually made available, or users are given the option to make their own updates.

The same types of network basemaps are used online for services such as MapQuest, Bing Maps, Here Maps, and Google Maps (see *Hands-On Application 8.3: Online Mapping and Routing Applications and Shortest Paths*), which allow you to perform geocoding and obtain directions for the shortest path between points. **Street View** allows Google Maps (or Google Earth) users to examine 360 degrees of location photography—on a street, say, as if you'd just stopped your car right there to take a look around. With this function, you can examine canned photography of an address or destination when you're planning a stop (**Figure 8.10**).

Cars equipped with special cameras capable of capturing a 360-degree view travel the nation's roads, taking in imagery along the way (see **Figure 8.11** and *Hands-On Application 8.5: Examining Google Street View*). Google is even extending their Street View into places inaccessible to motor vehicles by attaching the same type of camera equipment to bicycles and checking out the hiking and biking trails.

With this level of data availability, and geospatial networks being integrated into so many different applications, mapping (and providing accurate and up-to-date basemaps) has become big business. Companies such as Tele Atlas and HERE produce these maps, and their products are often what you're accessing on an online service or through a vehicle navigation system. Next time you access one of these kinds of services, look for the copyright data somewhere on the map to see which company is producing the map data you're using. HERE, for instance, sends teams out in high-tech

FIGURE 8.11

An example of a car capturing Google Street View images. *[AP Photo/dda, Michael Kappeler]*

cars to collect data on the location and attributes of new roads, housing developments, and other items for updates of maps (as well as photography equipment to produce items such as the Street View scenes).

Hands-On Application 8.5

Examining Google Street View

Google Street View is a very useful application that gives you a look at what the street network would look like if you were driving or riding past. Go to Google Maps at **www.google.com/maps**, and then enter a particular address—try typing in the address of the Rock and Roll Hall of Fame in Cleveland, Ohio (1100 Rock and Roll Blvd., Cleveland, OH). Zoom in until you can see the distinctive building of the Rock Hall and its surrounding streets. Below the zoom controls on the lower right-hand side of the map you'll see an orange icon that looks like a person (called the "pegman"). Grab that with the mouse and place it on top of the street in front of the Rock Hall (E. 9th Street). Streets that have been covered by Google Street View will be highlighted in dark blue. Place the icon on one of the available streets and the view will shift to photography of that area. Use the mouse to move the view about for 360-degree imagery of the area. You'll also see some white lines and arrows superimposed on the road—clicking on them will move you along the street and you'll see the imagery change. Move around until you can get a good street-level view of the Rock and Roll Hall of Fame.

Expansion Questions

➤ How would the imagery from Street View be helpful if you were traveling to the Rock Hall as a destination (and had never been there before)? There are no parking areas available immediately adjacent to the Rock Hall—how could Google Street View be more helpful to you than a road map or an overhead view in finding somewhere to park?

➤ Examine some local areas near you using Street View. Which roads are available with Street View imagery and which are not (for instance, is Street View imagery of residential areas available or is it available only on main roads)?

Thinking Critically with Geospatial Technology 8.2

What Kind of Issues Come with Google Street View?

Go to Google Maps at **www.google.com/maps**, and then search for: Longworth House Office Building, which is one of three buildings in Washington, D.C. that is home to the offices of the U.S. House of Representatives. When the map appears, you'll see the Longworth House Office Building flanked by the other two House of Representatives buildings—the Rayburn House Office Building to the west and the Cannon House Office Building to the east. However, when you try to move the Street View pegman to view the roads, you'll see that the streets immediately surrounding the three buildings are largely unavailable to view in Street View, presumably because they are inaccessible for security purposes. However, most of the surrounding streets can be viewed using Street View, allowing you to look at the exteriors of shops,

residences, and other government buildings. Can the images being collected by Street View pose a security risk, not just for government areas, but for the private security of homes or businesses?

An article in the May 31, 2008, issue of the *Star Tribune* (available online at **http://www.startribune.com/north-oaks-to-google-maps-keep-out/19416279)** describes how the community of North Oaks, Minnesota, demanded that Google remove its Street View images of the privately owned roads in the area, citing laws against trespassing (which may have been the first instance of these types of demands). Is Street View encroaching too far on people's private lives? Keep in mind that you can view or record the same things quite legally by driving or walking down the same streets.

Chapter Wrap-Up

Networks, geocoding, and routing are all powerful tools in everyday use in geospatial technology. Today, with GIS, these concepts are used in a variety of applications and businesses. Today 911 operators can geocode the address that a call is coming from, and emergency services can determine the shortest route to a destination. Delivery services can use geocoding and routing applications to determine locations quickly and reduce travel time by using the shortest paths to get to them. Also, the use and development of these types of techniques in satellite navigation systems and smartphone apps represent rapidly changing technology that keeps getting better. For current examples of geospatial technologies used for networks and routing, check out *Geocoding and Shortest Paths Apps, as well as Geocoding and Shortest Paths in Social Media*.

Geospatial Lab Application 8.1 uses Google Maps as well as Google Earth Pro for investigating geocoding and shortest path uses. In the next chapter, we're going to start looking at a whole different aspect of geospatial technology—remote sensing. All of these overhead images that you can see on applications such as Google Maps or MapQuest have to come from somewhere to get incorporated into the program, and we'll start looking at the methods behind remote sensing in the next chapter.

Important note: The references for this chapter are part of the online companion for this book and can be found at **www.macmillanlearning.com/catalog/shellitoigt4e.**

Geocoding and Shortest Path Apps

Here's a sampling of available representative apps using geocoding and shortest paths for your smartphone or tablet. Note that some apps are for Android, some are for Apple iOS, and some may be available for both.

- **Footpath Route Planner:** An app that allows you to draw a circle around an area and convert the drawing to a street route for use in walking, running, or cycling
- **Google Maps:** An app for a mobile version of Google's online maps
- **Google Street View:** An app for a mobile version of Google Street View
- **MapMyRide:** An app that lets you map and track your biking route
- **Maps (from Apple for iOS):** Apple's mapping app for mobile devices
- **MAPS.ME:** An app that allows you to browse and use maps and routing while offline
- **Moovit:** An app that provides information about the best route to a destination via buses or trains
- **Sygic: GPS Navigation, Maps, Traffic, Gas Prices:** A mobile navigation app
- **Trucker Path:** An app designed for navigation for professional truck drivers to locate truck stops, weigh stations, nearby fuel prices, and other essentials.
- **Waze:** An app used for obtaining and sharing real-time local traffic and road conditions

Geocoding and Shortest Paths in Social Media

Here's a sampling of some representative Facebook, Twitter, and Instagram accounts, along with some blogs related to this chapter's concepts.

On Facebook, check out the following:

- **Google Maps:**
  www.facebook.com/GoogleMaps
- **MapQuest:**
  www.facebook.com/MapQuest

Become a Twitter follower of:

➤ **BatchGeo:** @batchgeo

➤ **Bing Maps:** @bingmaps

➤ **Google Maps:** @googlemaps

➤ **Google Maps Update:** @GoogleMapUpdate

➤ **MapQuest:** @MapQuest

 Become an Instagram follower of:

➤ **Google Maps:** @googlemaps

For further up-to-date info, read up on these blogs:

➤ **Bing Maps Blog** (the blog for Microsoft's Bing Maps): **http://blogs. bing.com/maps**

➤ **Google Maps News** (a blog about Google Maps): **https://blog.google/ products/maps/**

Key Terms

8.1 Geospatial Lab Application

Geocoding and Shortest Path Analysis

This chapter's lab will introduce you to the concepts of calculating a shortest path between stops along a network, as well as generating directions for the path using Google Maps. You'll also be performing geocoding using Google Earth Pro and examining the geocoding results (as well as performing shortest path analysis using Google Earth Pro).

Objectives

The goals for you to take away from this lab are:

➤ To use Google Maps to create and alter shortest paths to account for route changes

➤ To utilize Google Earth Pro to geocode a series of addresses and then examine the results

➤ To familiarize yourself with the shortest path and directions functions of Google Earth Pro

Using Geospatial Technologies

The concepts you'll be working with in this lab are used in a variety of real-world applications, including:

➤ Emergency services, which utilize geocoding to determine the location of 911 callers, so that the nearest available aid can be quickly dispatched to the right place

➤ Commercial promotion and postal services, where traveling sales representatives and delivery drivers make use of routing information to take the shortest and most obstacle-free paths between locations

[Terry Moore / Stocktrek Images / Alamy Stock Photo]

Obtaining Software

The current version of Google Earth Pro (7.3) is available for free download at **https://www.google.com/earth/download/gep/agree.html**.

Important note: Software and online resources can change fast. This lab was designed with the most recently available version of the software at the time of writing. However, if the software or Websites have changed significantly between then and now, an updated version of this lab (using the newest versions) will be available online at **www.macmillanlearning.com/catalog/shellitoigt4e**.

Lab Data

Copy the folder **Chapter 8**—it contains a spreadsheet (in CSV format) called **VirginiaBeachLibraries.csv** featuring a series of addresses to be geocoded.

Localizing this Lab

The datasets in this lab focus on the locations of public libraries and the road network in Virginia Beach, Virginia. However, this lab can be modified to examine your local area. Use your local county library's Website (or the phone book) as a source of names and addresses of local libraries and use Microsoft Excel to create a file like the one in the **Chapter 8** folder (where each column has a different attribute of data). When you've entered the information into Excel, save the file in CSV format (it will be one of the file formats available to save to within Excel). If there are not enough local libraries around, use the addresses of other local venues, like coffee shops, pizza shops, or drugstores.

8.1 Google Maps' Shortest-Path Functions

1. To begin, we'll look at a simple shortest route calculation using Google Maps. Navigate to the Google Maps Website at **www.google.com/ maps**.

2. Type **Virginia Beach, VA,** and click the **Search** button.

3. When the map of Virginia Beach appears, select **Directions** from the pop-up menu that appears under where you searched.

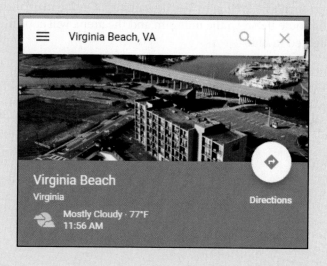

4. Let's start with a typical summertime scenario. You're on vacation in Virginia Beach, visiting the Old Coast Guard Museum (which faces the oceanfront boardwalk). You have tickets to see a show at Norfolk Scope Arena, located in nearby downtown Norfolk. You want to take the shortest route to get from your beachfront location to the arena. Use the following addresses for directions:

   a. Starting point: **2400 Atlantic Avenue, Virginia Beach, VA 23451** (this is the museum you're driving from)

   b. Ending point: **201 East Brambleton Avenue, Norfolk, VA 23510** (this is Norfolk Scope Arena)

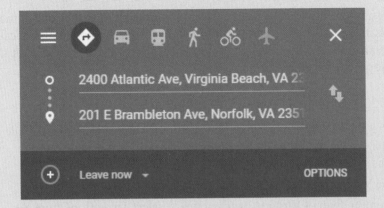

5. By default, Google Maps will use driving a car as the way to compute the shortest route (above the directions you'll see icons for other options such as traveling by bus or walking). On the map, the shortest driving route between the museum and the arena will appear, highlighted in blue.

> **Question 8.1** What is the approximate driving time between the two locations? What (in general) is the route that Google Maps suggests you take?

6. Select the option for walking between the museum and the arena (the icon is of a person on foot). A second path, shown with blue dotted lines, will appear.

> **Question 8.2** What is the new distance (and estimated walking time) of the route using this new path? Why is the walking route different from the driving route?

7. Return to the shortest-path driving distance. Zoom in closer to the starting point (the museum at 2400 Atlantic Avenue). Scroll the mouse over the blue shortest path line and you'll see a white circle appear along the path. This lets you change the calculated route to account for any variety of factors (such as your own travel preference, known congestion or construction areas, rerouting to avoid likely trouble spots or known delays, and so on).

8. Start by placing the mouse at the starting point and drag the circle up to state route 58 (also known as Laskin Road or Virginia Beach Boulevard—the other major east–west road just above and parallel to I-264). The route will change by first having you drive north on Atlantic Avenue, turning west on 58, then merging back onto I-264 again a little later on the route. Even though it's a small change, you'll see that the driving distance and estimated time have changed. Answer Question 8.3. When you're done, click the back arrow on your Web browser to return the path to how it originally was (if you made multiple changes to the route, you may have to click the back arrow a couple times to return the map to its original path).

> **Question 8.3** What is the new distance (and estimated driving time) of the route using this new path? What would account for this?

9. Scroll across the map until you see where the route crossed I-64 (about two-thirds of the way between the two points). I-64 is the major highway into the Hampton Roads area. Let's throw another change into the scenario—say, for instance, that there's heavy construction on I-264 west of the I-64 intersection. At the intersection, move the circle off the route and north onto I-64 far enough so that the remainder of the route to Norfolk Scope diverts your shortest path headed north onto I-64, and not so that it returns you to I-264 west (you may have to move other circles as well if Google Maps diverts your path back onto I-264).

> **Question 8.4** What is the new route (in terms of the general roads now taken)? What is the new distance (and estimated driving time) of the route using this new path?

10. Reset the route back to its original state. Now, change the route so you would have to travel south on I-64 instead of north (still avoiding I-264 for the whole route).

> **Question 8.5** What is the new route (in terms of the general roads now taken)? What is the new distance (and estimated driving time) of the route using this new path?

11. You can now close your Web browser.

8.2 Geocoding and Google Earth Pro

Before any type of shortest path could be calculated, the system first had to match the addresses of the starting and ending points to their proper location. The next section of this lab will introduce a method of doing this address matching (geocoding) to create a new point layer for use in Google Earth Pro.

1. Open the **VirginiaBeachLibraries.csv** file using Microsoft Excel (or a similar program). You'll see the 10 Virginia Beach libraries with their address information split into multiple fields (columns): Name, Address, City, State, and Zip. These last four fields will be used to geocode the libraries and then the Name field will be used to reference them.

2. Close the VirginiaBeachLibraries.csv file (don't save it if prompted).

3. Start Google Earth Pro (GEP), type Virginia Beach, VA, in the Search box, and then click on Search. Google Earth will switch to the view of Virginia Beach.

4. From the **File** pull-down menu, select **Import**. Navigate to the location of VirginiaBeachLibraries.csv and select that file to import into GEP.

5. In the Data Import Wizard that will open, select the radio button for **Delimited for the Field Type**, then select the radio button for **Comma** under the **Delimited** options. You'll see a preview of the data at the bottom of the Wizard, which should show each of the attribute fields split into their own column (as they were in the initial VirginiaBeachLibraries.csv file). Click **Next**.

6. In the second screen of the Data Import Wizard, make sure the box for **This dataset does not contain latitude/longitude information, but street addresses** has a checkmark next to it (because you will be geocoding addresses rather than plotting points based on their latitude and longitude coordinates). Click **Next**.

7. In the third screen of the Data Import Wizard, select the radio button for **Addresses are broken into multiple fields**.

8. In the options under **Select Address Field(s)**, be sure the following options are chosen (if they are not, you can select them from the pull-down menu next to each):

    a. Street field: **ADDRESS**

    b. City field: **CITY**

    c. State field: **STATE**

    d. Zip/postal code field: **ZIP**

    e. Country field: **N/A**

9. Click **Next**.

10. In the next screen of the Data Import Wizard, make sure that string is selected for the Type for each of the following fields: **NAME, ADDRESS, CITY,** and **STATE**. Make sure that **integer** is selected for **ZIP**.

11. Click Finish. When prompted by the question: **Do you want to apply a style template to the features you ingested?** click Yes. The Style Template Settings dialog box will appear. This will give you several options as to how you want your geocoded points displayed in GEP.

12. In the Style Template Settings dialog, select the **Name** tab. From the pull-down menu next to **Set name field**, choose **NAME** (this will tell GEP to use the attribute listed in the NAME field when labeling the geocoded points).

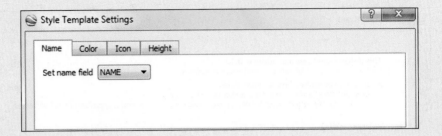

13. Next, click on the **Color** tab. Select the radio button next to **Use single color** and click on the **gray box** next to that. A new dialog box will appear that will allow you to choose whatever color you wish for the display of the geocoded points. Choose something appropriate and click **OK** to return to the Style Template Settings dialog box.

14. Next, select the **Icon** tab. Select the radio button next to **Use same icon for all features** and choose the **pushpin** from the pull-down menu next to that.

15. Last, select the Height tab. Select the radio button next to **Clamp features to ground.**

16. Click **OK.** You'll be prompted to save your style template—save it with the name **Chapter 8 style.** This will allow you to load these same options later if you do more geocoding on your own.

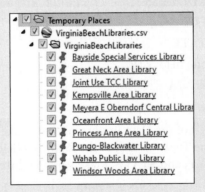

17. In the GEP Places box, you'll see an icon has been added called **VirginiaBeachLibraries.csv.** Expand this to see a folder under that called VirginiaBeachLibraries. Place a checkmark in both of these boxes and expand the folder option to see the labels for each of the libraries. In the main view, you will see the library addresses have been geocoded into points, which are shown with a pushpin in the color you selected.

18. To examine the results of the geocoding, locate the geocoded point for the Bayside Special Services Library and zoom in very closely so

you can see where it was placed in relation to GEP's crisp overhead imagery.

> **Question 8.6** Where is the geocoded point placed in relation to the actual library building?

19. Next, examine each of the other nine geocoded points and verify that they've been geocoded into the correct location. Zoom in closely to each one (a quick way to do this is to double-click on the library's name in the Places box to find the location and then zoom in closer from there) and answer Question 8.7. Return to the full view where you can see Virginia Beach and all 10 geocoded library points when you're done.

> **Question 8.7** For the other nine libraries in Virginia Beach, where were their geocoded points placed in relation to the library buildings themselves? Be specific in your answers.

8.3 Using Google Earth to Compute the Shortest Route Between Geocoded Points

1. The final step of the lab will investigate how to calculate shortest paths from the geocoded points in Google Earth. For instance, if you were driving between two of the geocoded libraries (such as the Great Neck Area Library and the Princess Anne Area Library), GEP can find you the shortest route. To start, zoom in on the Great Neck Area Library point.

2. Right-click on the **pushpin** representing the **Great Neck Area Library,** and select the **Directions from here** option. In the Search box, you'll see the Search option switch to say "Great Neck Area Library" with its latitude and longitude.

3. Next, locate the **Princess Anne Area Library,** right-click on its pushpin, and select the **Directions to here** option.

4. In the Search box, the information about the new Princess Anne destination will be added. A new shortest path from the Great Neck Area Library (its address will be listed as A) to the Princess Anne Area Library (its address will be listed as B) will be calculated and displayed in purple on the view. Different routes between the destinations will also be shown in the Search box. By clicking on one of them, the corresponding route will be displayed in the view.

> **Question 8.8** What are the suggested routes between Great Neck Area Library and the Princess Anne Area Library, and what is the distance (and estimated driving time) of each?

5. Locate both the Pungo-Blackwater Library and the Oceanfront Area Library.

> **Question 8.9** What are the suggested routes between the Pungo-Blackwater Library and the Oceanfront Area Library, and what is the distance (and estimated driving time) of each?

6. The geocoded points can also be treated by GEP just like any other destination. For instance, you can search for a location using GEP and find the shortest path to one of your geocoded points. For instance, search for Old Dominion University, 5115 Hampton Blvd, Norfolk, VA, 23529. GEP will put a placemark at this location.

7. If you wanted to find the shortest distance from Old Dominion University to the Joint Library operated by the City of Virginia Beach and Tidewater Community College (TCC), GEP can easily do this. To start, right-click on the newly added **placemark for Old Dominion University** and choose **Directions from here.**

8. Next, locate the **Joint Use TCC Library geocoded point**, right-click on it, and choose **Directions to here**.

> **Question 8.10** What are the suggested routes between Old Dominion University and the Joint Use TCC Library, and what is distance (and estimated driving time) of each?

Closing Time

This lab demonstrated several types of features associated with geospatial network data, including calculating different shortest paths and address matching with geocoding. Chapter 9 will switch gears and introduce some new concepts dealing with remote sensing of Earth (and the roads built on top of it). There's no need to save any data in this lab. Exit Google Earth Pro by selecting **Exit** from the **File** pull-down menu.

Remotely Sensed Images from Above

9

Where Aerial Photography Came from, UAS, Color Infrared Photos, Orthophotos, Oblique Photos, Visual Image Interpretation, and Photogrammetric Measurements

Whenever you take a picture with a digital camera, the image is captured and stored as a file on a memory card. Taking a picture at a birthday party or on vacation is actually a form of remote sensing—acquiring data without being in direct contact with the subject. For instance, you don't shove the camera right on top of the birthday cake to take a picture of it, you stand back a distance. In essence, the camera is "sensing" (taking the picture) while you're some distance ("remotely") away from the target. In the same way, your eyes function as "remote sensing" devices—when you're looking at this book, you don't place your eyeball directly on the page. Instead, your eyes are "sensing" the data, and they're "remotely" located several inches away from the page. Both a camera and your eyes are performing "remote sensing," as they're acquiring information or data (in this case, visual information) from a distance away without actually making contact with the item.

However, when it comes to **remote sensing** in geospatial technology, things are a little more specific. The data being acquired is information about the light energy being reflected off of a target. In this case, your eyes can still function like remote sensing devices, as what you're actually seeing is light being reflected off objects around you processed by your eyes. A digital camera does the same thing by capturing the reflection of light from whatever you're taking a picture of. In 1826, French inventor Joseph Niépce took the first photograph, capturing an image of the courtyard of his home. Photography has come a long way since those days (when it took Eight hours to expose the film), to the point where digital cameras are now commonplace, and it's difficult to find a cell phone without a built-in camera. In many ways, you can think of a camera as the original device for

remote sensing the process of collecting information related to the reflected or emitted electromagnetic energy from a target by a device in an aircraft or spacecraft at a considerable distance from that target remote sensing—it captures data from the reflection of visible light nearly instantaneously at a distance from the target.

People's fascination with the possibilities of remote observation led quickly to the development of **aerial photography**. Photographers have been capturing images of the ground from the sky for over 150 years. Initially they used balloons and kites to achieve the necessary height, but developing technologies of flight created new opportunities—airplanes, rockets, and, most recently, satellites, have enabled us to capture images of Earth's surface from greater and greater distances. When we're dealing with remote sensing in geospatial technologies, we're concerned with acquiring information from a platform on some type of aircraft or spacecraft. Currently, these types of imagery are obtained in both the public and private sectors.

Whether for military reconnaissance, surveillance, studies of the landscape, planning purposes, or just for being able to say, "Hey, I can see my house," people have been acquiring images of Earth from above for a long time. Today's aerial photographs are used for all manner of applications—for instance, aerial imagery serves as a base source for the creation and continuous updating of geospatial data and maps, such as the US Topo series of maps made by the USGS (see Chapter 13) and maps of road network data (see Chapter 8). Some of the extremely crisp high-resolution imagery you see on Google Earth is taken from aerial photography sources. In this chapter, we'll focus on how aerial images are captured and analyzed, and we'll learn to get used to looking at Earth from above.

aerial photography taking photographs of objects on the ground from an airborne platform

How Did Aircraft Photography Develop?

French photographer Gaspard-Félix Tournachon (known as "Nadar") took the first aerial photograph. In 1858, Nadar strapped himself into the basket of a hot-air balloon and captured an image of a landscape outside Paris. This first aerial photograph no longer exists, but this was clearly the beginning of something big. In 1860, aerial photographs were taken of Boston by James W. Black from a balloon above the city (**Figure 9.1**). However, development times, problems with landing at a prearranged location, and interference with the photographic plates from the balloon's gas all contributed to the view that balloons were not a practical platform for capturing aerial photographs.

By the end of the nineteenth century, photographic equipment was attached to kites (or a series of kites) and used to photograph the landscape remotely—while the photographer was safely on the ground. A famous example of kite photography comes from 1906, when photographer George Lawrence used a set of kites 1000 feet above San Francisco Bay to capture an aerial image of the city in the wake of the 1906 earthquake; he called it "San

FIGURE 9.1

An aerial photograph of Boston in 1860, taken by James W. Black. This is the earliest surviving aerial photograph taken from a balloon in the United States. *[Library of Congress/Corbis/VCG via Getty Images]*

Francisco in Ruins" (**Figure 9.2**). These early images taken from the air set the stage for the use of aircraft as platforms for aerial photography.

The history of modern aviation started in 1903, with the first flight of the Wright Brothers at Kitty Hawk, North Carolina. In 1908, a cameraman took the first photograph from an airplane as he flew with Wilbur Wright near Le Mans, France. During World War I, aerial photography from an airplane was commonly used for mapmaking and the planning of military tactics. Aerial photography of the battlefields allowed for the patterns of trenches to be determined, enemy positions to be observed, troop

FIGURE 9.2

"San Francisco in Ruins," an aerial photo taken in 1906 by George Lawrence, who suspended a remotely operated camera from a train of 17 kites, tethered 1000 feet over San Francisco Bay. *[Library of Congress, Prints and Photographs Division (LC-USZC4-3870 DLC)]*

FIGURE 9.3

World War I aerial photography showing trench formations. *[IWM via Getty Images]*

movements and encampments to be plotted, and the location of artillery or supplies to be found. Countless aerial photos were shot and developed over the course of the war, and aerial photography opened up new areas in military intelligence. See **Figure 9.3** for an example of trench formations captured by an aerial camera.

Aerial photography continued to be extremely important for tactical planning and reconnaissance during World War II. Both the Allies and the Axis forces depended heavily on aerial photography for obtaining vital information about locations prior to military action, and it became a key reconnaissance technique. For instance, a V-2 rocket base in Peenemunde, Germany, was the subject of British aerial photography in June of 1943, and bombing strikes destroyed the base less than two months later. Similarly, aerial reconnaissance was used in obtaining photographs of German dams and the subsequent "dam-busting" bombing missions to destroy them. A huge amount of aerial photography was performed during the war, and a good source of historical imagery is the National Collection of Aerial Photography (NCAP), which hosts a large archive of European World War II aerial photos (see *Hands-On Application 9.1: World War II Aerial Photography Online* for a glimpse of these historic photos).

Post–World War II aerial photography remained critical for military applications, especially for spy planes. The 1950s saw the development of the U-2 spy plane, capable of flying at an altitude of 70,000 feet, traveling at over 400 mph, and photographing the ground below. The U-2 was used for taking aerial photos over the Soviet Union, but also enabled the acquisition of imagery showing the placement of missile launchers at key points in Cuba during the Cuban Missile Crisis of October 1962 (**Figure 9.4**).

Spy plane technology continued to advance with the development of the Lockheed SR-71 Blackbird, which could fly at an altitude of over 80,000

Hands-On Application 9.1

World War II Aerial Photography Online

The National Collection of Aerial Photography is a great online resource for viewing aerial photography of Europe during World War II. Visit **http://ncap.org.uk/galleries**, which contains many different galleries of aerial imagery from several countries. Check out some of the countless historic aerial photos that have been archived online.

Expansion Questions

➤ A highlighted section of the National Collection of Aerial Photography showcases the "detective work" that has uncovered new information about historic events from the aerial archive. Look at **http://ncap.org.uk/news/detective-work-uncovers-hidden-secrets-aerial-reconnaissance-archive-tara**. What kinds of images linked to historic events have been found, and what kind of information has been gathered from them?

➤ Click on the Features section of the Website for access to several showcased historic aerial photos (and lessons on working with aerial photography). Select the option for The D-Day Landings and examine the aerial photos in this section. What aspects of the actions of the D-Day invasion have been captured through aerial photography? What particular features of the invasion can you interpret from the images?

➤ Also in the Features section, select the option for Camouflage, Concealment, and Deception. It's hard to hide something as big as an airfield or a warship from overhead cameras—but look closely at the images in this section. How were airfields, warships, oil refineries, and rocket launching sites concealed from aerial photos?

FIGURE 9.4

Aerial imagery of Cuba showing missile facilities, shot during the Cuban Missile Crisis of 1962.

[U.S. Air Force Photo]

FIGURE 9.5

FIGURE 9.5

The SR-71 Blackbird in action. *[USAF/Judson Brohmer]*

feet and travel at more than three times the speed of sound (see **Figure 9.5**), thus enabling it to go higher and faster than the U-2. The Blackbird was an essential aerial photography plane through the Cold War and beyond, until it was retired in the late 1990s. Today, digital aerial photography (among many other types of remotely sensed data) is also obtained by unmanned aircraft systems.

 ## What Are Unmanned Aircraft Systems?

UAS unmanned aircraft systems—a reconnaissance aircraft that is piloted from the ground via remote control

sUAS small unmanned aircraft systems that weigh less than 55 pounds

UAS (unmanned aircraft systems) are aircraft that are piloted from the ground via remote control. Sometimes these types of aircraft are also referred to as *drones* or *UAVs* (unmanned aerial vehicles). If the UAS weighs less than 55 pounds, it may sometimes be referred to as **sUAS** (small unmanned aircraft systems). As the name implies, UAS are not flown with a pilot in the cockpit—instead, the pilot remains behind on the ground and controls the UAS from there. UAS have been used for military operations to collect aerial imagery for reconnaissance as well as being weaponized for both imaging and combat strikes. Military UAS range from smaller drones such as the Predator to larger ones such as the RQ-4 Global Hawk (which is planned to eventually replace the U-2 for aerial imagery reconnaissance— see **Figure 9.6**).

Unmanned aircraft systems have increasingly branched out from military uses to civilian and commercial uses as well. Today, commercial, public sector, and research use of UAS has become big business and an integral part of the geospatial world. You can even get a bachelor's degree in UAS operations from several universities around the United States. The equipment

FIGURE 9.6

An RQ-4 Global Hawk
UAS (unmanned aircraft
system). *[U.S. Air Force photo/
Stacey Knott]*

onboard UAS collects remotely sensed aerial imagery and video, and UAS
can be launched and used on demand as the user needs it for a variety of
different applications. For instance, border patrols and port security use
UAS for viewing areas remotely and get a "bird's-eye view" for increased
monitoring and safety at these places. Police are using UAS for collecting
imagery to aid in fighting crime and for increased reconnaissance of locations. Farmers are utilizing UAS for collecting imagery to use for crop and
irrigation monitoring. Researchers and universities own UAS and fly them
for imagery data collection to aid in projects (see **Figure 9.7**).

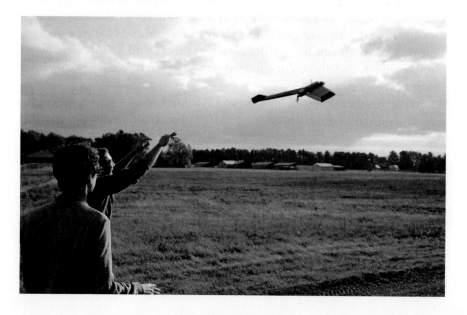

FIGURE 9.7

A UAS used for collecting
imagery for university
research. *[Zack Borst]*

Today, UAS are becoming more affordable and there are a multitude of companies (including firms such as 3DR, DJI, Parrot, senseFly, and Yuneec) that produce UAS that can be used for recreational, research, or commercial purposes. Civilian drones range from ones that you can easily carry in one hand to larger, more robust models, and can be purchased relatively inexpensively. As a result, recreational use of UAS is becoming more and more common, with the ability to launch a drone, control its flight by remote control (or sometimes even with an app on your smartphone or tablet), and quickly take high-resolution imagery of the ground from the sky (see **Figure 9.8**). UAS can be outfitted with cameras and sensors to collect photos, videos, and also color-infrared imagery, and be outfitted with lidar equipment (see Chapter 13).

With this growing proliferation of UAS (particularly sUAS), there are a number of rules governing who can fly UAS, what kind of permits are required, and where they can be flown. The U.S. Federal Aviation Administration (FAA) regulates the use of UAS for civilian use, and there are a variety of places where you're not permitted to fly UAS (see *Hands-On Application 9.2: No-Fly Zones for UAS* for more information).

If a realtor wants to use a drone to take a picture of a house to put in a listing, the drone can simply be flown above the house, take a few pictures, and land. However, when you fly a UAS to collect remotely sensed imagery

Hands-On Application 9.2

No-Fly Zones for UAS

There are several locations around the United States where it is illegal to fly a UAS, such as national parks, military installations, and within five miles of an airport. To examine spatially what areas are off-limits to flying UAS, go to **https://app.airmap.io**. Airmap is very useful to check before you take your UAS out for a flight to see what areas are restricted from flying over. Airmap allows you to toggle buffers for recreational and commercial airports on and off to show which flight areas around them are restricted. There are also controls for showing the various types of controlled airspace as well as temporary flight-restricted areas, national parks, protected marine areas, and other places you shouldn't be flying a UAS.

Expansion Questions

➤ Use the search box to look at the no-fly zones and restricted areas for Washington, D.C. (*Note:* you may have to zoom out some distance to find non-restricted airspace.) In what areas in and around D.C. are you permitted to fly a UAS?

➤ Use the search box to look at the no-fly zones and restricted areas for your local area. What specific places in your own area restrict flying UAS?

Thinking Critically with Geospatial Technology 9.1

How Can UAS Affect Your Privacy?

In the United States, the FAA (as of August 2016) puts very stringent regulations on the use of UAS for work, research, and commercial purposes. If you're going to fly a UAS of less than 55 pounds (a sUAS) for purposes other than recreation (for instance, to do a bridge survey, map out a bike trail, put together imagery of a park, or any type of non-recreational usage), there are a lot of requirements to meet (a list of which is available online at **https://www.faa.gov/uas/getting_started/fly_for_work_business**). You'll need to pass a piloting test, always fly the UAS below a certain altitude, always keep the UAS in full view while flying, not fly the UAS over moving vehicles, stay 500 feet away from people while flying, and, perhaps most importantly, you're not permitted to fly the UAS over another person without their consent (or a waiver). This last regulation in particular goes a long way toward maintaining people's privacy in relation to being in a photo or a video captured by a UAS.

However, if you're flying your UAS for strictly recreational purposes, those previous rules don't apply. With a recreational drone you can fly it as you wish and while the FAA has posted safety guidelines including "never fly over groups of people" or "never fly over stadiums or sporting events") items such as flying over or imaging a person or private property are not forbidden (see the FAA recreational drone information here: **https://www.faa.gov/uas/getting_started/fly_for_fun**). How can these very different rules affect one's privacy? Should there be more stringent regulations of recreational UAS, similar to those that apply to flying for work or business?

over a larger area for a project, the data collection process isn't that simple. The flight path of the UAS is plotted out well in advance and programmed with a series of waypoints. If the UAS is equipped with an autopilot feature (and many commercial-ready models are), after launch the UAS will fly itself along the path, make turns where necessary, and collect imagery as it goes. A significant amount of overlap is taken in each photo so that the images can be post-processed together (using a software program such as the popular Pix4D) to form a large seamless mosaic image of the study area, and can also be used to build a 3D model of the ground imaged by the UAS (see Chapter 14 for more about 3D rendering of images). Whether from a UAS or a regular piloted aircraft, photography is used extensively for many different kinds of imagery data collection.

What Are the Different Types of Aerial Photos?

When you're flying in a plane and you look out the window, you're looking down on the landscape below. If you had a camera strapped on the bottom of the plane and took a picture straight down from under the plane, you'd be capturing a **vertical photo**. Take a look back at Figure 9.4—the Cuban landscape as seen by the U-2 is photographed as if you were looking down from the plane. Many aerial photos are vertical (although there may be some minor tilting from the camera). The spot on the ground directly beneath the camera is referred to as the **nadir**.

Aerial photos can be **panchromatic** or color. Panchromatic imagery only captures the visible portion of light in its entirety. As a result, panchromatic aerial photos will be grayscale (that is, composed from a range of gray color values that exist between black and white—like the Cuban aerial photo). Color imagery captures the three main bands of visible light—red, green, and blue—and the colors are composited together in the digital imagery (or film) as a true color composite (we'll discuss much more about light and composites in Chapter 10). For more about examining color and panchromatic aerial photos, see *Hands-On Application 9.3: Viewing Aerial Photography Online.*

A color infrared or **CIR photo** is a distinctive type of aerial color photo. A CIR image is captured with a special type of film that's sensitive to infrared light. Infrared energy is invisible to the naked eye, but special film processes can capture it. To enable us to see infrared energy reflection, a special filter is applied to the film—the end result is a CIR photo, wherein near-infrared (NIR) reflection is displayed in the color red, red reflection is displayed in the color green, and green reflection is displayed in the color blue. Blue reflection is not shown in the image, as it is blocked by the filter and displayed as black (**Figure 9.9**).

vertical photo an aerial photo in which the camera is looking directly down at a landscape

nadir the location on the ground that lies directly below the camera in aerial photography

panchromatic black-and-white aerial imagery

CIR photo color infrared photo—a photo in which infrared reflection is shown in shades of red, red reflection is shown in shades of green, and green reflection is shown in shades of blue

Hands-On Application 9.3

Viewing Aerial Photography Online

To look at panchromatic and color aerial photographs of an area over time, go to the Historic Aerial Photography Viewer for the City of Greenville, South Carolina, at **http://gis.greenvillesc. gov/historicalimagery**. Aerial photography of Greenville is available for viewing going back to 1955, and the site also contains scanned maps of selected areas from 1920. To get started, use the search tools on the right-hand side of the screen to find street number 150 and street name Cleveland Park Drive. This is an address for the Greenville Zoo and the surrounding Cleveland Park area. Zoom and pan the view until you can see the entire zoo and park area, then use the time slider to move back and forth through time to see aerial photography of the area from 2015, 1997, 1989, 1979, 1965, and 1955.

Expansion Questions

➤ From examining the Cleveland Park area imagery from 1955 through 2015, what significant changes have occurred in the landscape there over the 60-year stretch? What new developments have been built and which features have changed from developed back to a natural environment?

➤ Another area in Greenville to examine using the viewer is Springwood Cemetery, located at 410 N Main St. From examining aerial photography from 1955 through 2015, what significant changes and developments to the area can be seen?

CIR imagery can be very useful in environmental studies. For instance, the relative health of areas of vegetation can be observed very clearly in CIR imagery. Examine the CIR aerial photo of Burley, Idaho, in **Figure 9.10**. The bright red sections of the image are indicative of areas with a high amount of NIR reflection (because NIR reflection is being shown with the color red

FIGURE 9.9

How the actual reflection of blue, green, red, and NIR light appears in its corresponding colors in a CIR image.

Observed colors	Actual wavelength
Black	Blue
Blue	Green
Green	Red
Red	Near infrared

FIGURE 9.10

A CIR aerial image of Burley, Idaho. *[NASA Airborne Science Program image acquired by NASA ER-2 aircraft 26 August 1983]*

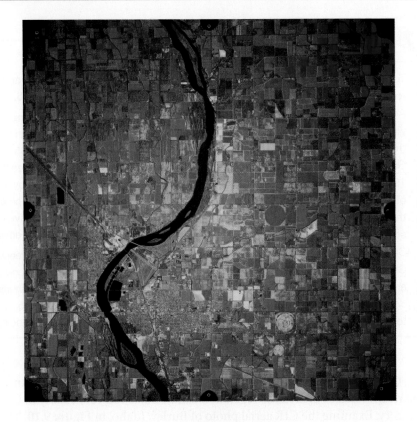

on the image)—probably grass, trees, or healthy agricultural fields (all elements that would reflect large amounts of NIR light).

The river appears black because water absorbs all types of light, except the color blue (which is then blocked in the CIR photo and shows up as black). Developed areas appear as a white or cyan, as do fields of bare earth that have either been harvested and plowed over, or have been prepared for crops but not yet sown with seed. For more about examining CIR aerial photos, see *Hands-On Application 9.4: Examining CIR Photos*.

No matter what type of aerial photo is being used, the big problem is that the photo cannot be used as a map. Even though it shows all the features you might need (like roads or footprints of buildings) you can't rely on the photo to be a map for one reason: The photo doesn't have the same scale at every spot on the image. As an aircraft flies over the surface, some terrain features will be closer to the camera and some will be further away—and thus, some areas will have a larger or smaller scale than others. The center of the photo is referred to as its **principal point**. In aerial photos, you'll see tall objects (such as terrain relief, towers, or buildings) have a tendency to "lean" away from this center point toward the edges of the photo—this effect is called **relief displacement**. See **Figure 9.11** for an example of this—if you

principal point the center point of an aerial photo

relief displacement the effect seen in aerial imagery where tall items appear to "bend" outward from the photo's center toward the edges

Hands-On Application 9.4

Examining CIR Photos

To look at the differences between color and color-infrared images, check out the Vermont Interactive Map Viewer (also used back in *Hands-On Application 1.4*) at **http://maps.vermont.gov/vcgi/html5viewer**. After the viewer opens, use the search box to type in the address of the University of Vermont's Office of Admissions: 194 South Prospect Street, Burlington, VT 05401-3596. A list of potential matching addresses will appear in the list. Choose the one that is the closest (or the exact) match. The view will shift over to the University of Vermont's campus. Then from the Layers options on the left-hand side of the screen, put a checkmark in the box that says All Imagery, then a checkmark in Color Imagery (and if the Buildings and Facilities layer is turned on, you'll want to turn that off to clear up the view). Zoom around the campus with the color imagery on to get a feel for the area, then from the Layers option, place a checkmark in the option for Color Infrared Imagery and remove the checkmark for Color Imagery. Take a look around the campus and also toggle back and forth between the two imagery options to see how things change.

Expansion Questions

➤ From examining the campus area and its surroundings, how are the following items represented in CIR imagery: buildings, rivers, grass, and fields? Why do areas in each of these categories appear in these colors in CIR imagery, as opposed to regular color imagery? You will also see some areas in green—what do these areas represent?

➤ Move south from your initial starting point at the Office of Admissions and you'll come to Gutterson Fieldhouse and the tennis courts and practice field just south of that, and then the large expanse of land for the Burlington Country Club just south of that. Why do the grass of the practice field and the tennis courts appear so differently in the CIR imagery? Also, what accounts for the differences in the landscape you can see in the country club in the CIR imagery?

look at the area labeled "Principal Point," it's like looking straight down at that area. However, if you look outward from the principal point, you'll see that the tall buildings seem to be "leaning" away from the center of the photo (rather than looking straight down on all the buildings in the image).

Regular aerial photos do not have uniform scale. **Orthophotos**, however, do. To create an orthophoto, a process called **orthorectification** is performed on a regular aerial photo. This process removes the effects of terrain and relief displacement and enforces the same scale across the whole photo. A regular aerial photo can't be used as a map because it doesn't have uniform scale everywhere, but an orthophoto can. Even with uniform scale, some of the side appearances of objects can still be seen because when the photo was taken, some of the sides of the building may have been hidden due to relief displacement and not visible in the photo. This effect is removed in a **true orthophoto**, which gives the appearance of looking straight down on all objects in the image. This is achieved by using several images of the area for filling in the missing information while keeping constant scale in the final photo.

orthophoto an aerial photo with uniform scale

orthorectification a process used on aerial photos to remove the effects of relief displacement and give the image uniform scale

true orthophoto an orthophoto in which all objects look as if they're being seen from directly above

FIGURE 9.11

The effect of relief displacement in an aerial photo. *[Michael H/Stone/Getty Images]*

DOQ a Digital Orthophoto Quad—orthophoto that covers an area of 3.75 minutes of latitude by 3.75 minutes of longitude, or one-fourth of a 7.5 minute USGS quad

Orthophotos are used to create special products called Digital Orthophoto Quads, or DOQs. A **DOQ** is orthophotography that covers 3.75 minutes of latitude and 3.75 minutes of longitude of the ground, the same area as one-fourth of a 7.5 minute USGS topographic map (see Chapter 13 for more about topographic maps). DOQs (sometimes also referred to as DOQQs, or Digital Orthophoto Quarter Quads, as they cover one-quarter of the size of a topographic map) have been georeferenced so that they can be matched up with other geospatial datasets (see Chapter 3 for more information on georeferencing). Also note that the USGS produces certain other DOQ products that cover the area of the entire quad.

NAIP the National Agriculture Imagery Program, which produces orthophotos of the United States

A similar resource is produced by the National Agriculture Imagery Program (**NAIP**), which is administered through the United States Department of Agriculture's Farm Service Agency. NAIP images are aerial photos of the United States that are taken on a regular basis, processed into DOQs (and also as mosaicked images of whole counties), and made publicly available. NAIP began in 2003 and continues to produce orthophotographic images today.

oblique photo an aerial photo taken at an angle

In an **oblique photo**, the camera is tilted so that it's not positioned directly at the nadir, but rather at an angle. Take another look at Figure 9.2—the image of San Francisco taken from the kites—for an example of an oblique photo. Unlike a vertical photo, you can see the sides of some of the buildings, and you're really seeing the city from a perspective

FIGURE 9.12

Oblique imagery of the Lincoln Memorial, Washington, D.C., as shown in Bing Maps' "Bird's eye" view feature.

view instead of an overhead one. Oblique images allow for very different views of a target than those taken from straight overhead and can capture other types of information (such as the sides of buildings or mountains). By examining different oblique photos of a single building, law enforcement agencies can obtain detailed information (the locations of entrances, exits, windows, and so on), which often provide vital information in deciding whether to send officers into a potentially dangerous situation. Companies such as EagleView make extensive collections of high-resolution oblique aerial imagery to sell commercially. You can view oblique images of several locations online using Microsoft's free Bing Maps utility, if its "Bird's Eye" imagery is available (see **Figure 9.12** and *Hands-On Application 9.5: Oblique Imagery on Bing Maps* for examples of oblique images created on Bing Maps).

How Can You Interpret Objects in an Aerial Image?

Understanding what you're looking at when you're viewing Earth from the sky instead of from ground level requires a completely different skill set. Objects that you're used to looking at head-on have a whole different appearance from directly above. For example, **Figure 9.13** shows two different views of the Stratosphere Hotel and Casino in Las Vegas, Nevada— one showing an overhead aerial view and the other an oblique photograph showing details of the tower and its buildings. Obviously, the complex looks completely different from the oblique angle than from directly above.

Hands-On Application 9.5

Oblique Imagery on Bing Maps

Bing Maps is a great source of online oblique imagery supplied via Pictometry. Like Google Earth and Google Maps, Bing Maps allows you to view aerial or satellite images of areas, but it also gives you the option to view high-resolution oblique images of areas (where oblique imagery is available). To start, go to **www.bing.com/maps** and type the name of a location (try Atlantic City, New Jersey). In the Atlantic City map, zoom in until you can see the ocean and boardwalk. From the pull-down menu in the upper left, select the option for "Bird's Eye." The view will switch to oblique photography of Atlantic City. You can zoom in and out and pan the imagery (when you move past the edge of a photo, a new tile of oblique imagery should load).

Reposition the images so that you can see the boardwalk and oceanfront at an oblique angle, and then move down the boardwalk to see the casinos, piers, and developments. In addition, Bing Maps provides oblique imagery from multiple angles—click on one of the arrows pointing downward to the left and right of the compass rose (around the red and white arrow symbol) to rotate the direction and load oblique imagery of the same location but shot from a different angle. You can use this function to examine a spot from multiple viewpoints. You can also click on the red and white arrow symbol to reset the orientation of the view. Once you get used to navigating and examining oblique images of the New Jersey shore, search for your local area with Bing Maps to see if Bird's Eye imagery is available to view your home, school, or workplace.

Expansion Questions

➤ What advantages are there for examining the Atlantic City boardwalk from oblique views rather than the usual overhead view? What specific features of the boardwalk can be seen in the oblique images that cannot be seen in a vertical perspective?

➤ If Bird's Eye imagery of your local area is available, examine it using Bing Maps. What specific features of your local area can be seen in the oblique images that cannot be seen from an overhead view?

If you look closely, you can see where areas in the two photos match up, such as the tower on the right-hand side of the image and the remainder of the hotel and casino on the left. However, if you're only given the aerial image and asked what building this is (of all possible structures in the world), it'll be a whole different ballgame.

From the aerial image alone, you'll have to search for clues to determine what's really being shown. For instance, based on the size of the complex (roughly a city block) you'll probably guess that it's some sort of hotel, resort, entertainment attraction, or museum. By looking at its location amidst other large buildings and multiple-lane roads, you may guess that the building's in a big city. On closer examination, you might make out a different colored object (the light blue) among the main complex and determine that it's a swimming pool, which narrows the chances further on the building being some sort of hotel or resort.

(a)

(b)

FIGURE 9.13

The Stratosphere Hotel
and Casino in Las Vegas,
Nevada: (a) an overhead
aerial view of the building
and (b) an oblique image
showing the tower
and the hotel layout.
*[DigitalGlobe/ScapeWare3d/Getty
Images]*

Although the height of the tower on the right-hand side of the image isn't really discernible (because the image was taken looking down onto the tower itself), the huge shadow cast by the tower, stretching upward, is visible. Based on the length and shape of the shadow being cast, you may determine that the object casting it is a tower (or a similar very large, very tall object). Putting all of these clues together (city block–sized hotel/resort in a big city with a massive tower in front of it), with the help of some outside information, like a few travel books and maybe a quick Web search, you should be able to determine pretty quickly that you're looking at the Stratosphere Hotel in Las Vegas.

When you try to interpret features in an aerial photo or a satellite image (for instance, objects in developed areas or physical features in a natural landscape), you act like a detective searching for clues in the image to figure out what you're really looking at. As in the Stratosphere Hotel example, clues such as the size and shape of objects or the shadows they cast help to determine what you're really looking at. **Visual image interpretation** is the process of identifying objects and features in an aerial (or other remotely sensed) image based on a number of distinct elements:

➤ **Pattern:** This is the physical arrangement of objects in an image. How objects are ordered (or disordered) will often help to interpret an image. A large array of cars set up as if for inspection in a parking lot around a relatively small building will probably be a car dealership rather than some other type of shopping area. Evenly spaced rows of jet airplanes suggest a military base, while a haphazard arrangement of aircraft could be a group of statics on display at an aircraft or military museum.

➤ **Site and association:** Site deals with the location characteristics of an item, while association relates an object in an image to other nearby

**visual image
interpretation** the process
of examining information to
identify objects in an aerial (or
other remotely sensed) image

pattern the arrangement
of objects in an image, used
as an element of image
interpretation

site and association the
information referring the
location of objects and their
related attributes in an image,
used as elements of image
interpretation

features in the image. For example, a football field itself has enough distinctive features to identify it, but the related phenomena you could see in the image (the number of bleachers, the number and distribution of the seats, and perhaps the amount of nearby parking) would help in determining if you're looking at a field used by a high school team, a 1-AA college team, or a professional NFL team.

size the physical dimensions (length, width, and area on the ground) of objects, used as an element of image interpretation

➤ **Size:** This describes the length, width, and area on the ground of objects in the image. The relative size of objects in an image can offer good clues in visual image interpretation. For instance, the average length of a car is about 15 feet. If a car is present in an image, you can gain information about other objects by comparing their length with that of the car—you can tell, for instance, if a structure near the car is the size of a house or a shopping center. The sizes of some features remain the same, whatever the image. If baseball diamonds or football fields are present in an image, elements in them (the 90 feet between bases, or the 100 yards between goal lines) can always be compared with other objects in the image to calculate their relative sizes.

shadow the dark shapes in an image caused by a light source shining on an object, used as an element of image interpretation

➤ **Shadow:** This is the dark shape cast by an object with a source of light shining on it. Shadows help to provide information about the height or depth of the objects that are casting them. For instance, in the Stratosphere Hotel example, the height of the tower itself was partially hidden due to the nature of being photographed from directly above, but we can figure out that it must be very tall because of the shadow it casts. Shadows can help us to identify objects that are virtually unidentifiable from a viewpoint directly above them. From directly above, a set of railings looks like a thin, almost invisible thread around a park or in front of a building, and telephone poles look like dotted lines. However, if you look down at them from directly above late on a sunny afternoon, railings and telephone poles can be identified immediately by their shadows.

shape the distinctive form of an object, used as an element of image interpretation

➤ **Shape:** This is the particular form of an object in an image. The distinctive shapes of objects in an aerial photo can help us to identify them. A baseball field has a traditional diamond shape, a racetrack has a distinctive oval shape (and even an abandoned horse racetrack may still show evidence of the oval shape on the landscape), and the circular shape of crops may indicate the presence of center-pivot irrigation in fields.

texture repeated shadings or colors in an image, used as an element of image interpretation

➤ **Texture:** This refers to the differences of a certain shading or color throughout parts of the image. The texture of objects can be identified as coarse or smooth. For instance, different types of greenery can be quickly distinguished by their texture—a forest of trees and a field of grass may have the same tone, but the trees appear very rough in an image, while grass will look very smooth. The texture of a calm lake will look very different from the rocky beach surrounding it.

➤ **Tone:** This is the particular grayscale (black to white) or intensity of color in an image. The tone of an object can convey important information about its identity. In the Stratosphere example, the light-blue color of the swimming pool made it easy to identify. Similarly, a wooden walkway extending into a sandy beach will have different tones that help in distinguishing them from one another in a photo.

tone the grayscale levels (from black to white), or range of intensity of a particular color discerned as a characteristic of particular features present in an image, used as an element of image interpretation

These same elements can be used to identify items in satellite images as well. **Figure 9.14** is an overhead view of an area containing several objects that all add up to represent one thing. We'll apply these elements to this image to determine what is being shown.

➤ Pattern: The three main objects are arranged in a definite diagonal line.

➤ Texture: The area around the three objects is of very fine texture, compared with some of the rougher textures nearby, especially in the east.

➤ Tone: The area around the three objects is much lighter than the area of rougher texture (and that area's probably a town or city of some sort, given the pattern of the objects in this range). Given the tone and texture of the ground around the three objects, it's likely not smooth water or grass or concrete, therefore sand is a likely choice.

➤ Site and association: The three objects are located in a large sandy area, adjacent to a city, with what resembles roads leading to the objects.

FIGURE 9.14

An overhead view of an area containing several different objects. *[Space Imaging Europe/Science Source]*

➤ Size: Comparing the size of the objects to the size of the buildings indicates that the objects are very large—even the smallest of the three is larger than many buildings, while the largest is greater in size than blocks of the city.

➤ Shadow: The shadows cast by the objects indicate that not only are they tall, but they have a distinctive pointed shape at their top.

➤ Shape: The objects are square at the base, but have pointed, triangular tops, making them pyramid shaped.

Bringing all of these elements of interpretation together, we can find three pyramids of different sizes (although still very large), in a definite fixed arrangement, in sand near a busy city. All of these clues combine to identify the Great Pyramids at Giza, just outside Cairo, Egypt. Aerial images often contain clues that help to identify the objects they contain, or at least narrow down the options to a handful. Some brief research among collateral material (like groupings of large desert pyramids in that particular arrangement) should nail down the identification.

Visual image interpretation skills enable the viewer to discern vital information about the objects in a remotely sensed image. The elements of interpretation can be used to identify, for instance, the specific type of fighter jet on a runway based on the size and shape of its features, wingspan, engines, and armaments. A forestry expert examining a stand of trees in a remotely sensed image might use visual image interpretation to identify the species of trees based on their features.

How Can You Make Measurements from an Aerial Photo?

photogrammetry the process of making measurements using aerial photos

Once positive identification of objects in an image has been made, these objects can be used for making various types of measurements. **Photogrammetry** is the process of obtaining measurements from aerial photos. Photogrammetric techniques can be used for determining things like the height and depth of objects in an aerial photo. For instance, the heights of buildings and the lengths of visible features in a photo can be calculated. There are a lot of possible photogrammetric measurements—the following are two simple examples of how photogrammetric measurements can be made.

photo scale the representation used to determine how many units of measurement in the real world are equivalent to one unit of measurement on an aerial photo

As every map has a scale (discussed in Chapter 7), so every photo has a **photo scale**, which is listed as a representative fraction. For example, in a 1:8000 scale aerial photo, one unit of measurement in the photo is equivalent to 8000 of those units in the real world. Using the photo scale, you can determine the real-world size of features. Say you measure a section of railroad track in an aerial photo as a half-inch, and the photo scale is 1:8000. Thus, 1 inch in the photo is 8000 inches in the real world, and therefore a measurement of 0.5 inches in the photo will be 4000 inches (0.5 times 8000), or 333.33 feet.

The scale of a photo relies on the focal length of the camera's lens and the altitude of the plane when the image is captured. The problems start when you try to make measurements from an orthophoto if you don't know the photo's scale. If you don't know how many real-world units are equal to one aerial photo unit, you can't make accurate measurements. The photo scale of a vertical photo taken over level terrain can be determined by using a secondary source that has a known scale, as long as an item is visible in both the photo and the secondary source and as long as you can measure that item in both. A topographic map (see Chapter 13) with a known scale will make a good secondary source because it will contain many of the features that will also be clearly seen in an aerial photo (such as a road, or a section of road, where the beginning and end are visible). When you make these types of measurements, it's important to remember that a regular aerial photo will not have the same scale everywhere in the photo, whereas an orthophoto will have uniform scale throughout the image.

By being able to make the same measurement on the map (where the scale is known) and on the aerial photo (where the scale is unknown), you can determine the photo scale. This is the way it works: Your photo could be 1:6000 or 1:12000 or 1:"some number." Just like the representative fraction (RF) discussed in Chapter 7, the photo scale can be written as a fraction—1:24000 can be written as 1/24000. So, assume your unknown photo scale is the RF—this is equal to the distance measured on the photo, known as the photo distance (PD), divided by the real-world distance measured on the ground, known as the ground distance (GD), or:

$$RF = \frac{PD}{GD}$$

Say you have a 1:12000 scale map showing an oceanfront boardwalk, and you also have an orthophoto of unknown scale of the same area, showing the same features. For the photo to be useful, you have to determine its scale. You can find the same section of boardwalk in both the map and the photo. By measuring the boardwalk section on the map, you find it is 0.59 inches. However, that's not the ground distance, or how long that section of the boardwalk is in the real world—because of the map scale, every 1 inch measured on the map translates into 12,000 inches in the real world. Therefore, 0.59 inches on the map is actually 7080 inches in the real world. This measure is the GD variable in the equation above (the ground distance):

$$RF = \frac{PD}{7080 \text{ in}}$$

You can measure the same section of boardwalk on the photo and find that it's 1.77 inches. This measurement is the PD variable in the equation (the photo distance):

$$RF = \frac{1.77 \text{ in}}{7080 \text{ in}}$$

Doing some quick division, you find that RF is equal to 1/4000

$$RF = \frac{1}{4000}$$

Therefore the photo scale is 1:4000. One unit measured on the photo is equal to 4000 units in the real world.

A second type of measurement that you can make from the elements found in an aerial photo is to calculate the height of an object in the photo simply by examining its shadow. At first blush, you might think you'll need to know all sorts of other information—where the photo was taken, what time of day it was taken, the date on which it was taken—all variables related to the relative location of the Sun and how the shadows would be cast. Chances are you won't be able to get your hands on a lot of this type of information easily, so you're probably thinking there must be a better way to do this.

You'd be right—photogrammetric measurements give you a simple way of determining the heights of objects in a photo from their shadows, without needing all that other data. They rely on three things: (1) knowing the scale of the photo (which we just figured out); (2) being able to see, fully and on level ground, the shadows of all objects (from the top of each object) whose height you want to measure; and (3) already knowing the height of one object with a shadow you can measure. If you have all three, measuring heights is a snap.

Let's take that hypothetical 1:4000 photo from the last example and assume that it's got a number of large hotels casting shadows on the boardwalk. You know the height of one building (115 feet), and its shadow in the photo (from the base to the top) is 0.10 inches. You can use this information to calculate the angle of the Sun, which is casting the shadows in the photo, as follows:

$$\tan a = \frac{h}{L}$$

In this equation, a is the angle of the Sun, h is the real-world height of the object, and L is the real-world length of the shadow. From basic trigonometry, the tangent of a right angle ($\tan a$) is equal to its opposite value (h) divided by its adjacent value (L). See **Figure 9.15** for a diagram that shows how this works. You already know the height, h (115 feet). The length, L, can be found by taking the length of the shadow measured in the photo (0.10 inches) and multiplying it by the photo scale (1:4000). The 0.10 inches on the photo is 400 inches in the real world, or 33.33 feet. This means that if you want to check your calculations by getting into a time machine and traveling back to the boardwalk on the day the aerial photo was taken, you'll find the shadow of the building will be 33.33 feet long. Plugging these numbers into the formula, we find that:

$$\tan a = \frac{115\,ft}{33.33\,ft}$$

and that the tangent of angle a (or "$\tan a$") is equal to 3.45.

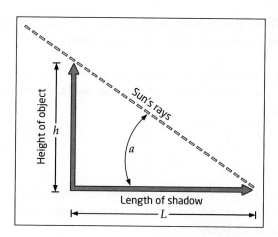

FIGURE 9.15

The relationship between the height of an object (*h*), the length of its shadow (*L*), and the angle of the Sun's rays.

Now, because an aerial photo represents a single snapshot in time, we can assume that the angle of the Sun is going to remain the same across the photo for all of the buildings casting shadows. Therefore, the measure for "tan *a*" will be the same value when applied to all heights we're trying to determine. We can use this information to calculate the height of any object in the photo casting a shadow that we can measure. Say you measure the shadow cast by a second building in the photo on the boardwalk and find it to be 0.14 inches. The 1:4000 photo scale tells us that the length of this shadow in the real world will be 560 inches, or 46.67 feet. The only thing that's unknown now is the actual height of the building. Using the equation

$$3.45 = \frac{h}{46.67 \text{ ft}}$$

and solving for *h*, we can find that the height of the new building is about 161 feet. This type of calculation can be applied to other objects casting shadows, meaning we can quickly determine the height of multiple objects in a single photo. Be cautious, however, when you employ these techniques—to determine height from shadows, the object must be straight up and down, and it must cast a full shadow on level ground.

Chapter Wrap-Up

Aerial photography has a wide variety of uses and is an integral part of geospatial technology. Aerial photos are used for interpretation, photogrammetry, and as additional data sources for other aspects of geospatial technology, such as GIS. Despite its myriad uses and applications, aerial photography is just one part of remote sensing. The next chapter will delve into how the whole remote sensing process actually works. First, however, see *UAS Apps* for some mobile apps related to this chapter's concepts, as well as *Aerial Imagery in Social Media* for several blogs, Facebook pages, Twitter accounts, and YouTube videos related to aerial photography.

This chapter's lab will put you to work with several remotely sensed images and help you to apply elements of visual image interpretation to items in the images.

Important note: The references for this chapter are part of the online companion for this book and can be found at **www.macmillanlearning. com/catalog/shellitoigt4e.**

UAS Apps

Here's a sampling of available representative UAS apps for your smartphone or tablet. Note that some apps are for Android, some are for Apple iOS, and some may be available for both.

➤ **Airmap for Drones:** An app used for examining flight conditions or restrictions in your area

➤ **B4UFly:** An app for examining and monitoring flight restrictions or special requirements for your location before flying a UAS

➤ **DJIGo 4:** An app used for flying and capturing imagery from a DJI drone

➤ **Pix4D Capture:** A mobile app used to supplement the Pix4D drone mapping software

Aerial Imagery in Social Media

Here's a sampling of some representative Facebook, Twitter, and Instagram accounts, along with some YouTube videos and blogs related to this chapter's concepts.

 On Facebook, check out the following:

➤ **3DR (UAS company):**
https://www.facebook.com/3drobotics

➤ **Airmap:**
https://www.facebook.com/airmapio

➤ **DJI (UAS company):**
https://www.facebook.com/DJI

➤ **EagleView (bird's eye imagery):**
https://www.facebook.com/EagleView-101026084310

- ➤ **Parrot (UAS company):**
  https://www.facebook.com/Parrot
- ➤ **Pix4D (UAS mapping software):**
  https://www.facebook.com/Pix4D/
- ➤ **senseFly (UAS company):**
  https://www.facebook.com/senseFlyDrones
- ➤ **Yuneec (UAS company):**
  https://www.facebook.com/YuneecAPV

Become a Twitter follower of:

- ➤ **3DR:** @3DRobotics
- ➤ **Airmap:** @AirMapIO
- ➤ **EagleView:** @EagleViewTech
- ➤ **DJI:** @djiglobal
- ➤ **Parrot:** @Parrot
- ➤ **Pix4D:** @pix4d
- ➤ **senseFly:** @sensefly
- ➤ **sUASnews (news about UAS):** @sUASnews
- ➤ **Yuneec:** @YuneecAviation

Become an Instagram follower of:

- ➤ **3DR:** @3DRobotics
- ➤ **DJI:** @djiglobal
- ➤ **EagleView:** @eagleview_tech
- ➤ **Parrot:** @parrot_official
- ➤ **Pix4D:** @pix4Dgram
- ➤ **Yuneec:** @yuneecapv

On YouTube, watch the following videos:

- ➤ **3DRobotics** (this is the YouTube channel for 3DR, showcasing several different uses of its UAS): **https://www.youtube.com/user/3DRobotics**
- ➤ **DJI** (this is the YouTube channel for DJI, featuring the many applications of the DJI brand of UAS drones): **https://www.youtube.com/user/djiinnovations**
- ➤ **EagleView** (this is the YouTube channel for EagleView, showcasing high-resolution aerial photography applications): **https://www.youtube.com/user/myEagleView**

➤ **Jarlath O'Neill-Dunne** (a YouTube channel from the Director of the University of Vermont's Spatial Analysis Laboratory, highlighting many of its UAS applications and projects): **https://www.youtube. com/user/joneildunne**

➤ **Parrot** (this is the YouTube channel featuring videos from the Parrot company and its UAS): **https://www.youtube.com/user/parrot**

➤ **Pix4D** (this is the YouTube channel for Pix4D, demonstrating how its software is used with UAS in a variety of ways): **https://www.youtube. com/user/pix4dsoftware**

➤ **senseFly** (this is the YouTube channel for senseFly, showing off what the senseFly UAS drones can do): **https://www.youtube.com/user/ senseFly**

➤ **Yuneec Electric Aviation** (this is the YouTube channel for Yuneec, featuring many different applications of its UAS): **https://www. youtube.com/user/YuneecAPV**

For further up-to-date info, read up on these blogs:

➤ **Altitude** (news and blog posts from DJI): **http://www.dji.com/ altitude**

➤ **EagleView Blog** (a blog about developments at EagleView): **http:// blog.eagleview.com**

➤ **Plane-ly Spoken** (a blog about aviation regulations, including UAS): **www.planelyspokenblog.com**

Key Terms

aerial photography (p. 330)
CIR photo (p. 338)
DOQ (p. 342)
nadir (p. 338)
NAIP (p. 342)
oblique photo (p. 342)
orthophoto (p. 341)
orthorectification (p. 341)
panchromatic (p. 338)
pattern (p. 345)
photo scale (p. 348)
photogrammetry (p. 348)
principal point (p. 340)

relief displacement (p. 340)
remote sensing (p. 329)
shadow (p. 346)
shape (p. 346)
site and association (p. 345)
size (p. 346)
sUAS (p. 334)
texture (p. 346)
tone (p. 347)
true orthophoto (p. 341)
UAS (p. 334)
vertical photo (p. 338)
visual image interpretation (p. 345)

Visual Imagery Interpretation

This chapter's lab will get you started thinking about what objects look like from the sky rather than the ground. You'll be examining a series of images and applying the elements of visual image interpretation we discussed in the chapter—think of it like being a detective and searching the image for clues to help you figure out just what it is you're looking at. You should be able to figure out exactly what you're looking at, or at least narrow down the possibilities to a short list before you turn to support data for help.

Although the items you'll be examining will all be large and prominent—buildings or other very visible features—the application of visual image interpretation elements for these simple (and hopefully, fun) examples will help get you started with looking at the world from above.

Objectives

The goals for you to take away from this exercise are:

➤ To think of how objects look from an aerial perspective

➤ To apply the elements of visual image interpretation (as described in the chapter) in order to identify objects in the images

Using Geospatial Technologies

The concepts you'll be working with in this lab are used in a variety of real-world applications, including:

➤ Military intelligence, which relies on knowing the locations of items on the ground—interpretation of objects in aerial reconnaissance photographs is often a key component in planning military operations

➤ Forest resource management, which utilizes aerial photo interpretation to monitor the health and development of different types of trees by careful observation of foliage and the height and density of the tree canopy

Obtaining Software

There is no special software used in this lab, aside from whatever program your computer uses to view graphics and images. You may find Google Earth useful during the lab, however.

[Agencja Fotograficzna Caro/Alamy]

Lab Data

Copy the folder **Chapter 9**—it contains a series of PNG images (.png files) in which you'll find items that you'll be attempting to interpret. These are numbered, and tend to increase in difficulty as the numbers go up.

Localizing This Lab

The images used in this lab were taken from a variety of locations around the United States. You can use Google Earth, Google Maps, or Bing Maps to locate good overhead views of nearby areas (showing prominent developed or physical features) to build a dataset of local imagery that you can use for visual image interpretation.

9.1 Applying Elements of Visual Image Interpretation

Take a look at each of the images (there are questions related to each further down) and try to determine just what it is you're looking at. Although everyone's application of the elements may vary, there are some guidelines that you may find useful.

Several images may contain multiple items, but all of them work together to help define one key solution.

1. We'll start with a sample image (a digital copy of this sample image is also in the **Chapter 9** folder labeled "sample"):

2. First, look at the image as a whole for items to identify:

 a. The central object is obviously some sort of structure—and a large one at that, judging from the shadow being cast. The relative size of the structure (when compared with the size of several of the cars visible throughout the image) suggests that it's probably a building and not a monument of some kind.

 b. It's set on a body of water (the tone of the water is different from the tone of the nearby concrete or greenery). Thus, the site can be fixed.

 c. One of the most notable features is the large triangular shape of the main portion of the building—note too that its texture is very different from the rest of the building (the three white sections).

 d. The shape of the entranceway is also very distinctive, being large and round, with a couple of concentric circles.

3. Second, take the interpretation of the initial items and start looking for specifics, or relationships between the items in the image:

 a. There's something about the pattern of those concentric circles at the entrance plaza that's striking—two outer rings and a third inner ring with some sort of a design placed in it. Judging from the relative size of a person (pick out shadows near the plaza or on the walkway along the water), the plaza is fairly large. The pattern sort of resembles a giant vinyl record album (and the concept solidifies further with the shape of the large curved feature that follows around the plaza on its right—that looks like the arm of an old record player).

 b. The texture of the triangular portions of the building suggests that they could be transparent, as if that whole portion of the structure were made of glass.

c. There's no parking lot associated with the building, so you'd have to park somewhere else and walk there. Further, there are no major roads nearby, again indicating that this is not some sort of destination that you just drive up to and park in front. The large body of water gives us a clue that this building is in a large city.

4. Third, put the clues together and come up with an idea of what the object or scene could be:

    a. The building's distinctive enough to be a monument, but probably too large to be one. It could be some sort of museum, casino, shopping center, or other attraction. However, the building's too small to be a casino or resort center, and lacks the necessary parking to be a casino, office, or shopping center.

    b. The very distinct "record" motif of the front plaza seems to indicate that music (and more specifically older music, given the whole "vinyl record" styling) plays a large part in whatever the building is.

    c. From the clues in the image, this could be some sort of music museum, like the Rock & Roll Hall of Fame, which is located in Cleveland, Ohio, right on the shore of Lake Erie.

5. Fourth and finally, use a source to verify your deduction, or to eliminate some potential choices.

    a. By using Google Earth to search for the Rock & Roll Hall of Fame, the identification will be confirmed. An image search on Google will also turn up some non-aerial pictures of the site to verify your conclusion.

Not all images will be this extensive, and not all may make use of all elements—for instance, there may be one or two items in the image that will help you make a quick identification. However, there are enough clues in each image to figure out what you're looking at—or at least narrow the choices down far enough that some research using other sources may be able to lead to a positive identification.

For instance, by putting all of the clues together, you could start searching for large museums dedicated to music—add the triangular glass structures, and the Rock & Roll Hall of Fame will certainly come up.

9.2 Visual Image Interpretation

1. To examine an image, open the **Chapter 9** folder and double-click on the appropriate image (.jpg) file. Whatever program you have on your computer for viewing images should open with the image displayed. You may find it helpful to zoom in on parts of an image as well.

2. For each image, answer the questions presented below, then explain which of the elements of visual image interpretation led you to this

conclusion and how they helped you to decide. Writing "shadow and shape helped identify traits of the building" will be unacceptable. However, writing something that answers the question, "What was so special about the shadows in the scene or the shape of items that helped?" will be much better.

3. When you identify the items, be very specific (for instance, not just "a baseball stadium," but "Progressive Field" in Cleveland, Ohio).

   Important note: All images used in this exercise are from areas within the United States.

4. You may want to consult some outside sources for extra information, such as Websites, search engines, books, or maps. Lastly, when you think you finally have the image properly identified, you may want to use something like Google Earth or Bing Maps to obtain a view of the object in question to help verify your answer.

> **Question 9.1** Examine image 1. What (specifically) is being displayed in this image? What elements of visual image interpretation lead you to draw this conclusion?

> **Question 9.2** Examine image 2. There are many similar objects in this image, but you'll notice some differences too. What (specifically) is being displayed in this image, and what is its location? What elements of visual image interpretation lead you to draw this conclusion?

> **Question 9.3** Examine image 3. What (specifically) is being displayed in this image? What elements of visual image interpretation lead you to draw this conclusion?

> **Question 9.4** Examine image 4. What (specifically) is being displayed in this image? What elements of visual image interpretation lead you to draw this conclusion?

> **Question 9.5** Examine image 5. What (specifically) is being displayed in this image? What elements of visual image interpretation lead you to draw this conclusion?

> **Question 9.6** Examine image 6. What (specifically) is being displayed in this image? What elements of visual image interpretation lead you to draw this conclusion?

Question 9.7 Examine image 7. There are several different items in this image, but they all add up to one specific thing. What, specifically, is this image showing, and what is its location? What elements of visual image interpretation lead you to draw this conclusion?

Question 9.8 Examine image 8. There are several items visible in this image, but there's one that's more prominent than the others. What, specifically, is this item? What elements of visual image interpretation lead you to draw this conclusion?

Question 9.9 Examine image 9. What (specifically) is being displayed in this image? What elements of visual image interpretation lead you to draw this conclusion? For instance, it's obviously an island, but what types of features does this island have (or not have) that may aid in identifying what is being shown here?

Question 9.10 Examine image 10. What (specifically) is being displayed in this image? What elements of visual image interpretation lead you to draw this conclusion?

Closing Time

Now that you've gotten your feet wet with some visual image interpretation and have hopefully gotten used to viewing Earth from above, Chapter 10 will delve into satellite imagery, and its exercise will have you doing (among other things) some more interpretation tasks based on satellite capabilities.

Once you've identified all of the images and explained your application of the elements of image interpretation, you can close the **Chapter 9** folder and any other programs or resources you have open.

How Remote Sensing Works

Electromagnetic Energy, the Remote Sensing Process, Spectral Reflectance, NDVI, Digital Imagery, and Color Composites

When you use Google Earth to get an overhead view of your house, Google had to get those images from somewhere, and a number of steps had to be taken before that image of your house can be acquired and finally turned into the picture you can view on the computer screen. In Chapter 9, we discussed how what's really being captured via **remote sensing** is reflected light—this is a form of electromagnetic energy, and a sensor is measuring the amount of energy that reflects off a target on the ground. In addition, in remote sensing the sensor will be on an airborne or spaceborne platform that will be a considerable distance away from a target on the ground below. Some remote sensing devices measure other forms of energy emitted by objects on the ground (heat, or the thermal properties of items, for instance). This chapter, however, focuses on measuring the reflection of electromagnetic energy.

If remote sensing measures the reflectance of electromagnetic energy, there needs to be a source for that energy. Luckily, we have one nearby—the Sun. The Sun is almost 93 million miles away from Earth, and sunlight takes about 8.3 minutes to reach Earth. This solar energy radiates through space at the speed of light (almost 300 million meters per second), and passes through and interacts with Earth's atmosphere before it reaches Earth's surface. The energy then interacts with objects on Earth's surface, and some of the energy is reflected back into space. That reflected energy passes back through the atmosphere, and is picked up and recorded by a sensor, either on an airborne or spaceborne platform (**Figure 10.1**).

remote sensing the process of collecting information related to the electromagnetic energy reflected or emitted by a target on the ground, using a device a considerable distance away from the target on board an aircraft or spacecraft

FIGURE 10.1

The remote sensing process: Energy radiates from the Sun, interacts with the atmosphere, interacts with items on the ground, then some energy reflects upward to be measured by the sensor.

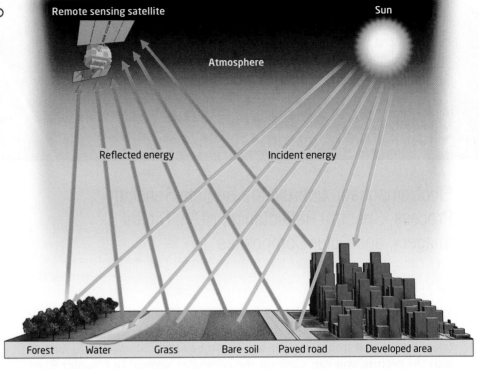

These measurements made at the sensor are being captured as remotely sensed data. Thus, the remotely sensed imagery you see on Google Earth and similar sites is actually the reflection of energy off targets on the ground that has been measured, processed, and turned into imagery to view. We'll go through each of these stages in more detail to see how you can start with energy from the Sun and end up with a crisp satellite image of a target on the ground (also see *Hands-On Application 10.1: Viewing Remotely Sensed Imagery Online* for some finished products of remote sensing).

Note that this process (and this chapter) describes passive remote sensing—where the sensor doesn't do anything but simply measure and record reflected or emitted energy. In active remote sensing, the sensor generates its own energy, casts it at a target, and then measures the rebound, echo, or reflection of that energy. Radar is a good example of active remote sensing—a device on a plane throws radar waves (microwaves) at a target, those waves interact with the target, and the device measures the backscatter of the returning radar waves. Lidar (see Chapter 13) is a similar kind of active remote sensing technique that shoots a laser beam at a target and uses the beam's reflection to measure the heights of objects on the ground.

Hands-On Application 10.1

Viewing Remotely Sensed Imagery Online

An excellent tool for viewing different types of remotely sensed data in your Web browser (without using a separate program such as Google Earth) is Zoom Earth. To get started, go to **Zoom.Earth**. When Zoom Earth starts, you can pan around Earth and zoom in on an area (or search by location). To get started, search for New York City. Zoom Earth will change the view to show New York and its surrounding environs. By zooming out, you'll see a different type of broader-scale imagery loaded, and as you zoom in, finer resolution imagery will load that will show you more detail. By clicking on Map Labels, overlays of roads, names of streets, and other information will be added to the image for

reference. You can also select a particular month, day, and year from which to view imagery.

Expansion Questions

➤ While you're looking around Manhattan, search for Governors Island, turn on the Map Labels and zoom in closely—what kinds of detail are you able to discern in the image? What types of details are still unclear from the imagery?

➤ When you're done looking around Manhattan, type your home location into the search box (or click on Locate Me to change to your estimated current location). What kinds of details can be seen as you zoom in closely?

What Is Remote Sensing Actually Sensing?

The information that is being sensed is the reflection of electromagnetic (EM) energy off a target. There are two ways of thinking of light energy: as a particle or as a wave. For our purposes, we'll stick to the concept of light energy as a wave. When the Sun radiates light energy, it radiates in the form of waves. Think of it as sitting on the beach and watching the ocean waves come in and crash on the shore. The distance between the crests of two waves is called the **wavelength** (λ)—some wavelengths are very long, and some are very short. If it's a calm day at the ocean, you see waves less frequently than on a rough day—there will be a longer distance between waves (and therefore a longer wavelength). If it's a rough day on the ocean, you'll see waves a lot more frequently, and the distance between waves will be shorter (that is, a shorter wavelength).

wavelength the distance between the crests of two waves

As we can observe, there's a definite connection between wavelengths and the frequency of the waves themselves. With a longer wavelength, you'll see waves far less frequently, and with a shorter wavelength, you'll see the waves a lot more frequently (**Figure 10.2**). If you sit on the beach and count the waves, you'll count a higher number of waves when the wavelength between them is small, and you'll count fewer waves when the wavelength between them is longer.

Waves of energy work on a similar principle. Energy travels at a constant speed—that is, the speed of light (c), which is about 300 million meters

The relationship between wavelength and frequency: Long wavelengths equate to low frequencies, whereas short wavelengths equate to high frequencies.

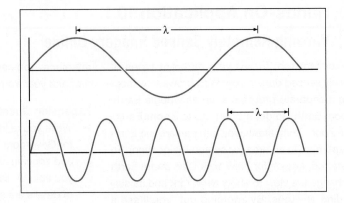

per second. If you multiply the frequency by the wavelength, the result is always equal to the speed of light. When the value for the wavelength of light energy is high, the frequency is low, and if the value for the frequency is high, the value for the wavelength has to be lower to make the product of the two values always come out to the same number.

Electromagnetic energy at different wavelengths also has different properties. Light energy with very long wavelengths will have different characteristics from light that has very short wavelengths. The **electromagnetic spectrum** is used to examine the properties of light energy in relation to wavelength. The values of the wavelengths are usually measured in extremely small units of measurement called **micrometers** (μm): A micrometer is one-millionth of a meter (roughly the thickness of a single bacterium). Sometimes an even smaller unit of measurement is used, called a **nanometer** (nm), which is one-billionth of a meter. Forms of electromagnetic energy with very short wavelengths are cosmic rays, gamma rays, and x-rays, while forms of electromagnetic energy with very long wavelengths are microwaves and radio waves (**Figure 10.3**).

It's important to remember that the Sun radiates many types of energy, but they're invisible to the human eye. When you put a slice of cold pizza in

electromagnetic spectrum the light energy wavelengths and the properties associated with them

micrometer a unit of measurement equal to one-millionth of a meter, abbreviated μm

nanometer a unit of measurement equal to one-billionth of a meter, abbreviated nm

The electromagnetic spectrum, showing the wavelengths of energy and the properties with which they correspond. The visible light portion of the spectrum occurs at wavelengths between 0.4 and 0.7 micrometers.

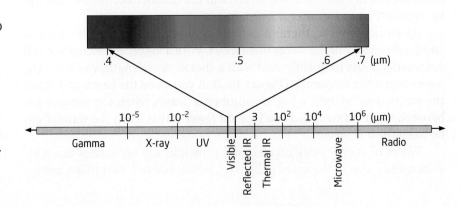

the microwave oven to heat it, you don't actually see the microwaves bombarding the pizza. In the same way, you can't see the x-rays that a doctor uses in an exam or the radio waves that are picked up by your car's antenna. All this energy is out there, but you just can't see it. This is because the composition of your eyes (which are structured to respond to reflected energy) is only sensitive to the reflected light that has a wavelength between 0.4 and 0.7 micrometers. This portion of the electromagnetic spectrum is called the **visible light spectrum** or visible spectrum (see Figure 10.3 to see what portion of the whole electromagnetic spectrum is occupied by the visible light spectrum). If your eyes were sensitive to the reflection of slightly shorter wavelengths of energy, you'd be able to see ultraviolet (**UV**) light (as bees can), and if you could see longer wavelengths of energy, you'd be able to see infrared light. However, the human eye is only sensitive to electromagnetic energy in this very small range of wavelengths.

The visible spectrum can be broken down further to the colors that it comprises. Shorter visible light wavelengths (between 0.4 and 0.5 micrometers) represent the blue portion of the spectrum. Medium visible light wavelengths (between 0.5 and 0.6 micrometers) represent the green portion of the spectrum. Longer visible light wavelengths (between 0.6 and 0.7 micrometers) represent the red portion of the spectrum. When you see a bright red shirt, your eyes are actually sensing the reflection of light with a wavelength measurement somewhere between 0.6 and 0.7 micrometers.

Each of these narrow portions of the electromagnetic spectrum that represent a particular form of light is referred to as a **band** of energy in remote sensing. For example, the shorter wavelengths of 0.4 to 0.5 micrometers are referred to as the **blue band** of the electromagnetic spectrum, because the properties of those wavelengths of light correspond to the characteristics of blue energy. Wavelengths between 0.5 and 0.6 micrometers make up the **green band** of the electromagnetic spectrum and wavelengths between 0.6 to 0.7 micrometers make up the **red band**.

Keep in mind that while the human eye is restricted to seeing only these narrow bands of the electromagnetic spectrum, remote sensing devices have the capability to sense energy from other sections of the spectrum. For instance, the infrared (**IR**) light portion of the electromagnetic spectrum covers wavelengths between 0.7 and 100 micrometers. A sensor could be tuned to measure only the reflection of infrared energy between 0.7 and 0.9 micrometers, and even though this energy is not visible to the human eye, the sensor would have no difficulty measuring this narrow band of reflected energy. As we'll see, an enormous amount of information can be gained from examining these other forms of energy as they are reflected off objects. In remote sensing of infrared light, there are a few wavelength ranges that are utilized much more than the others:

➤ 0.7 to 1.3 micrometers: Bands in this portion, referred to as "near infrared" (**NIR**), are frequently used in satellite remote sensing (see Chapter 11) and in color infrared photography (see Chapter 9).

visible light spectrum the portion of the electromagnetic spectrum with wavelengths between 0.4 and 0.7 micrometers

UV ultraviolet; the portion of the electromagnetic spectrum with wavelengths between 0.01 and 0.4 micrometers

band a narrow range of wavelengths that may be measured by a remote sensing device

blue band the range of wavelengths between 0.4 and 0.5 micrometers

green band the range of wavelengths between 0.5 and 0.6 micrometers

red band the range of wavelengths between 0.6 and 0.7 micrometers

IR infrared; the portion of the electromagnetic spectrum with wavelengths between 0.7 and 100 micrometers

NIR near infrared; the portion of the electromagnetic spectrum with wavelengths between 0.7 and 1.3 micrometers

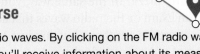

Hands-On Application 10.2

Wavelengths and the Scale of the Universe

The Scale of the Universe is an online tool for comparing items of different sizes—everything from the size of the known universe all the way down to sub-atomic particles. It's a fun visualization tool for measuring sizes, including making comparisons of the wavelengths of energy used in remote sensing with other items. To get started, go to **http://htwins.net/scale2**. Click on the Start button to begin. You'll start with the size of a human relative to other objects. As you move the scroll bar at the bottom of the screen to the right you'll see the scale increase and show larger objects, while moving the scroll bar to the left will decrease the scale and allow you to view smaller items. By moving the bar slightly to the right you'll see the wavelengths for FM radio waves. By clicking on the FM radio waves icon, you'll receive information about its measurement (these radio wavelengths are about 1 meter). Move the scroll bar to the left to start examining items at a smaller scale, including the wavelengths used in remote sensing.

Expansion Questions

➤ What is the wavelength of infrared energy measured as and what is it compared to in size?

➤ What is the wavelength of red light measured as and what is it compared to in size?

➤ What is the wavelength of ultraviolet measured as and what is it compared to in size?

SWIR shortwave infrared; the portion of the electromagnetic spectrum with wavelengths between 1.3 and 3.0 micrometers

TIR thermal infrared; the portion of the electromagnetic spectrum with wavelengths between 3.0 and 14.0 micrometers

➤ 1.3 to 3.0 micrometers: Bands of this portion, referred to as "shortwave infrared" (**SWIR**) or "middle infrared" (MIR), are also utilized by various other satellite sensors. For instance, measurements of bands of SWIR energy are used regularly in studies related to the water content of plants.

➤ 3.0 to 14.0 micrometers: Bands of this portion are referred to as "thermal infrared" (**TIR**), and are widely used for measuring heat sources and radiant heat energy.

For a quick example of examining the size of these wavelengths that we deal with in remote sensing, check out *Hands-On Application 10.2: Wavelengths and the Scale of the Universe* for a very neat online tool that will help demonstrate just how small the wavelengths we'll be working with really are.

What Is the Role of the Atmosphere in Remote Sensing?

Earth's atmosphere acts like a shield around the planet, and the energy from the Sun has to pass through it before it reaches Earth's surface. As a result, a significant amount of the Sun's electromagnetic energy never actually makes it to the ground—and if it never makes it to the ground, it's never going to be reflected back to the remote sensing device (and will never be part of the whole remote sensing process). Earth's atmosphere contains a variety of gases (including carbon dioxide and ozone) that serve to absorb electromagnetic energy of numerous wavelengths. For instance, ozone absorbs wavelengths of energy that correspond to ultraviolet light; these are "UV rays" that can be harmful to humans. Thus, a considerable amount of ultraviolet gets trapped in the atmosphere and doesn't pass through to Earth. Very short wavelengths and some very long wavelengths are also absorbed by the atmosphere.

Those wavelengths of energy that pass through the atmosphere (rather than being absorbed) are referred to as **atmospheric windows**. Remote sensing of the wavelengths that make up these "windows" is possible since they are the ones that transmit most of their energy through the atmosphere to reach Earth. The visible light wavelengths of the electromagnetic spectrum (wavelengths between 0.4 and 0.7 micrometers) are examples of atmospheric windows. Most of the energy at these wavelengths is not absorbed by the atmosphere and instead is transmitted to Earth, where it reflects off objects and surfaces to be viewed by our eyes as color (and viewed also by cameras and satellite sensors). Other windows used in remote sensing include portions of the infrared and thermal infrared sections of the spectrum (see **Figure 10.4** for a diagram of the atmospheric windows). Sensors are set up to measure the energy at the wavelengths of the windows. It won't do any good whatsoever to have a sensor measuring wavelengths (such as the entire ultraviolet portion of the spectrum) that are completely absorbed by the atmosphere, as there will be no reflectant energy to measure.

While some light energy is absorbed by the atmosphere, other light energy is scattered by it. Scattering is caused by particles, such as dust and water, in the atmosphere—it is always present, but its effects are always unpredictable. Light hitting these particles will be absorbed and redirected back out in random directions, causing scattering. There are three types of scattering that can occur: The first of these is **Rayleigh scattering**, which occurs when the particles causing the scattering are significantly smaller than the wavelengths being affected. Keeping in mind that the visible-light

atmospheric windows
those wavelengths of electromagnetic energy in which most of the energy passes through Earth's atmosphere

Rayleigh scattering
scattering of light caused by atmospheric particles smaller than the wavelength being scattered

FIGURE 10.4

The absorption of various wavelengths by different gases in Earth's atmosphere and the atmospheric windows used in remote sensing.

wavelengths are smaller than one-millionth of a meter, Rayleigh scattering is caused by molecules in the atmosphere. In addition, shorter wavelengths are scattered to a much greater degree than longer wavelengths. This explains why the sky is blue during the day: The shorter blue wavelengths are scattered far more than the longer red wavelengths. However, at sunset and sunrise, sunlight has a longer way to travel to reach Earth, and thus with all of the blue wavelengths already scattered, we see the reds and oranges of longer wavelengths as the Sun goes out of view.

The second is **Mie scattering**, which occurs when the particles causing the scattering are roughly the same diameter as the wavelengths they're scattering (this would include things like dust or smoke). The third is **non-selective scattering**, where what's causing the scattering—say, water droplets or clouds—is larger than the wavelengths being scattered and scatters all wavelengths equally. Non-selective scattering also explains why we see clouds as white—the visible primary colors of blue, green, and red are all scattered the same, and equal levels of red, green, and blue will produce the color white.

Mie scattering scattering of light caused by atmospheric particles the same size as the wavelength being scattered

non-selective scattering scattering of light caused by atmospheric particles larger than the wavelength being scattered

What Happens to Energy When It Hits a Target on the Ground?

After so much energy has been absorbed and scattered by the atmosphere, many of the wavelengths that correspond to the windows finally make it to Earth's surface and interact with targets there. One of three things can happen to that energy (on a wavelength-by-wavelength basis)—either it can be transmitted through the target, absorbed by the target, or reflected off the target. **Transmission** occurs when a wavelength of energy simply passes through a surface to interact with something else later. Think of light passing through the windshield of a car—most of the light will pass through the glass to affect things in the car rather than the windshield itself.

transmission when light passes through a target

Absorption occurs when the energy is trapped and held by a surface rather than passing through or reflecting off it. Think of walking across a blacktop parking lot during a hot summer day—if you don't have something on your feet, they're going to hurt from the heat in the pavement. Because the blacktop has absorbed energy (and converted it into heat), it's storing that heat during the day. Absorption also explains why we see different colors. For instance, if someone is wearing a bright green shirt, there's something in the dyes of the shirt's material that is absorbing all other colors (that is, the portions of the visible light spectrum) aside from that shade of green, which is then being reflected to your eyes.

absorption when light is trapped and held by a target

Different surfaces across Earth have different properties that cause them to absorb energy wavelengths in various ways. For instance, we see a clear lake as a blue color—it could be dark blue, lighter blue, maybe even some greenish blue, but it's nonetheless a shade of blue. We see it this way because water strongly absorbs all electromagnetic wavelengths except for the range of 0.4 to 0.5 micrometers (and some lesser absorption past the edges of that range), which corresponds to the blue portion of the spectrum. Thus, other wavelengths of energy (such as the red and infrared portions) get absorbed and the blue portion gets reflected.

What remote sensing devices are really measuring is the reflectance of energy from a surface—the energy that rebounds from a target to be collected by a sensor. Just as different objects and surfaces absorb energy in different ways, so different objects and surfaces reflect energy differently. The total amount of energy that strikes a surface (the **incident energy**) can be calculated by adding up the amounts of energy that were transmitted, absorbed, and reflected per wavelength by a particular surface as follows:

incident energy the total amount of energy (per wavelength) that interacts with an object

$$I = R + A + T$$

where I is the incident energy (the total amount of energy of a particular wavelength) that strikes a surface and is made up of R (reflection) plus A (absorption) plus T (transmission) of that particular wavelength. Because remote sensing is focused on the reflection of energy per wavelength, let's change that equation to focus on calculating what fraction of the total incident energy was reflected (and to express this value as a percentage instead of a fraction, we'll multiply by 100):

$$\rho = (R/I) \times 100$$

where ρ is the portion of the total amount of energy composed by reflection (rather than absorption or transmission) per wavelength. This final value is the **spectral reflectance**—the percentage of total energy per wavelength that is reflected off a target, makes its way toward the sensor, and is utilized in remote sensing.

spectral reflectance the percentage of the total incident energy that was reflected from that surface

Thinking Critically with Geospatial Technology 10.1

How Does Remote Sensing Affect Your Privacy?

The remote sensing processes described in this chapter are systems that measure energy reflection and turn it into images without interfering with what's happening below them. For instance, when an aircraft flies overhead to take pictures or a satellite crosses more than 500 miles above your house, it collects its imagery and moves on. This data collection process doesn't physically affect you or your property in the slightest, and all of this is done without your explicit permission and usually without your knowledge. If you can see your car in your driveway on Google Earth, then it just happened to be parked there when the airplane or satellite collected that image. Does this unobtrusive method of data collection affect your privacy? Is the acquisition of images via remote sensing an invasion of your personal space? In what ways could this use of geospatial technology intrude on someone's life? Also, a lot of this kind of imagery is being acquired by private companies—is this more or less intrusive than a government agency doing the same thing?

How Can Spectral Reflectance Be Used in Remote Sensing?

Everything on Earth's surface reflects energy wavelengths differently from everything else. Green grass, bare soil, a parking lot, a sandy beach, and a large lake—all reflect portions of the visible, near-infrared, and middle-infrared energy differently (these items emit thermal energy differently as well). In addition, some things reflect energy differently depending on different conditions. As an example of examining these remote sensing concepts, let's look at something simple—measuring reflectance of energy from the leaves on a tree. During late spring and summer, a tree canopy will be dense and full, as the leaves increase in number, stay on the trees, and are a healthy green color. The human eye sees healthy leaves as green because the chlorophyll in the leaves is reflecting the green portion of the spectrum and absorbing the red and blue portions. What can't be seen by the human eye (but can be seen by remote sensing instruments) is that, due to their structure, healthy leaves also very strongly reflect near-infrared energy. Thus, measurement of near-infrared energy is often used (in part) as an indicator of the relative health of leaves and vegetation.

However, leaves aren't always going to stay green. As autumn progresses, leaves go through a senescence process in which they lose their chlorophyll. As this happens, the leaves reflect less green energy and begin to absorb less (and thus, reflect more) red and blue energy, causing the leaves to appear in other colors, such as yellow, orange, and red. As the leaves turn brown and fall off the trees, they have a lot more red reflection, causing their brownish appearance (and there will also be less reflectance of near-infrared energy from the tree).

If a sensor were able to measure all of these energy wavelengths simultaneously for different objects, it would give the person examining the data the ability to tell objects apart by examining their reflectance values in all of these wavelengths. If you were to chart the spectral reflectance values against the wavelengths being measured for each item, you would find that each of these things would have a different line on the chart. This is referred to as an item's **spectral signature** (also called a spectral reflectance curve), as each set of charted measurements will be unique to that item, just like your handwritten signature on a piece of paper is different from everyone else's. Remote sensing analysts can use these spectral signatures to distinguish items in an image and to tell the difference between different types of plants or minerals (see **Figure 10.5** on page 371 for comparisons of various spectral signatures).

One useful application of remotely sensed imagery is its use in assessing the health of green vegetation (such as fields, grass, and foliage). When vegetation is very healthy, it will have a strong reflection of near-infrared energy and absorption of red energy. As the vegetation becomes more stressed, less near-infrared energy will be reflected and more red energy will be reflected

spectral signature a unique identifier for a particular item, generated by charting the percentage of reflected energy per wavelength against a value for that wavelength

FIGURE 10.5

Four spectral signatures, generated for different items by charting the percentage of reflection on the *y*-axis and wavelengths on the *x*-axis.

rather than absorbed. An image can be processed to measure the health of various kinds of vegetation. The Normalized Difference Vegetation Index (**NDVI**) is a means of doing so—the process takes a remotely sensed image and creates a new NDVI image from it, containing values related to the relative health of the vegetation in the image. As long as the sensor can measure the red and near infrared portions of the electromagnetic spectrum, NDVI can be calculated.

NDVI is computed using the measurements for the red and near-infrared bands. The formula for NDVI is:

$$\frac{(NIR - Red)}{(NIR + Red)}$$

NDVI returns a value between –1 and +1—the higher the value (closer to 1), the healthier the vegetation is at the area being measured. Low values indicate unhealthy vegetation or a lack of biomass. Very low or negative values indicate areas where nothing is growing (like pavement, clouds, or ice). **Figure 10.6** shows an example of NDVI being calculated for both a healthy vegetation source (one that reflects a high amount of near-infrared energy and less red energy), resulting in a high NDVI value (0.72) and an unhealthy source (which reflects less near-infrared energy and more red energy), which results in a lower NDVI value (0.14).

NDVI can be calculated by a variety of remote sensing devices, as long as their sensors can measure the red and near-infrared bands (which several of the specific sensors discussed in the next two chapters can do). NDVI gives a quick means of assessing the health of various forms of vegetation and is used in a variety of environmental applications, both local and global—from crop monitoring to surveying the planet's natural resources (see **Figure 10.7** for an example of NDVI).

NDVI Normalized Difference Vegetation Index; a method of measuring the health of vegetation using near-infrared and red energy measurements

FIGURE 10.6

An example of NDVI calculation for healthy (left) and unhealthy (right) vegetation. *[Adapted from NASA/Robert Simmon]*

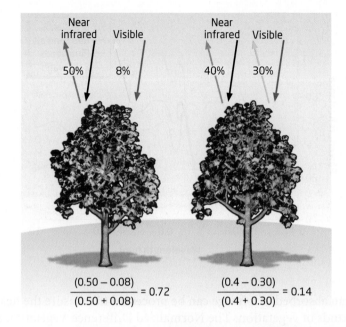

FIGURE 10.7

An NDVI image of Africa, showing the health (or presence) of vegetation on the continent. *[Adapted from NASA/University of Maryland]*

How Do You Display a Digital Remotely Sensed Image?

Once the digital data is collected, the next step is to transform it into something that can be viewed on a computer screen, especially since many of these bands of energy are only visible to the sensor and not to the human eye. Sensors can only "see" a certain area of the ground at a time and will record measurements within that particular area. The size of this area on the ground is the sensor's **spatial resolution**: It's the smallest unit of area the sensor can collect information about. If a sensor's spatial resolution is 30 meters, each measurement will consist of a 30 meter × 30 meter area of ground, and the sensor will measure thousands of these 30 meter × 30 meter sections at once. No matter how many sections are being measured, the smallest area the sensor has information for is a block that is 30 meters × 30 meters. Keep in mind that the spatial resolution of the image is going to limit the kind of visual information you can obtain from it. In an image with 4-meter resolution, you may be able to discern that an object is a large vehicle, but you won't be able to go into any details about it. An image with 0.5-meter resolution, however, will contain much more detailed information about the vehicle.

We'll discuss more about spatial resolution in Chapter 11, but for now, each one of these blocks being sensed represents the size of the area on the ground having its energy reflection being measured by the sensor. Owing to the effects of the atmosphere, there's not a straight one-to-one relationship between the reflectance of energy and the radiance being measured at the satellite's sensor. The energy measurements at the sensor are converted to a series of pixels, with each pixel receiving a **brightness value**, or BV (also referred to as a *digital number*, or *DN*), for the amount of radiance being measured at the sensor for that section of the ground. This measurement is made for each wavelength band being sensed.

Brightness values are scaled to fit a range specified by the sensor. We'll discuss these ranges in more detail in Chapter 11 (with regard to a sensor's radiometric resolution), but for now, the scale of values in a range represents the energy being recorded—from a low number (indicating very little radiance) to a high number (indicating a lot of radiance)—and measured as a number of bits. For instance, an 8-bit sensor will scale measurement values between 0 (the lowest value) and 255 (the highest value). These values represent the energy being recorded in a range from 0 to 255 (these values are related to the amount of energy being measured). For instance, all values in a band of an image could possibly only be in a range such as 50 to 150, with nothing reaching the maximum value of 255. Other types of sensors can have larger ranges, such as 0–2047 or 0–65535.

Each wavelength band that is being sensed is assigned a BV in this range for each pixel in the image. Brightness values can be translated to a grayscale to be viewed on a computer screen. With the range of values

spatial resolution the size of the area on the ground represented by one pixel's worth of energy measurement

brightness values the energy measured at a single pixel according to a pre-determined scale; also referred to as digital numbers (DNs)

8-bit imagery a digital image that carries a range of brightness values from 0 to 255

panchromatic imagery black and white imagery formed by viewing the entire visible portion of the electromagnetic spectrum

multispectral imagery remotely sensed imagery created from the bands collected by a sensor capable of sensing several bands of energy at once

hyperspectral imagery remotely sensed imagery created from the bands collected by a sensor capable of sensing hundreds of bands of energy at once

color gun equipment used to display a color pixel on a screen through the use of the colors red, green, and blue

FIGURE 10.8 ———o

Pixels of a sample 8-bit image, with their corresponding brightness values and grayscale shading.

associated with **8-bit imagery**, values of 0 represent the color black and values of 255 represent the color white. All other integer values between 0 and 255 are displayed in shades of gray, with lower numbers being darker shades and higher numbers being lighter shades. In this way, any of the wavelength band measurements (including the wavelengths normally invisible to the human eye) can be displayed in grayscale on a computer screen (see **Figure 10.8**). For instance, a near-infrared image will have the brightness value for each pixel representing the near-infrared measurements at the sensor for that area.

If a sensor is measuring the visible portion of the spectrum and treating the entire 0.4- to 0.7-micrometer range as if it were one band, the result will be black-and-white **panchromatic imagery**. However, remote sensing devices are capable of sensing several wavelengths of the electromagnetic spectrum simultaneously. Some widely used remote sensing devices can sense several different bands at once—U.S. government satellite sensors are often capable of sensing 7, 10, or even 36 bands at the same time. Imagery created by sensing multiple bands together is referred to as **multispectral imagery**. Some technology can simultaneously sense over 200 bands of the electromagnetic spectrum, producing **hyperspectral imagery**.

When multiple bands of imagery are available, they can be combined so that the image may be viewed in color, rather than grayscale. Displaying imagery in color requires three bands to be displayed simultaneously, each in a separate color other than gray. This is simplifying the technical aspects a bit, but a display monitor (like a computer screen) is equipped with devices that can draw a pixel on the screen in varying degrees of red, green, and blue. These are often referred to in remote sensing as **color guns** (a name that refers back to equipment in old CRT monitors and tube televisions for displaying colors in red, green, and blue). If an 8-bit image is drawn using the red color gun, the values of 0 to 255 will not be in a range

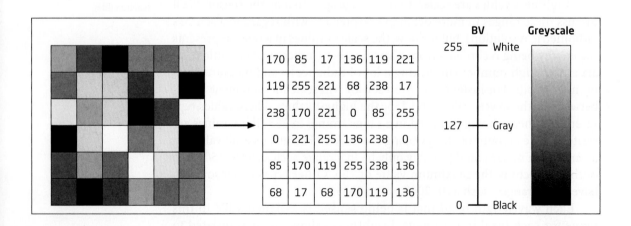

of black, gray, and white, but instead will be in a range of the darkest red, medium red, and very bright red. The same goes for the blue and green color guns—each gun is able to display an image in its respective color on a range of 0 to 255.

A **color composite** can be generated by displaying the brightness values of one band of imagery in the red gun, a second band of imagery in the green gun, and a third band of imagery in the blue gun. Each pixel will then be drawn using brightness values as the intensity of the displayed color. For instance, a pixel can be displayed using a value of 220 in the red gun, 128 in the green gun, and 75 in the blue gun, and the resulting color shown on the screen will be a combination of these three color intensities. By using the three primary colors of red, green, and blue and combining them to varying degrees using the range of values from 0 to 255, many other colors can be created—using the 256 numbers in the 0–255 range, with three colors, there are 16,777,216 possible color combinations (see **Figure 10.9** for examples of color formation).

To create colors, different values can be assigned to the red, green, and blue guns. **Table 10.1** shows some examples of ways to combine values together to form other colors (keeping in mind that there are millions of possible combinations of these values). See also *Hands-On Application 10.3: Color Tools Online: Color Mixing* for an online tool that can be used to create colors from various 0 to 255 values.

The color for each pixel in an image corresponds to particular values of red, green, and blue being shown on the screen. For a remotely sensed image, those numbers are the brightness values of the band being displayed

color composite an image formed by placing a band of imagery into each of the three color guns (red, green, and blue) to view an image in color rather than grayscale

FIGURE 10.9

Colors that can be formed by adding different colors together (left) and subtracting different colors from each other (right).

TABLE 10.1	Examples of colors generated through 8-bit combinations of red, green, and blue.		

Red Value	Green Value	Blue Value	Color Formed
255	255	0	Yellow
255	0	255	Magenta
0	255	255	Cyan
255	128	0	Orange
0	64	128	Dark Blue
255	190	190	Pink
255	255	115	Bright Yellow
115	76	0	Brown
0	197	255	Sky Blue
128	128	128	Gray
255	255	255	White
0	0	0	Black

Hands-On Application 10.3

Color Tools Online: Color Mixing

The Color Tools Website provides several great resources related to color. The Color Mixer tool available at **www.colortools.net/color_mixer.html** allows you to input values (between 0 and 255) for red, green, and blue and see the results. On the Website, type a value into the box in front of the "/255" for each of the three colors and check out the result. Try the values listed in Table 10.1 as well as some of your own to see how different numbers create different colors.

Expansion Questions

➤ What colors do the following combinations generate? (Try to figure out the result first, then use the Color Mixer to verify your results.)

- R: 0, G: 128, B: 255
- R: 175, G: 175, B: 175
- R: 85, G: 85, B: 128
- R: 20, G: 100, B: 50
- R: 120, G: 90, B: 10

with that color gun. Any of the bands of an image may be displayed with any of the guns, but as there are only three color guns, only three bands can be displayed as a color composite at any time (**Figure 10.10**).

For example, the near-infrared band values could be displayed using the red color gun, the green band values could be displayed using the green color gun, and the red band values could be displayed using the blue color gun. Using this arrangement, a brown pixel can be seen on the screen. On closer examination, it's discovered that the red gun (which has the

FIGURE 10.10

Combining the display of three 8-bit imagery bands, each being displayed in a different color gun, to form a color composite image. *[NASA Images acquired by Landsat 5]*

near-infrared band) has a value of 150, the green gun (which has the green band) has a value of 80, and the blue gun (which has the red band) has a value of 20. These values correspond to the brightness values of their respective bands and combine together to form that pixel color. If you change the arrangement of bands and guns, you'll see different colors displayed on the screen, but the actual brightness values will not change. **Table 10.2** shows how the display color will change depending on which band is being displayed with which gun.

Note that while the pixel being displayed on the screen will change color depending on what band is being displayed with a different gun, none of the brightness values of that pixel will be altered. Different band combinations can reveal different information about objects in an image due to the varying brightness values of the bands for those items.

TABLE 10.2 Different combinations of for various bands displayed with different color guns from different colors on the screen.

Red Value	Green Value	Blue Value	Color Formed
NIR band (150)	Green band (80)	Red band (20)	Brown
Red band (20)	Green band (80)	NIR band (150)	Deep blue
Green band (80)	NIR band (150)	Red band (20)	Green
Red band (20)	NIR band (150)	Green band (80)	Paler green

true color composite an image arranged by placing the red band in the red color gun, the green band in the green color gun, and the blue band in the blue color gun

false color composite an image arranged when the distribution of bands differs from placing the red band in the red color gun, the green band in the green color gun, and the blue band in the blue color gun

standard false color composite an image arranged by placing the near-infrared band in the red color gun, the red band in the green color gun, and the green band in the blue color gun

Several different types of composites can be formed using the various bands and guns. Whenever a composite has the red band displayed in the red gun, the green band displayed in the green gun, and the blue band displayed in the blue gun (the way natural colors would be shown), a **true color composite** is formed. A true color composite appears as you would expect a remotely sensed image to appear if you were looking down at the ground from an airplane (that is, the trees would be green and the water would be blue). Whenever the distribution of bands to guns deviates from this arrangement, a **false color composite** is generated. For instance, all four of the combinations described in Table 10.2 would result in false color composites as none of them display the red band in the red gun, the green band in the green gun, and the blue band in the blue gun. See **Figure 10.11** for a comparison of true color and false color composites.

With all of these different possible combinations, there is a **standard false color composite** that is used in many remote sensing studies. In this standard false color arrangement, the near-infrared band is displayed in the red gun, the red band is displayed in the green gun, and the green band is displayed in the blue gun. The false color composite image is similar to the CIR photos discussed back in Chapter 9. Objects that have a strong reflection of near-infrared energy (like healthy plants, grass, or other green vegetation) appear in a variety of bright reds, while water (which has very little reflection of the near-infrared, red, and green energy) appears as black. With composites, multispectral imagery can be displayed in color, regardless of what bands are being used, and these composites can be used for many different types of image analysis (see *Hands-On Application 10.4: Comparing True Color and False Color Composites* for online methods of comparing how objects on the ground are shown in true color and false color composites).

FIGURE 10.11

Two views of the Cleveland area: (a) a true color composite, and (b) a false color composite. *[NASA/Purdue Research Foundation]*

(a)

(b)

Hands-On Application 10.4

Comparing True Color and False Color Composites

An excellent online resource for examining Landsat imagery in different band combinations from different dates is the Esri ChangeMatters tool. Open your Web browser and go to **http://changematters.esri.com/compare** to get started. ChangeMatters will show you three remotely sensed images—two Landsat images from different dates and then an NDVI comparison of the two. By default, the initial images should be those of Mount St. Helens (from 1975, 2000, and an NDVI change between the two dates).

Start by selecting the dates 2005–2010 from the Select Dates pull-down menu (if a warning message appears, just close it and continue). This will change the first image to an image from a 2005 date and the second image to that taken from a 2010 date. Next, choose the option for Natural Color from the Select Image Map pull-down menu. This will change both the 2005 and 2010 images to true color composites (note that the colors may be a little different in each one). You can zoom in on one of the two images to examine some of the landscape features (such as Spirit Lake, the volcano itself, or the vegetated areas surrounding it) and all three images will zoom to the same extent. To examine the same area in a false color composite, choose the option for Infrared from the Select Image Map pull-down menu. Check out the same features in a false color arrangement to see how they appear differently.

In the Search box, you can type the name of a location (such as Cleveland, Ohio) and the images will switch to Landsat images from 2005 and 2010 (and the NDVI change comparison) of Cleveland. Take a close look at the urban areas of Cleveland and the surrounding non-urban areas, as well as Lake Erie in both the Natural Color and Infrared settings. In addition, use the Search box to view your own local area and examine it on both Natural Color and Infrared imagery from the two dates.

A second Esri resource for examining remotely sensed imagery in different band combinations is Landsat Explorer, available online here: **http://landsatexplorer.s3-website-us-west-2.amazonaws.com**. With this application, you can view Landsat imagery in different band combinations that are commonly used for analysis of different features on the landscape. For instance, clicking on the "wrench" icon (the "Renderer" option in the upper left corner) will allow you to view Earth in different composites. By selecting the option for "Natural Color" you'll see Earth in a true color composite, while selecting "Color Infrared" will allow you to see Earth in a standard false color composite. Check out the different band combination options (including Agriculture, Shortwave Infrared, Geology, and Bathymetric) as well as the Panchromatic option for a sharper 15-meter resolution image and the Vegetation Index and Moisture Index options for other images derived from Landsat imagery.

Expansion Questions

➤ How did all the features change in appearance (from Mount St. Helens and your own local area) when you switched from the true color composite (Natural Color) to the false color composite (Infrared), and why?

➤ Check out the parts of Cleveland, Ohio, around Lake Erie, both rural and urban. How do the lake and areas around the lake appear in true color and false color composites, and why?

➤ In the second application, why do features like vegetation appear bright red in the Color Infrared option, bright green in the Agriculture option, and dark green in the Natural Color option?

Chapter Wrap-Up

The basics described in this chapter cover the process of remote sensing and how a digital image can be formed from the measurement of electromagnetic energy reflection. Remote sensing provides a valuable technique for a variety of applications, including monitoring the health of vegetation and mapping land cover. See *Remote Sensing Apps* for some representative apps for your smartphone or tablet and *Remote Sensing in Social Media* for some Facebook pages, Twitter accounts, blogs, and YouTube videos related to remote sensing and this chapter's concepts. In the next chapter you'll be looking at how these concepts can be applied to perform remote sensing from a satellite platform in space. Chapters 11 and 12 will examine different types of remote sensing platforms and look at some of the uses to which their imagery is put, including assessment of air pollution, wildfire monitoring, and measurements of water quality.

The lab for this chapter uses a free program called MultiSpec to examine a digital remotely sensed image as well as different band combinations and color composites.

Important note: The references for this chapter are part of the online companion for this book and can be found at **www.macmillanlearning.com/catalog/shellitoigt4e.**

Remote Sensing Apps

Here's a sampling of available representative remote sensing apps for your smartphone or tablet. Note that some apps are for Android, some are for Apple iOS, and some may be available for both.

➤ **Flat Earth HD:** An app for viewing global remotely sensed imagery of Earth

➤ **Images of Change:** A NASA app that compares satellite imagery (and sometimes regular photography) of environmental features and changes from two time periods

➤ **iNSIGHT Color Mixing:** An app that allows you to mix the primary colors of red, green, and blue to different degrees and generate new colors

Remote Sensing in Social Media

Here's a sampling of some representative Twitter accounts, along with some YouTube videos and blogs related to this chapter's concepts.

On Facebook, check out the following:

➤ **Google Earth Outreach:**
www.facebook.com/earthoutreach

Become a Twitter follower of:

➤ **Google Earth Outreach:** @earthoutreach
➤ **USGS Land Cover** (the USGS Land Cover Institute):
@USGSLandCover

On YouTube, watch the following videos:

➤ **Google Earth Outreach** (a Google channel on YouTube showcasing how geospatial technology, including remotely sensed imagery, is applied in numerous areas of study): **www.youtube.com/user/EarthOutreach**
➤ **Monitoring Forests from the Ground to the Cloud** (a Google Earth Outreach video showing how remotely sensed imagery is used for monitoring deforestation): **www.youtube.com/watch?v=ymKHb3WJLz4**
➤ **San Francisco-True&FalseColorImagery-Landsat7.mov** (a video showing the differences between true color and false color composites using Landsat 7 bands): **www.youtube.com/watch?v=9pVF8LK_KMs**

For further up-to-date info, read up on this blog:

➤ **ArcGIS Resources—Imagery Category** (blog concerning topics related to working with remotely sensed imagery): **http://blogs.esri.com/esri/arcgis/category/subject-imagery**

Key Terms

8-bit imagery (p. 374)
absorption (p. 368)
atmospheric windows (p. 367)
band (p. 365)
blue band (p. 365)
brightness value (p. 373)
color composite (p. 375)
color guns (p. 374)
electromagnetic spectrum (p. 364)
false color composite (p. 378)
green band (p. 365)
hyperspectral imagery (p. 374)
incident energy (p. 369)
IR (p. 365)
micrometers (p. 364)
Mie scattering (p. 368)
multispectral imagery (p. 374)
nanometer (p. 364)
NDVI (p. 371)

NIR (p. 365)
non-selective scattering (p. 368)
panchromatic imagery (p. 374)
Rayleigh scattering (p. 367)
red band (p. 365)
remote sensing (p. 361)
spatial resolution (p. 373)
spectral reflectance (p. 369)
spectral signature (p. 370)
standard false color composite (p. 378)
SWIR (p. 366)
TIR (p. 366)
transmission (p. 368)
true color composite (p. 378)
UV (p. 365)
visible light spectrum (p. 365)
wavelength (p. 363)

10.1 Geospatial Lab Application

Remotely Sensed Imagery and Color Composites

This chapter's lab will introduce you to some of the basics of working with multispectral remotely sensed imagery through the use of the MultiSpec software program.

Objectives

The goals to take away from this exercise are:

➤ To familiarize yourself with the basics of the MultiSpec software program

➤ To load various bands into the color guns and examine the results

➤ To create and examine different color composites

➤ To compare the brightness values of distinct water and environmental features in a remotely sensed satellite image in order to create basic spectral profiles

Using Geospatial Technologies

The concepts you'll be working with in this lab are used in a variety of real-world applications, including:

➤ Environmental science, which uses satellite imagery to derive signatures for various classes of land cover and to create accurate land cover maps

➤ Urban planning, where satellite imagery collected over time can be used to study the expansion (and sometimes the contraction) of cities and to evaluate proposals for commercial, industrial, and residential development

[NASA/GSFC]

Obtaining Software

The current version of MultiSpec (3.6) is available for free download at **https://engineering.purdue.edu/~biehl/MultiSpec**.

Important note: Software and online resources can change fast. This lab was designed with the most recently available version of the software at the time of writing. However, if the software or Websites have changed significantly between then and now, an updated version of this lab (using the newest versions) will be available online at **www.macmillanlearning.com/catalog/shellitoigt4e.**

Lab Data

Copy the folder **Chapter 10**—it contains a Landsat 8 OLI/TIRS satellite image (called "cle.img") from 9/27/2014 of Cleveland, Ohio. This file shows a subset of a larger Landsat satellite image. We will discuss more about Landsat imagery in Chapter 11, but the OLI and TIRS imagery bands refer to the following portions of the electromagnetic (EM) spectrum in micrometers (μm):

➤ Band 1: Coastal (0.43 to 0.45 μm)

➤ Band 2: Blue (0.45 to 0.51 μm)

➤ Band 3: Green (0.53 to 0.59 μm)

➤ Band 4: Red (0.64 to 0.67 μm)

➤ Band 5: Near infrared (0.85 to 0.88 μm)

➤ Band 6: Shortwave infrared 1 (1.57 to 1.65 μm)

➤ Band 7: Shortwave infrared 2 (2.11 to 2.29 μm)

➤ Band 8: Panchromatic (0.50 to 0.68 μm)

➤ Band 9: Cirrus (1.36 to 1.38 μm)

➤ Band 10: Thermal infrared 1 (10.60 to 11.19 μm)

➤ Band 11: Thermal infrared 2 (11.50 to 12.51 μm)

Keep in mind that Landsat 8 imagery has a 30-meter spatial resolution (except for the panchromatic band, which is 15 meters). Thus, each pixel you will examine covers a 30 meter by 30 meter sized area on the ground.

Localizing This Lab

Although this lab focuses on a section of a Landsat scene from northeast Ohio, Landsat imagery is available for free download via LandsatLook at **https://glovis.usgs.gov**. This site will provide raw data that will have to be imported into MultiSpec and processed to use in the program. Information for using this free Landsat data for use in MultiSpec is available at **https://engineering.purdue.edu/~biehl/MultiSpec/tutorials/MultiSpec_Tutorial_5.pdf**.

10.1 MultiSpec and Band Combinations

1. Start **MultiSpec**. It will start with an empty text box (which will give you updates and reports of processes in the program). You can minimize the text box for now.

2. To get started with a remotely sensed image, select **Open Image** from the **File** pull-down menu. Alternatively, you can select the **Open** icon from the toolbar:

3. Navigate to the **Chapter 10** folder, select **cle.img** as the file to open, and click **Open**.

4. A new dialog box will appear that allows you to set the display specification for the "cle" image.

5. Under Channels, you will see the three color guns available to you (red, green, and blue). Each color gun can hold one band (see the Lab Data section for a listing of which bands correspond to which parts of the electromagnetic spectrum). The number listed next to each color gun represents the band being displayed with that gun.

6. Display **band 5** in the red color gun, **band 4** in the green color gun, and **band 3** in the blue color gun.

7. Accept the other defaults for now and click **OK**.

8. A new window will appear and the "cle" image will begin loading. This might take a minute or two to load completely.

9. For the best results, maximize both MultiSpec and the window containing the "cle" image. Use the bars at the edge of the window to move around the image. You can zoom in or zoom out by using the zoom tools on the toolbar:

The large mountain icon will zoom in and the small mountain icon will zoom out.

10. Zoom around the image, paying attention to some of the city, landscape, and water areas.

> **Question 10.1** What wavelength bands were placed into which color guns?

> **Question 10.2** Why are the colors in the image so strange compared to what we're normally used to seeing in other imagery (such as Google Earth or Google Maps)? For example, why is most of the landscape red?

> **Question 10.3** In this color composite, what colors are the water, vegetated areas, and urban areas displayed as in the image?

11. Reopen the image, this time using **band 4** in the red gun, **band 3** in the green gun, and **band 2** in the blue gun (referred to as a 4-3-2 combination). Once the image reloads, pan and zoom around the image, examining the same areas you just looked at.

> **Question 10.4** What kind of composite did we create in this step? How are the bands being displayed in this color composite in relation to their guns?

> **Question 10.5** Why can we not always use this kind of composite (from Question 10.4) when analyzing satellite imagery?

12. Reopen the image yet again, this time using **band 7** in the red gun, **band 5** in the green gun, and **band 3** in the blue gun (referred to as a 7-5-3 combination). Once it reloads, pan and zoom around the image, examining the same areas you just looked at.

> **Question 10.6** Once again, what kind of composite was created in this step?

> **Question 10.7** How are vegetated areas being displayed in this color composite (compared with the arrangement in Question 10.4)? Why are they displayed in this color?

10.2 Examining Color Composites and Color Formations

1. Reopen the image one more time, returning to the 5-4-3 combination (**band 5** in the red gun, **band 4** in the green gun, and **band 3** in the blue gun). You can close the other images, as we'll be working with this one for the rest of the lab.

2. Zoom and move around the image to find and examine Burke Lakefront Airport as follows:

3. Zoom in to the airfield. From its shape and the pattern of the runways, you should be able to clearly identify it in the Landsat image.

4. Examine the airfield and its surroundings.

> **Question 10.8** Why do the areas in between the runways appear red?

5. Open a new image, this time with a 4-5-3 combination. Examine Burke Lakefront Airport in this new image and compare it to the one you've been working with.

> **Question 10.9** Why do the areas in between the runways now appear bright green?

6. Open another new image, this time with a 4-3-5 combination. Examine Burke Lakefront Airport in this new image and compare it with the others you've been working with.

> **Question 10.10** Why do the areas in between the runways now appear blue?

7. At this point, only keep the 5-4-3 image open and close the other two.

10.3 Examining Specific Brightness Values and Spectral Profiles

Regardless of how the pixels are displayed in the image, each pixel in each band of the Landsat 8 image has a specific brightness value set in the 0–65536 range. By examining those pixel values for each band, you can chart a basic "spectral profile" of some features in the image.

1. Zoom in to the area around Cleveland's waterfront and identify the urban areas (in the image these will mostly be the white or cyan regions).

2. From the **Window** pull-down menu, select **New Selection Graph**. Another (empty) window (called Selection Graph) will open in MultiSpec.

3. In the image, locate a pixel that's a good example of an urban or developed area. The cursor will change to a cross shape, so give the pixel one more click.

   Important note: Zoom in so that you are only selecting one pixel with the cursor.

4. A chart will appear in the Selection Graph window that will show graphically the BVs for each band at that particular pixel (see the first page of this lab for the portions of the electromagnetic spectrum that match up with each band). The band numbers are on the *x*-axis and the BVs are on the *y*-axis. The chart can be expanded or maximized as necessary to allow you to examine the values better.

5. The Selection Graph window now shows the data that can be used to compute a simplified version of a spectral profile for an example of the particular urban land use pixel you selected from the image.

6. For the next question, you'll be required to find a pixel that's a good example of water and another pixel that's a good example of vegetation. You'll also be required to translate the data from the chart to a spectral profile for each example. In drawing the profiles from the information on the chart, keep two things in mind:

   a. First, the values at the bottom of the Selection Graph window represent the numbers of the bands being examined. On the chart below, the actual wavelengths of the bands are plotted out, so be very careful to make sure you properly match up each band with its respective wavelength. Note that bands 10 and 11 are not being charted, because they measure emitted thermal energy rather than reflected energy like bands 1 through 9. Also note that band 8 (panchromatic) is not charted either.

   b. Second, the values on the *y*-axis of the Selection Graph window are BVs, not the percentage of reflection as seen in a spectral signature diagram. There are a number of factors involved with transforming BVs into the actual percent reflectance values (a BV and the percent reflectance don't have an exact one-to-one ratio, as there are other factors that affect the measurement at the sensor, such as atmospheric effects), but in this simplified example, you'll just chart the BV for this spectral profile.

7. Examine the image for a good example of a water pixel and a vegetation pixel.

8. Plot a spectral profile diagram for the water and vegetation pixels you chose on the following diagram (remember to plot values as calculated from the BVs).

9. Examine your two profiles and answer Questions 10.11 and 10.12.

> **Question 10.11** What information can you gain from the spectral profile for water about the ability of water to reflect and absorb energy (that is, what types of energy are most reflected by water, and what types of energy are most absorbed by water)?

> **Question 10.12** What information can you gain from the spectral profile for vegetation about the ability of vegetation to reflect and absorb energy (that is, what types of energy are most reflected by vegetation, and what types of energy are most absorbed by vegetation)?

10. Exit MultiSpec by selecting **Exit** from the **File** pull-down menu. There's no need to save any work in this exercise.

Closing Time

Now that you have the basics of working with remotely sensed data and examining the differences in color composites, the next two chapters will build on these skills. The Chapter 11 lab will return to using MultiSpec for more analysis of Landsat satellite imagery.

Images from Space

Satellite Remote Sensing, Satellite Orbits, Sensor Resolutions, the Landsat Program, High-Resolution Satellite Sensors, Small Satellites, and Using Satellites for Monitoring

How many times have you seen something like this in a movie? The heroes are racing against the clock to stop the villains from setting off their doomsday weapon. In the tense mission-control center, the heroes' computer expert taps a few keys, a spy satellite swings into view, and suddenly the monitors fill up with crystal-clear images from space of the villains' remote hideout. The imagery's crisp enough for the heroes to run some facial recognition software and pick out the international arms dealer selling the weapons of mass destruction (WMDs), as well as read the license plates of the nearby cars. Rest assured, the members of the elite counter-intelligence strike team will show up on cue to save the world, probably watched by their teammates at the home base via a live remotely sensed satellite video feed.

From this kind of scenario, you might get the impression that remote sensing satellites are magical things that can do anything—rotate into whatever orbital position is required to track people, zoom in (from outer space) to make out the details of someone's license plate, read lips from space, or send live video feeds to the good guys in the field. Certainly, best-selling books and blockbuster movies can help to perpetuate the idea that remote sensing (and anything involving satellites) is some sort of utopian technology that can do anything. The goal of this chapter is to examine how satellite remote sensing really works and to have a look at what it can and can't do in the real world.

The first man-made satellite to achieve orbit around Earth was Sputnik, launched in 1957 by the Union of Soviet Socialist Republics (USSR). It was quickly followed by the U.S. launch of Explorer I the next year. These initial satellite launches established that satellites carrying equipment (or, in the case of Sputnik II, animals) could be placed into orbit and that

NASA the National Aeronautics and Space Administration, established in 1958; it is the U.S. government's space exploration and aerospace development branch

Corona a U.S. government remote sensing program, which utilized film-based camera equipment mounted in a satellite in Earth's orbit; Corona was in operation from 1960 to 1972

the equipment could then be put to use. As a result, **NASA** (the National Aeronautics and Space Administration) was established in 1958 to head up the U.S. space and aeronautics program. Satellites were first used for remote sensing reconnaissance and surveillance in 1960, when the **Corona** program was initiated.

A Corona satellite had an onboard camera that took pictures of areas on the ground. When the roll of film was used up, it would be ejected from the satellite and collected by a U.S. plane. After the film was developed, it would then be interpreted by agents to provide intelligence information about locations around the world (see **Figure 11.1** for an example of Corona imagery). Corona and its counterpart—the USSR's Zenit satellite—provided numerous images from orbit, with Corona itself providing an estimated 800,000 remotely sensed satellite images. Corona ceased operation in 1972, although similar satellite reconnaissance programs continued, such as the Gambit and Hexagon programs, which remained active until 1984. Corona's imagery was declassified in 1995 but satellite remote sensing has continued, and a series of satellites are now used by the U.S. government for acquiring digital imagery of areas around the world.

FIGURE 11.1

Corona imagery of Dolon Airfield in the former USSR (taken in 1966).
[National Reconnaissance Office]

Beyond government surveillance, satellites are used today for a wide variety of purposes, constantly collecting data and imagery of the world below. How does satellite imaging compare with the aerial photography we discussed in Chapter 9? Using a satellite instead of an aircraft has several apparent advantages. Satellites are constantly orbiting Earth and taking images—there's no need to wait for a plane to fly over a particular area of interest. Satellites can image a much larger area than a single aerial photograph can. Also, remote sensing satellites provide global coverage and are not restricted to geographic boundaries or constraints the way aircraft are, because satellites orbit hundreds of miles above Earth's surface.

How Do Remote Sensing Satellites Collect Data?

Remote sensing satellites are placed in a specific orbital path around Earth. One type of orbit used in remote sensing is **geostationary orbit,** in which satellites rotate at the same speed as Earth. Because a geostationary satellite takes 24 hours to make one orbit, it's always in the same place at the same time. For instance, although it's not a remote sensing system, satellite television utilizes a geostationary orbit—the satellite handling the broadcasting is continuously in orbit over the same area, which is why a satellite reception dish always needs to be pointed at the same part of the sky. Likewise, the WAAS and EGNOS satellites discussed in Chapter 4 are in geostationary orbit so that they can provide constant coverage to the same area on Earth. There are some remote sensing satellites that use geostationary orbit to collect information about the same area on Earth's surface all the time (such as a weather satellite).

geostationary orbit an orbit in which an object precisely follows the direction and speed of Earth's rotation and is therefore always directly above the same point on Earth's surface

Many remote sensing satellites operate in **near-polar orbit,** a north-to-south path wherein the satellite moves close to the North and South Poles while it makes several passes a day about Earth. For instance, the Landsat 8 satellite (a U.S. satellite that we'll discuss in more detail shortly) completes a little more than 14 orbits per day (each orbit takes about 99 minutes)—see **Figure 11.2** for Landsat 8's near-polar orbit. Also, check out *Hands-On Application 11.1: Examining Satellite Orbits in Real Time* for two methods of tracking satellite orbits in real time.

near-polar orbit an orbital path close to the North and South Poles that carries an object around Earth at an unvarying elevation

While a satellite in continuous orbit passes over Earth, it will be imaging the ground below it. However, during an orbit, a satellite can only image a certain area of ground at one time—it can't see everything at once. The **swath width** is the measurement of the width of ground the satellite can image during one pass. A Landsat satellite's sensor has a swath width of 185 kilometers, meaning that a 185-kilometer-wide area on the ground is imaged as Landsat flies overhead. As satellites orbit, Earth is rotating beneath them, meaning that several days may elapse before an orbital path (and thus,

swath width the width of the ground area the satellite is imaging as it passes over Earth's surface

FIGURE 11.2

The near-polar orbit of Landsat 8.

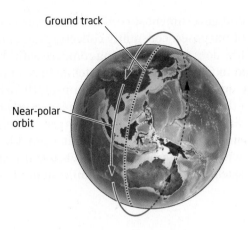

Ground track

Near-polar orbit

the swath of ground being imaged) takes a satellite over nearby geographic areas. For example, a Landsat satellite may pass over the middle of the United States on Day 1 of its orbit, pass hundreds of miles to the west on Day 2, and not return to the path next to the first one until Day 8 (**Figure 11.3**).

FIGURE 11.3

An example of the paths covered for one section of the United States during the 16-day revisit cycle of a Landsat satellite.

Ground track on Day 1
Day 2
Day 3
Day 4
Day 5
Day 6
Day 7
Day 8
Day 9
Day10
Day11
Day12
Day13
Day14
Day15
Day16

Hands-On Application 11.1

Examining Satellite Orbits in Real Time

Think about all the jobs that satellites are doing—communication, TV transmission, weather monitoring, military applications, and GPS—and then add remote sensing to that list. When you consider how useful these tasks are, you can see the need for many, many satellites in orbit. To get started looking at the orbital paths of these satellites, go to **http://www.n2yo.com**, which is the Website for a tracking app for satellites. It will open with the current location of the International Space Station (ISS) plotted on a map, but can track the path of a multitude of different remote sensing satellites. From the Satellites on Orbit pull-down menu, choose the option for More Categories, and then choose the option for Earth resources. A new list of satellites will appear that you can sort alphabetically by clicking on the option for Name. By clicking on the name of a satellite, a new Webpage will appear with information about the satellite, as well as a map of its current location, and when the satellite will next be in orbit over your current location.

Expansion Questions

➤ What is the current position of the following remote sensing satellites: Landsat 7, Landsat 8, Sentinel-2A, Sentinel-2B, and SPOT 7?

➤ For each of the five satellites whose current position you found, when is the next time that they will pass over your current location in orbit?

After a certain amount of time, the satellite will completely pass over all areas on Earth and again resume its initial path. For instance, each one of the Landsat satellites is set up in an orbital path so that it will pass over the same swath on Earth's surface every 16 days. Thus, when Landsat's orbit carries it to the swath where it's collecting imagery of your current location, it will be another 16 days before it's able to image your location again. A **sun-synchronous orbit** occurs when a satellite's orbit always takes it over the same swath of Earth's surface at the same local time. As a result, images collected from the satellite will have similar sun-lighting conditions and can be used for comparing changes to an area over time. For instance, Landsat 8 has a sun-synchronous orbit, allowing it to cross the Equator at the same time every day.

The satellites' onboard sensors have different ways of actually scanning the ground and collecting information in the swath being imaged. In **along-track scanning**, a linear array scans the ground along the satellite's orbital path. As the satellite moves, the detectors in the array collect the information from the entire swath width below them. This type of sensor is referred to as a "pushbroom." Think of pushing a broom in a straight line across a floor—in this case, the broom itself is the sensor, and each bristle of the broom represents a part of the array that senses what is on the ground below it. A second method is **across-track scanning**; here, a rotating mirror moves back and forth over the swath width of the ground to collect information. This type of sensor is referred to as a "whiskbroom." Once again, think of sweeping a floor with a broom, swinging the broom back and forth

sun-synchronous orbit an orbital path set up so that the satellite always crosses the same areas at the same local time

along-track scanning a scanning method using a linear array to collect data directly on a satellite's path

across-track scanning a scanning method using a rotating mirror to collect data by moving the device back and forth across the width of the satellite's swath

FIGURE 11.4

The operation of across-track and along-track scanners.

across a path on the floor—the broom is the sensor, and the bristles represent the detectors that collect the data on the ground below them. Satellite systems usually use one of these types of sensors to obtain the energy reflectance measurements from the ground (see **Figure 11.4** for examples of the operation of both sensor types).

After a satellite's sensors have collected data, the data need to be off-loaded from the satellite and sent to Earth. Because data collection is continuous, this information needs to get to people on the surface so that it can be processed and utilized as it becomes available. The Earth Resources Observation Science (**EROS**) Center located outside Sioux Falls, South Dakota, is one of the downlink stations in the United States that receives data from numerous remote sensing satellites. Satellite data can be sent directly to a receiver at a station or transmitted to other tracking and data relay satellites, which then send the data to Earth. Once received on the ground, the data can be processed into imagery.

EROS the Earth Resources Observation Science Center, located outside Sioux Falls, South Dakota, that serves (among many other things) as a downlink station for satellite imagery

What Are the Capabilities of a Satellite Sensor?

A satellite sensor has four characteristics that define its capabilities: spectral resolution, spatial resolution, radiometric resolution, and temporal resolution. A sensor's first parameter is its **spectral resolution**, which refers to the bands and their wavelengths measured by the sensor. For instance, a sensor on the WorldView-4 satellite can measure four bands of energy simultaneously

spectral resolution the bands and wavelengths measured by a sensor

Thinking Critically with Geospatial Technology 11.1

What Effect Does Satellite Remote Sensing Have on Political Borders?

When imagery is acquired by satellite remote sensing for any part of the globe, the satellite sensors effectively ignore political boundaries. After the 2010 earthquake in Haiti, imagery of the country was quickly acquired by satellites and made available for rescue and recovery efforts. When satellite images of nuclear facilities in North Korea or Iran are shown on the news, the satellite's sensors were able to image these places from orbit. Satellites can acquire global imagery of events ranging from the 2016 Olympics in Rio de Janeiro to parades in Beijing; and in all cases, the political boundaries of these places are being ignored by the "eye in the sky" that is looking down from Earth's orbit. Satellites are not restricted by no-fly zones or other restrictions involving a country's airspace, and they are capable of acquiring imagery from nearly everywhere. Given these imaging capabilities, what effect does acquisition of images via satellites have on political borders? There are many areas that can't be approached from the ground without causing international tensions, but these same places can easily be viewed from space. How are political boundaries affected by these types of geospatial technologies?

(wavelengths in the blue, green, red, and near-infrared portions of the spectrum), and the sensor onboard the Landsat 8 satellite can sense in seven different wavelengths at the same time. A sensor with a higher spectral resolution can examine several finer intervals of energy wavelengths (for example, the MODIS sensor onboard the Terra satellite—see Chapter 12—can sense in 36 different bands at once). Being able to sense numerous smaller wavelengths in the near-infrared portion of the spectrum gives much finer spectral resolution than just sensing one wider wavelength of near-infrared energy.

A sensor is defined by which bands it can measure. A **panchromatic sensor** measures one wide range of wavelengths, usually the entire visible portion as one band (or sometimes the visible portion and part of the infrared spectrum) as one single band. This will produce the black, white, and grayscale panchromatic imagery (see page 374). A **multispectral sensor** will measure several bands of energy—for instance, the MODIS sensor mentioned above is an example of a multispectral sensor that measures 36 different bands and will produce multispectral imagery (see page 374). A **hyperspectral sensor** is capable of sensing hundreds of very fine band wavelengths at once and will produce hyperspectral imagery (see page 374).

In Chapter 10, we discussed the concept of resolution in the context of **spatial resolution**, with the area on the ground represented by one pixel in a satellite image. A sensor's spatial resolution will affect the amount of detail that can be determined from the imagery—a sensor with a 1-meter spatial resolution has much finer resolution than a sensor with spatial resolution of 30 meters (see **Figure 11.5** for examples of the same area viewed by different spatial resolutions). A satellite's sensor is fixed with one spatial

panchromatic sensor a sensor that can measure one range of wavelengths

multispectral sensor a sensor that can measure multiple wavelength bands simultaneously

hyperspectral sensor a sensor that can measure hundreds of different wavelength bands simultaneously

spatial resolution the ground area represented by one pixel of satellite imagery

FIGURE 11.5

The same area on the landscape as viewed by three different sensors, each with a different spatial resolution. *[NASA Marshall Space Flight Center]*

pan-sharpening the technique of fusing a higher-resolution panchromatic band with lower-resolution multispectral bands to improve the clarity and detail seen in the image

radiometric resolution the degree of a sensor's ability to determine fine differences in a band of energy measurements

resolution for the imagery it collects—a 30-meter resolution sensor can't be "adjusted" to collect 10-meter resolution images.

If two bands (one with a higher resolution than the other) are used by the sensor to image the same area, the higher-resolution band can be used to sharpen the resolution of the lower-resolution band. This technique is referred to as **pan-sharpening** because the higher-resolution band used is a panchromatic band acquired by a panchromatic sensor. For example, a satellite (like Landsat 8) that senses the blue, green, and red portions of the electromagnetic spectrum at 30-meter resolution could also have a sensor equipped to view the entire visible portion of the spectrum and part of the near-infrared spectrum as a panchromatic band at 15-meter resolution. This panchromatic image can then be fused with the 30-meter resolution color imagery to create a higher-resolution color composite. Many satellites with high spatial resolution capabilities sense in a panchromatic band that allows for pan-sharpening of the imagery in the other bands.

As we mentioned in Chapter 10, a sensor scales the energy measurements into several different ranges (referred to as *quantization levels*) to assign brightness values to a pixel. The sensor's **radiometric resolution** refers to its ability to measure fine differences in energy. For instance, a sensor with 8-bit radiometric resolution places the energy radiance values on a scale of 0 (lowest radiance) to 255 (highest radiance). The wider the range of measurements that can be made, the finer the sensor's radiometric resolution is. The radiometric resolution (and the resulting number of levels) is a measure of the number of the sensor's bits of precision. To calculate the number of levels, take the number of bits and apply it as an exponent to the number 2. For instance, a 6-bit sensor would have 64 levels (2^6) ranging from a value of 0 to a value of 63. An 8-bit sensor would have 256 levels (2^8) ranging from 0 to 255. An 11-bit sensor would have 2048 levels (2^{11}) ranging from 0 to 2047.

The finer a sensor's radiometric resolution, the better it can discriminate between smaller differences in energy measurements. For example, a 2-bit sensor (2^2 or four values) would have to assign every pixel in an image a brightness value of 0, 1, 2, or 3, which would result in most pixels having the same value, and it would be difficult to distinguish between items or to create effective spectral signatures. However, an 11-bit sensor would assign a value anywhere from 0 to 2047, creating a much wider range of value levels and allowing much finer distinctions to be made between items. See **Figure 11.6** for examples of differing effects of radiometric resolution on an image when performing remote sensing.

The sensor's **temporal resolution** refers to how often the sensor can return to image the same spot on Earth's surface. For instance, a sensor with 16-day temporal resolution will collect information about the swath containing your house and not return to image it again until 16 days have passed. The finer a sensor's temporal resolution, the fewer days it will take between return times. A way to improve a sensor's temporal resolution is to use **off-nadir viewing** capability, in which the sensor is not fixed to sense what's directly below it (the nadir point). A sensor that is capable of viewing places off-nadir can be pointed to image locations away from the orbital path. Using off-nadir viewing can greatly increase a sensor's temporal resolution, as the sensor can image a target several times during its orbits, even if the satellite isn't directly overhead. This capability is a great help in monitoring conditions that can change quickly or require rapid responses (such as mitigating disasters resulting from hurricanes, tornadoes, or widespread fires).

 ## What Is a Landsat Satellite, and What Does It Do?

This chapter and Chapter 10 have mentioned a satellite system called Landsat but, until now, haven't really explained what it is, how it operates, or why it's so important. What would eventually become the U.S. **Landsat** program was conceived in the late 1960s as a resource for doing remote sensing of

FIGURE 11.6

The effect of different levels of radiometric resolution on an image (simulated 2-bit, 4-bit, and actual 8-bit imagery of one band). *[NASA]*

temporal resolution the length of time a sensor takes to come back and image the same location on the ground

off-nadir viewing the capability of a satellite to observe areas other than the ground directly below it

Landsat a long-running and ongoing U.S. remote sensing project that launched its first satellite in 1972

FIGURE 11.7

The history of the Landsat program launches. *[Courtesy of the U.S. Geological Survey]*

MSS the Multispectral Scanner aboard Landsat 1 through 5

TM the Thematic Mapper sensor onboard Landsat 4 and 5

Landsat 7 the seventh Landsat mission, launched in 1999, which carries the ETM+ sensor

ETM+ the Enhanced Thematic Mapper sensor onboard Landsat 7

Landsat 8 the latest Landsat mission, launched in 2013

Earth from space. The program's first satellite, Landsat 1, was launched in 1972 and the program continues today with the most recent launch, Landsat 8, in 2013 (see **Figure 11.7** for a graphical timeline of Landsat launches).

Landsat 1's sensor was the Landsat **MSS** (Multispectral Scanner), which had a temporal resolution of 18 days. The MSS would continue to be part of the Landsat missions through Landsat 5. The MSS was capable of sensing four bands (red and green, plus two bands in the near infrared) at 79-meter spatial resolution and 6-bit radiometric resolution. As the name implies, MSS is an example of a multispectral sensor, which can measure multiple wavelengths at once.

Landsats 4 and 5 were also equipped with a new remote sensing device called the **TM** (Thematic Mapper). The TM was also a multispectral sensor that allowed for sensing in seven bands in the electromagnetic spectrum simultaneously, including the three visible bands, a band of NIR, two bands of SWIR, and a thermal band (see **Table 11.1** for specifics about these bands and their wavelengths). The TM bands had improved radiometric resolution over MSS of 8 bits while the spatial resolution of the multispectral bands was improved to 30 meters (the thermal band had a spatial resolution of 120 meters, however) and had a swath width of 185 kilometers. TM also had an improved temporal resolution of 16 days.

Landsat 7, launched in 1999, carried neither the MSS nor the TM; instead, it carried a new sensor called the **ETM+** (Enhanced Thematic Mapper Plus). This new device senses the same seven bands and their wavelengths as TM at the same swath width as well as the same radiometric and spatial resolutions (although the thermal band's spatial resolution was improved to 60 meters). It also added a new band 8—this was a panchromatic band (covering wavelengths of 0.52–0.90 micrometers) with a spatial resolution of 15 meters. With this improved spatial resolution, the panchromatic band could be used to pan-sharpen the quality of the other imagery. Landsat 7's temporal resolution was still 16 days, but it could collect about 475 images per day.

The latest Landsat mission is **Landsat 8** launched on February 11, 2013 (See **Figure 11.8** on page 404). Landsat 8 currently collects up to 725 images per day, with its instruments covering a 185-kilometer swath

TABLE 11.1	The sensors from the various Landsat satellites and their remotely sensed bands and wavelengths (all wavelength values are listed in micrometers, µm).				
Bands and Sensors	OLI / TIRS (Landsat 8)	ETM+ (Landsat 7)	TM (Landsat 4, 5)	MSS 1, 2, 3 (Landsat 1, 2, 3)	MSS 4, 5 (Landsat 4, 5)
Band 1	Coastal (0.43–0.45)	Blue (0.45–0.52)	Blue (0.45–0.52)		Green (0.5–0.6)
Band 2	Blue (0.45–0.51)	Green (0.52–0.60)	Green (0.52–0.60)		Red (0.6–0.7)
Band 3	Green (0.53–0.59)	Red (0.63–0.69)	Red (0.63–0.69)		NIR (0.7–0.8)
Band 4	Red (0.64–0.67)	NIR (0.77–0.9)	NIR (0.77–0.9)	Green (0.5–0.6)	NIR (0.8–1.1)
Band 5	NIR (0.85–0.88)	SWIR (1.55–1.75)	SWIR (1.55–1.75)	Red (0.6–0.7)	
Band 6	SWIR (1.57–1.65)	Thermal (10.4–12.5)	Thermal (10.4–12.5)	NIR (0.7–0.8)	
Band 7	SWIR (2.11–2.29)	SWIR (2.09–2.35)	SWIR (2.09–2.35)	NIR (0.8–1.1)	
Band 8	Pan (0.5–0.68)	Pan (0.52–0.9)			
Band 9	Cirrus (1.36–1.38)				
Band 10	Thermal (10.6–11.19)				
Band 11	Thermal (11.5–12.51)				

(See **Figure 11.9** on page 404 for an example of Landsat 8 imagery). Like Landsat 7, Landsat 8 also has a 16-day temporal resolution. However, by having both Landsat 7 and 8 on the job, one of the two satellites can obtain an image of anywhere on the globe 8 days after the other.

Landsat 8's sensor is the next iteration of Landsat technology, called the **OLI** (Operational Land Imager), which has capabilities similar to ETM+. OLI senses visible and infrared bands at 30-meter resolution, a panchromatic band at 15-meter resolution, and two new 30-meter bands aimed at monitoring clouds and coastal zones. A thermal band is not part of OLI; instead, a separate instrument on the satellite, called **TIRS** (the Thermal Infrared Sensor), collects two different bands of thermal data listed as bands 10 and 11 for Landsat 8. This thermal data is acquired at 100-meter resolution but is then resampled to 30-meter resolution when delivered to users. See **Table 11.1** for a comparison of the characteristics of OLI and TIRS with previous Landsat sensors.

At the time of writing, Landsat 9 is currently in development with a tentative launch date of December 2020. Landsat 9 is expected to have

OLI the Operational Land Imager, the multispectral sensor onboard Landsat 8

TIRS the Thermal Infrared Sensor, the instrument that acquires thermal imagery onboard Landsat 8

FIGURE 11.8

Landsat 8 in action. *[USGS/ NASA Landsat]*

FIGURE 11.9

Landsat 8 images showing the Virgin Islands before and after Hurricane Irma in 2017. *[NASA]*

sensors similar to OLI and TIRS onboard and also to have radiometric and temporal resolution characteristics similar to Landsat 8. In addition, Landsat 10 is currently aimed at launching in the mid-to-late 2020s, but nothing definitive has been set concerning the sensors or capabilities of Landsat 10 as it is still in its conceptual stage.

The bands that were chosen for Landsat sensors were (and still are) key to a variety of studies. The blue band is used for examining bathymetry due to its lack of absorption by water. The green, red, and NIR bands are used to monitor vegetation (and the red and NIR bands are also used to examine NDVI, as discussed in Chapter 10). The shorter of the SWIR bands is useful when examining the moisture content of plants and soils, while the wavelengths of the longer SWIR band are useful in distinguishing various rock and mineral types. The thermal bands enable the monitoring of heat energy in a variety of settings.

Thinking Critically with Geospatial Technology 11.2

What If There Is No Landsat 9 or Landsat 10?

The Landsat program, which has been providing global coverage since 1972, is the longest running remote sensing program in existence. However, the U.S. remote sensing mission currently relies on the Landsat 7 satellite (launched in 1999 and now impaired since 2003 due to the SLC error) and also on Landsat 8, which was launched in 2013. Despite Landsat 5's successful 28-year tenure, the satellite missions are not intended to last for that long—a Landsat satellite has about 5 years of prime operating shelf life, though the satellites can last longer than that. At present, Landsat 7's mission is only expected to last until somewhere around 2021, when it will run out of fuel.

With only one Landsat satellite launched during the twenty-first century, the long-term future of the Landsat program has been a question. In recent years, federal monies have been appropriated to clear a path for Landsat 9 to be launched by NASA and have its ground operations done by USGS. It takes several years to develop, build, and launch a Landsat satellite, and as of the time of writing (summer 2017), Landsat 9 is being designed with its earliest tentative launch date being the end of 2020. Also at the time of writing, requirements are being

gathered for what Landsat 10 might be and its tentative launch is sometime in the mid-to-late 2020s.

Since there was a 14-year gap between the launches of Landsat 7 and Landsat 8, and, at present, a minimum 7-year stretch between the launch of Landsat 8 and the proposed launch for Landsat 9 (and likely the same amount of time before a Landsat 10 launches), what might happen to data acquisition if there are gaps between successful Landsat satellites? This may sound like a doomsday scenario, but if Landsat 8 develops unfixable problems as Landsat 7 did or if the satellites somehow become unavailable for image acquisition, then what happens?

Will the United States have to purchase imagery from other sources or countries to make up the data gap until another Landsat mission can be launched? Are there other sources that the U.S. government could use to make up for the lack of Landsat imagery? Would the U.S. government be reliant on data from the Sentinel satellites (see page 409) or imagery from other countries? What impact would a lack of continuous Landsat imagery have on the numerous applications and studies previously described?

Each **Landsat scene** measures an area about 170 kilometers long by 183 kilometers wide and is considered a separate image. See **Figure 11.10** for an example taken from the Land Viewer Website which shows a single Landsat scene in relation in the geographic area it actually covers. One Landsat scene provides imagery for a portion of the coastlines of Maryland and Virginia – in order to get imagery of an entire state, you would need multiple Landsat scenes. Thus, if you want access to imagery of a particular area, you need to know which scene that ground location is in. Landsat uses a **Worldwide Reference System** that divides the entire globe into a series of Paths (columns) and Rows. For instance, imagery of Columbus, Ohio, extends from the scene at Path 19, Row 32, while imagery of Toledo would be found in Path 20, Row 31 (**Figure 11.11**).

In an area 170 kilometers long by 183 kilometers wide, a scene will often be much larger than the area of interest—for example, Path 19, Row 32 contains much more than just Columbus; it encompasses the entire middle section of the state of Ohio, so to examine only Columbus, you would

Landsat scene a single image obtained by a Landsat satellite sensor

Worldwide Reference System the global system of Paths and Rows that is used to identify what area on Earth's surface is present in which Landsat scene

FIGURE 11.10

A single Landsat 8 OLI scene (note the size of one scene in relation to coastlines of Virginia and Maryland).

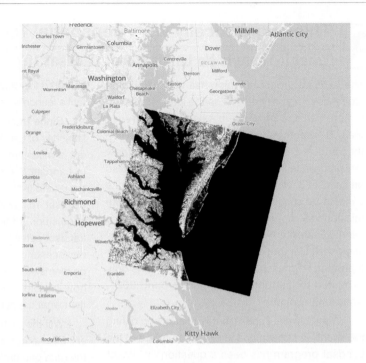

FIGURE 11.11

The total Landsat coverage for the state of Ohio. It requires nine Landsat scenes to cover the entire state.

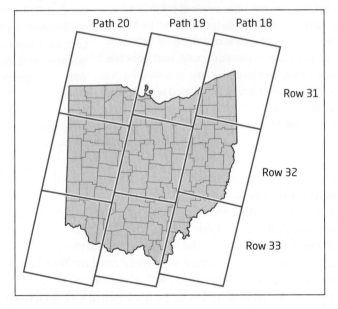

have to work with a subset of the image that matches the city's boundaries. Also, an area of interest may sometimes straddle two Landsat scenes—for instance, Mahoning County in Ohio is located entirely in Path 18, but part of the county is in Row 31 and part of it is in Row 32. In this case, the scenes would need to be merged together and then have the boundaries of the county used to define the subsetted area.

In 2008, the USGS announced that the entire Landsat archive would be made freely available. This USGS decision has allowed free worldwide access to a vast library of remote sensing data (approximately 6 million Landsat scenes) for everyone, which can only help with furthering research and education in remote sensing. The USGS maintains several online utilities for the free distribution of Landsat imagery including Earth Explorer and GloVis Next, while many other non-governmental services (such as Amazon Web Services and EOS Data Analytics) also offer free download of Landsat imagery. See *Hands-On Application 11.2: Viewing Landsat Imagery with LandsatLook* to view Landsat 8 imagery interactively online.

The Landsat program has not been without its issues and challenges. For instance, a Landsat satellite has a prime technology life of five years, but will often continue to be used far beyond that timespan—Landsat 5 was launched in 1984 and its successor, Landsat 6, was not launched until 1993. However, Landsat 6 failed to achieve orbit at launch and the satellite crashed into the Indian Ocean and was lost. This forced the Landsat 5 mission to continue on until the launch of Landsat 7 in 1999.

In 2003, Landsat 7's **SLC** (Scan Line Corrector) failed and could not be repaired. Because of the malfunctioning SLC, the ETM+ produces imagery that's missing certain data. Without a working SLC, an ETM+ image contains only about 75% of the data that should be there (the imagery contains what looks like diagonal black stripes of missing pixels). These gaps in ETM+ imagery are filled by substituting pixels from other Landsat 7 scenes of the same area taken on a different date, as well as by other methods. With

SLC the Scan Line Corrector in the ETM+ sensor; its failure in 2003 caused Landsat 7 ETM+ imagery to not contain all data from a scene

Hands-On Application 11.2

Viewing Landsat Imagery with LandsatLook

The USGS has a second utility for viewing and downloading Landsat imagery called the Landsat-Look Viewer, available at **http://landsatlook.usgs.gov/viewer.html**. To begin, from the set of tools in the upper left corner, choose the magnifying glass (the search tool) and a new box will open, allowing you to type the name of the area you want to view Landsat imagery for and the Viewer will zoom to that location. A pop-up window will allow you to select imagery years, months, and percentage of cloud cover and to also select the sensors you wish to see for imagery (such as Landsat 7 ETM+ and Landsat 8 OLI). By clicking on Show Images, LandsatLook will show you the available Landsat imagery for that location along with a grid denoting the path and row of the available scene.

In the pop-up window, you can slide the bar to view imagery of a specific date or use the arrows to advance between images. You can also adjust the transparency of the Landsat images to see how they match up with the base map.

Expansion Questions

➤ In what Path and Row is your local area located? What Landsat imagery is available via LandsatLook for that Path/Row combination for last month's date?

➤ How many Landsat images are available for your local area through LandsatLook? Are there any large-scale land-cover changes that you can see over time by viewing imagery from past and current years for your local area?

these types of issues emerging in Landsat 7, Landsat 5 continued to be used as a primary source of imagery. Landsat 5 continued collecting imagery (with many mechanical issues along the way) until it finally ended its mission in 2012 and was shut down in 2013. Landsat 5 was not likely intended to be the cornerstone of the program, but that's what it became, with its mission lasting for 28 years.

There's a strong, ongoing need for Landsat imagery that monitors large-scale phenomena, such as environmental effects of land-use changes and urban expansion. Long-term gaps in available current data and between the dates of imagery would have a huge negative impact on these types of studies. Being able to monitor large-scale conditions on Earth on a regular basis at 30-meter resolution with a variety of multispectral bands (including near-infrared, middle-infrared, and thermal characteristics) gives users a rich dataset for numerous applications.

With Landsat imagery stretching back for more than 45 years and providing a constant source of observations, long-term tracking of environmental phenomena (such as ongoing land-use or land-cover changes over a particular area) can be performed. For instance, the growth of urban areas over time can be observed every 16 days (cloud cover permitting, of course), enabling planners to determine the spatial dimensions of new developments across a region. **Figure 11.12** shows an example of utilizing

FIGURE 11.12 ⟶○

Multiple dates of Landsat imagery of Rondonia, Brazil, showing the forested areas (in dark green) and the growing patterns of deforestation over time (in lighter shades). *[NASA/Goddard Space Flight Center Scientific Visualization]*

Landsat imagery over time to track the patterns of deforestation in Brazil through examination of clear-cut areas and new urban areas built up in the forest. The color composite images show the rain forest in dark green shades, while the deep-cut patterns of forest clearing are clearly visible and the encroachment of development into the forest can be tracked and measured over time.

Beyond examining land-cover changes over time, the multispectral capabilities of Landsat imagery can be used for everything from measuring drought conditions and their effects on crops, to glacier features, to the aftermath of natural disasters. By examining the different band combinations provided by Landsat imagery, scientists can find clues to understanding the spread of phytoplankton in coastal waters, enabling environmentalists to try to head off potential algae blooms or red-tide conditions. Threats to animal habitats can be assessed through forest and wetland monitoring via Landsat imagery. Landsat images also allow scientists to track the movements of invasive species that threaten timber industries. *Hands-On Application 11.3: Google Earth Engine and Applications of Landsat Imagery* lets you dig further into the myriad uses of Landsat remote sensing.

Hands-On Application 11.3

Google Earth Engine and Applications of Landsat Imagery

Landsat imagery has been adopted for numerous environmental monitoring, measuring, and tracking applications. Google Earth Engine (which, despite the name, is a separate entity from regular "Google Earth") is a Web resource that will allow you to work with Landsat imagery applications. To get started, go to **https://earthengine.google.com**. Earth Engine has a wide variety of analytical tools for use with satellite imagery (such as image classification) but you need to apply to Google directly for access to these features. Despite that, Earth Engine has a number of useful items that are publicly available. One of them is the Timelapse feature—click on the Timelapse link in the upper right portion of the screen. This one part shows various places around Earth (such as Las Vegas, Dubai, or the Columbia Glacier) in a series of Landsat images between 1984 and 2016. You can pause the Timelapse imagery and also slowly step through it to view changes over time. You can also select the Search

for places option to type a location and see time-lapse Landsat imagery of that place from the last several years.

Also, click on the Case Studies link in the upper right corner and you'll be brought to a page with further discussion of the applications of Landsat (and at times other satellites') imagery (on topics such as global forest cover change, malaria mapping, and predicting locations of at-risk species) as well as several videos showcasing these topics.

Expansion Questions

➤ Using the Timelapse feature, where has growth been occurring in Las Vegas, Dubai, and your own local area as viewed by Landsat imagery from 1984 to 2016?

➤ Select one of the Case Studies presented on the Website. How has satellite imagery been used for this type of application?

Sentinel a satellite program operated by the European Space Agency

Other programs exist with similar capabilities to Landsat. For instance, the **Sentinel** program, operated by the European Space Agency and part of the Copernicus program, is a new series of satellites designed to acquire various types of remote sensing imagery. Each type of Sentinel satellite has different capabilities—for instance, the Sentinel-1 series of satellites use radar instruments for remote sensing while the Sentinel-3 series of satellites are geared toward remote sensing related to ocean topography and sea surface temperature.

Sentinel-2 a Sentinel mission, consisting of two satellites with characteristics similar to Landsat

MSI Multispectral Instrument, the multispectral sensor onboard the Sentinel-2 satellites

The **Sentinel-2** satellite missions are similar in capabilities to Landsat. At the time of writing, both Sentinel-2A and Sentinel-2B were in orbit with additional satellites (Sentinel-2C and Sentinel-2D) being readied. Sentinel-2 satellites operate with a 10-day revisit cycle and collect multispectral imagery using their **MSI** (Multispectral Instrument) sensor in 13 bands. MSI bands include visible, NIR, and SWIR and Cirrus bands similar to Landsat 8 OLI, but also have unique red-edge/vegetation detection and water vapor bands.

Sentinel-2 imagery is available in 10m, 20m, or 60m resolutions, depending on the bands being sensed (see **Table 11.2** for further detail about Sentinel-2 bands and their resolutions) with 12-bit radiometric resolution. While you can download Sentinel-2 imagery from the European

TABLE 11.2 The bands sensed by MSI onboard the Sentinel-2 series of satellites and the central wavelength of each band (all wavelength values are listed in micrometers, μm).

Band	Central Wavelength	Spatial Resolution
1	Coastal Aerosol (0.443)	60 m
2	Blue (0.490)	10 m
3	Green (0.560)	10 m
4	Red (0.665)	10 m
5	Red Edge (0.705)	20 m
6	Red Edge (0.740)	20 m
7	Red Edge (0.783)	20 m
8	NIR (0.842)	10 m
8a	NIR (0.865)	20 m
9	Water vapor (0.945)	60 m
10	Cirrus (1.375)	60 m
11	SWIR (1.610)	20 m
12	SWIR (2.190)	20 m

Hands-On Application 11.4

Viewing Sentinel-2 Imagery Online

A very useful online tool for viewing Sentinel-2 imagery is Land Viewer, available at **https://lv.eosda.com**. With Land Viewer, you'll be able to examine individual Sentinel-2 scenes and Landsat 8 scenes as well. To maximize your use of Land Viewer, you should first create a free account on the Website. When Land Viewer opens, click the option for Sentinel-2, then in the search box at the left of the screen, type a location to see imagery of, for instance, Miami, Florida. From there, several options for Sentinel-2 imagery will be available for you to view. Choose the one that's the most cloud-free and it will load in the main screen (the visible bands of Sentinel-2's MSI instrument have 10-meter resolution, therefore the image should be fairly crisp when compared to a Landsat image).

By choosing the Band Combinations button on the right (the two squares stacked on top of each other) you will be able to view your chosen Sentinel-2 image with different bands and color composite choices other than the default true color composite (see Table 11.2 on page 410 for a description of which parts of the electromagnetic spectrum each Sentinel-2 MSI band refers). Try Color Infrared (vegetation), Land/Water, and False Color (Urban) for some different composite options to view. Also try selecting a Landsat 8 OLI image for the same area to compare to the Sentinel-2 MSI imagery.

Expansion Questions

➤ How do each of the other band combination options (Color Infrared vegetation, Land/Water, and False Color urban) compare to the default Natural Color option in terms of what details you can discern from the image of Miami as well as how features are highlighted?

➤ How does a Sentinel-2 MSI Natural Color scene compare with a Landsat 8 Natural Color and Landsat 8 Pan-Sharpened scene in terms of detail?

Copernicus Open Access Hub (**https://scihub.copernicus.eu**) the USGS also archives and distributes Sentinel-2 imagery in conjunction with Landsat imagery. To take a look at Sentinel-2 imagery, see *Hands-On Application 11.4: Viewing Sentinel-2 Imagery Online*.

Although Landsat data has been widely used for numerous environmental studies, 30-meter resolution isn't nearly enough to do very detailed, high-resolution studies. At 30-meter resolution, broad land-use change modeling and studies can be performed, but a much finer spatial resolution would be required to pick out individual details in an image. High-resolution sensors have a different purpose than Landsat imagery—with a very fine spatial resolution, crisp details can emerge from a scene. Spy satellites or military surveillance satellites require this type of finer spatial resolution, but of course the imagery they acquire remains classified. Much of the non-classified and available high-resolution satellite imagery is obtained and distributed commercially rather than from government sources.

What Satellites Have High-Resolution Sensors?

There are several remote sensing satellites with high-resolution capabilities; this section describes a few of the more prominent ones. Much of the available high-resolution satellite imagery is distributed and sold by private companies from the satellites that they own and operate. For instance, the high-resolution satellite imagery seen in some images on Google Earth comes from a commercial provider. One thing to keep in mind is that prior to June 2014, U.S. government regulations did not permit the sale of imagery of less than 0.5-meter resolution to commercial customers (so imagery of finer resolution was resampled to this level before it was distributed). Commercial imagery after this date was permitted to be sold at its regular resolution (which is often much finer than 0.5 meters).

A common characteristic of satellites with high-resolution capabilities is a panchromatic band that allows pan-sharpening of other multispectral bands to provide very crisp and detailed image resolution—details of urban environments can be distinguished with little difficulty, although details of objects that are smaller than its resolution cannot be detected. Another common characteristic of these types of satellites is the ability to perform off-nadir viewing for satellite **tasking**, whereby the sensor can be directed on-demand to take imagery of specified locations up to several orbital paths away.

Two pioneering commercial launches for high-resolution satellite imagery were the **IKONOS** (launched in 1999) and **QuickBird** (launched in 2001) satellites. Both of these satellites have been decommissioned in recent years but remain cornerstones of early high-resolution commercial imagery. Both satellites featured off-nadir capabilities to reduce their revisit time to 1–3 days and a four-band (blue, green, red, and near-infrared) multispectral sensor with very high radiometric resolution (11 bits). The **IKONOS** sensors were capable of spatial resolution of 0.8-meter panchromatic and 3.2-meter multispectral imagery while QuickBird collected panchromatic imagery of 0.61-meter panchromatic imagery and 2.4-meter resolution multispectral imagery.

Improvements to the spatial resolution of commercial satellite imagery continued. In the United States, the premier commercial high-resolution satellite imagery provider is DigitalGlobe, which operates the WorldView series of satellites. In 2007, DigitalGlobe launched **WorldView-1**, which featured a single sensor, a panchromatic one, with 0.5-meter spatial resolution. The **GeoEye-1** satellite (also operated by DigitalGlobe), launched in 2008, has two sensors—a four-band multispectral (blue, green, red, and near-infrared) sensor with 1.65-meter spatial resolution and also a panchromatic sensor with 0.46-meter spatial resolution. In 2009, **WorldView-2**

tasking the ability to direct the sensors of a satellite on demand to collect imagery of a specific area

IKONOS a satellite launched in 1999 that features sensors with a multispectral spatial resolution of 3.2 meters and panchromatic resolution of 0.8 meters

QuickBird a satellite launched in 2001 by DigitalGlobe with sensors that have 2.4-meter multispectral resolution and 0.61-meter panchromatic resolution

WorldView-1 a satellite launched in 2007 by DigitalGlobe whose panchromatic sensor has 0.5-meter spatial resolution

GeoEye-1 a satellite launched in 2008 that features a panchromatic sensor with a spatial resolution of 0.46 meters

WorldView-2 a satellite launched in 2009 that features an eight-band multispectral sensor with 1.84-meter resolution and a panchromatic sensor of 0.46-meter resolution was launched and featured 0.46-meter panchromatic and 1.84-meter eight-band multispectral spatial resolution.

The highest spatial resolution imagery currently commercially available is from the sensors onboard **WorldView-3,** launched in 2014, which provides 0.31-meter panchromatic and 1.24-meter eight-band multispectral resolution. WorldView-3 also contains advanced multispectral capabilities with 8 additional shortwave infrared bands and an additional 12 CAVIS bands for monitoring clouds, aerosols, and water. This spatial resolution represents the finest commercial resolution on the market today, currently making World-View-3 the satellite with the sensors to beat when it comes to high-resolution imagery. (See **Figure 11.13** and **Figure 11.14** for examples of WorldView-3 imagery that demonstrates the kind of spatial resolution this sensor can provide.)

Launched by DigitalGlobe at the end of 2016, **WorldView-4** features both a multispectral sensor with spatial resolution of 1.24 meters and a panchromatic sensor with 0.31-meter resolution. At the time of writing, WorldView-4 was the most recent high-resolution satellite launch from DigitalGlobe.

Another example of high-resolution, commercially available imagery is the **SkySat** program, operated by Terra Bella. At present, seven SkySat satellites have been launched, with an additional 13 SkySats planned to be launched in the near future. A SkySat satellite features a panchromatic sensor with 0.9-meter resolution and a multispectral sensor with 2-meter resolution. The SkySat satellites contain video cameras capable of recording up to 90 seconds of video as well.

WorldView-3 a satellite launched in 2014 by DigitalGlobe that features shortwave infrared and CAVIS bands, plus a panchromatic sensor of 0.31-meter resolution and eight multispectral bands with 1.24-meter resolution

WorldView-4 a satellite launched in 2016 that features a panchromatic sensor of 0.31-meter resolution and 4 multispectral bands with 1.24-meter resolution

SkySat a high-resolution, small satellite program operated by Terra Bella

FIGURE 11.13

A WorldView-3 image of Big Pine Key (in the Florida Keys) from September 2017 after Hurricane Irma passed through the Keys.
[DigitalGlobe/ScapeWare3d/Getty Images]

FIGURE 11.14

A WorldView-3 image of Bay Drive (east of Key West, FL) from September 2017 after Hurricane Irma passed through. *[DigitalGlobe/ ScapeWare3d/Getty Images]*

small satellite a type of satellite that weighs between 220 and 1100 pounds

nanosatellite a type of satellite that weighs between 2.2 and 22 pounds

CubeSat a type of satellite with a weight of less than 3 pounds and a length of about 4 inches

Dove the name for the CubeSat series launched by Planet Labs

Flock-1 a constellation of high-resolution remote sensing CubeSats launched by Planet Labs in 2014

SPOT a high-resolution imagery satellite program operated by Airbus Defence and Space

A SkySat satellite is an example of a **small satellite,** a type of remote sensing satellite that is simpler in design and smaller in size (it weighs between 220 and 1100 pounds) as well as cheaper to produce. With these kinds of attributes, small satellites such as SkySat are becoming more common and more commercially viable to put in orbit.

Even smaller remote sensing satellites, such as **nanosatellites** (which weigh between 2.2 and 22 pounds), are being designed and launched today. The **CubeSat** design is a version of these nanosatellites, where a simple box-like remote sensing satellite weighing less than 3 pounds and measuring about 4 inches long is placed into orbit and used for capturing high-resolution imagery.

Planet Labs has designed the **Dove,** a CubeSat with panchromatic and visible sensors that can provide imagery of 3- to 5-meter resolution. While a single CubeSat would take a very long time to obtain imagery from everywhere on Earth, Planet Labs launches Doves in groups, referred to as "flocks." **Flock-1** (see **Figure 11.15**) was the first group of Dove satellites (released into orbit from the International Space Station in 2014) set up as constellation around Earth. Further Flock series of groups of Dove CubeSats have followed the launch of Flock-1, with each being a group of Dove CubeSats (for instance, Flock-2p consisted of 12 Dove satellites). At the time of writing, multiple Flocks have been launched by Planet Labs, including the 2017 launch of Flock-3p, which consisted of 88 Doves.

A different satellite program, **SPOT** (Satellite Pour l'Observation de la Terre), is currently owned and operated by Airbus Defence and Space. The SPOT series of satellites was initiated by CNES, the French space agency,

FIGURE 11.15

The Dove CubeSat satellites of Flock-1 being released from the International Space Station into Earth orbit.
[Source: Image courtesy of NASA]

and developed by Swedish and Belgian space agencies with the first satellite (SPOT 1) launched in 1986.

Airbus Defence and Space developed SPOT 6 and SPOT 7, launched in 2012 and 2014, respectively. SPOT 6 and SPOT 7 feature improved resolution capabilities (including a 1.5-meter panchromatic spatial resolution and 6-meter resolution for four multispectral bands) as well as the stereo capability for production of digital elevation models (see Chapter 13). At the end of 2014, Airbus sold SPOT 7 to the Azerbaijan National Aerospace Agency, which renamed it Azersky.

A related Airbus program is the **Pléiades** series of satellites (currently consisting of Pléiades-1A and Pléiades-1B), the first of which launched in 2011. Pléiades-1A and Pléiades-1B feature very high-resolution panchromatic (0.5-meter spatial resolution) and multispectral (2-meter spatial resolution of four bands, which can then be pan-sharpened to 0.5-meters) capabilities (see **Figure 11.16** for an example of Pléiades-1A's high-resolution imagery).

Pléiades a series of satellites operated by Airbus Defence and Space, which feature sensors with a multispectral spatial resolution of 2 meters and panchromatic resolution of 0.50 meters

How Can Satellites Be Used for Monitoring?

One of the advantages of the SPOT 6 and SPOT 7 (Azersky) satellites and the Pléiades series is that by having several satellites with similar capabilities in orbit, the amount of revisit time it takes for a location on the ground

FIGURE 11.16

Pléiades-1A images of Tacloban, Philippines, showing the area before and after the Haiyan typhoon in 2013. *[dpa picture alliance / Alamy Stock Photo]*

RapidEye a constellation of five satellites operated by Planet Labs that feature sensors with 5-meter multispectral sensors and a daily global revisit

to be seen by one of them is greatly reduced. For instance, Pléiades-1A can view a location on Day 1, then Pléiades-1B can image the area on Day 2. Having multiple satellites (such as the Pléiades twins) improves the overall temporal resolution of the satellites by being able to deliver imagery on a much faster basis.

This notion of using a constellation of satellites to be able to collect imagery of areas swiftly greatly enhances the ability to monitor areas on the ground. An example of this kind of high-resolution and quick-repeat capability is the **RapidEye** constellation owned by Planet Labs. RapidEye is a series of five satellites with identical capabilities (5-meter resolution in five multispectral bands). With five satellites in orbit (see **Figure 11.17**), RapidEye can deliver daily imagery of anywhere on Earth. With such high-resolution sensors and a daily imagery capability, RapidEye is an excellent example of using remote sensing as a monitoring tool for environmental conditions such as land use or forestry. With the RapidEye satellites combined with their constellations of Dove satellites, Planet Labs has the capability of daily imaging of the entire planet.

The satellites operated by DigitalGlobe, Terra Bella, Planet Labs, and Airbus Defence and Space are only a few prominent examples of high-resolution imagery sources. Other companies and governments have launched or are planning to launch their own satellites with similar capabilities. See *Hands-On Application 11.5: Viewing High-Resolution Satellite Imagery* for examples of all of these high-resolution images online.

FIGURE 11.17

The RapidEye
constellation of satellites
in orbit. *[BlackBridge]*

Using satellite imagery as a monitoring device has become more common now with the ever-increasing number of satellites and space agencies across the planet. Having high-resolution satellites that can be tasked to view specific off-nadir locations allows end users the capacity to capture imagery of events around the world rapidly, especially natural disasters such as floods, fires, tsunamis, or landslides. The **International Charter** "Space and Major Disasters" was put in place by the international community for just this purpose. The International Charter is an agreement among numerous government and commercial space agencies from around the world

International Charter an agreement among multiple international space agencies and companies to provide satellite imagery and resources to help in times of disaster

Hands-On Application 11.5

Viewing High-Resolution Satellite Imagery

While the data acquired by satellites such as WorldView-3 and Pléiades-1B is sold commercially, you can view samples of these satellite images at several places online. The Satellite Imaging Corporation's Website provides a means for viewing high-resolution imagery. Open your Web browser and go to **www.satimagingcorp.com**. Select the option for Gallery. From the next Web page, your available options include GeoEye-1, IKONOS, Pléiades-1A, Pléiades-1B, RapidEye, QuickBird, SkySat-1, SkySat-2, SPOT 6, SPOT 7, WorldView-1, WorldView-2, WorldView-3, and WorldView-4. Examine some examples of each sensor, keeping in mind the specifications of

each. You can zoom in on the sample imagery from each satellite to get a real sense of what can be seen via high-spatial-resolution satellites. Note that the images available on the Website are just crisp graphics, not the actual satellite data itself.

Expansion Questions

➤ For what kinds of analysis or interpretation are these high-resolution images being used?

➤ What current events have samples of high-resolution images available for viewing through the Website (and which satellite sensors acquired them)?

(including the United States, Canada, Russia, China, India, and many more) to provide satellite imagery and resources to aid in the event of a disaster emergency that activates the charter.

Utilizing satellite imagery for disaster management has extended throughout the international community. For instance, **UN-SPIDER** (United Nations Space-Based Information for Disaster Emergency and Response) is a program established in 2006 by the United Nations as a means to help get satellite imagery and other geospatial resources into the hands of disaster management personnel who need them to help aid in disaster mitigation and recovery.

An additional way of using satellite imagery resources in an emergency is to get them into the hands of everyone so that people can use their own geographic knowledge to help, regardless of where they are in the world. The notion of using satellite imagery to help crowdsource (see Chapter 1) disaster recovery is demonstrated by the Tomnod program set up by DigitalGlobe (the word "tomnod" means "big eye" in Mongolian). Through **Tomnod,** DigitalGlobe makes high-resolution satellite imagery available online and gives end users the ability to examine and make edits to the data to offer their own insights. For instance, when Malaysian Airlines Flight MH370 went missing in 2014, DigitalGlobe posted satellite imagery of the suspected areas where the plane may have gone down to Tomnod. Over two million people examined the hundreds of thousands of satellite images, marking things that they thought might have been evidence of the missing plane. Tomnod is active today with a variety of questions that rely on your time and expertise with satellite imagery to help answer (see *Hands-On Application 11.6: Crowdsourcing Satellite Imagery*).

UN-SPIDER the United Nations Space-Based Information for Disaster Emergency and Response program

Tomnod an online crowdsourcing initiative that applies satellite imagery for a variety of real-world issues

---

Hands-on Application 11.6

Crowdsourcing Satellite Imagery

The search for the missing Malaysian Airlines Flight MH370 is perhaps the most publicized usage of Tomnod, but it's far from the only thing that can be done with it. DigitalGlobe uses Tomnod to involve people from all over the globe in applying their time and knowledge to a variety of different global events. To check some of these out (or to even volunteer your own geographic knowledge), go to **www.tomnod.com**. On the Website, you'll see different ongoing activities for which Tomnod is being used. One example at the time of writing allowed users to view satellite imagery from Peru to help locate and preserve archeological sites. Take a look at some of the current and ongoing Tomnod campaigns.

Expansion Questions

➤ What kinds of current activities is Tomnod being used for to crowdsource information through the use of satellite imagery?

➤ What are some of the tips provided by Tomnod to help in using satellite imagery to identify various features of the campaign?

Chapter Wrap-Up

Satellite remote sensing is a powerful tool for everything from covert surveillance to the monitoring of land use and land cover. The next chapter delves into a new set of geospatial satellite tools—the Earth Observing System, a series of orbital environmental satellites that continuously gather data about global processes. *Geospatial Lab Application 11.1* returns to the subject of using MultiSpec, and it has you doing more with Landsat 8 imagery and the multispectral capabilities of the OLI sensor. Before you get to the lab, check out the resources in *Satellite Imagery Apps* for some representative mobile apps as well as *Satellite Imagery in Social Media* for Facebook pages and Twitter accounts related to many of this chapter's concepts as well as a few YouTube videos that show off various satellite remote sensing applications.

Important note: The references for this chapter are part of the online companion for this book and can be found at **www.macmillanlearning. com/catalog/shellitoigt4e**.

Satellite Imagery Apps

Here's a sampling of available representative satellite imagery apps for your smartphone or tablet. Note that some apps are for Android, some are for Apple iOS, and some may be available for both.

➤ **DigitalGlobe FirstLook:** An app designed for examining satellite imagery of events before and after emergency situations

➤ **NASA Earth as Art:** An app for viewing stunning imagery from Landsat satellites and other NASA remote sensing satellites

Satellite Imagery in Social Media

Here is a sampling of available social media outlets related to this chapter's concepts.

f On Facebook, check out the following:

➤ **European Space Imaging:**
   www.facebook.com/europeanspaceimaging

➤ **DigitalGlobe:**
   www.facebook.com/DigitalGlobeInc

> ➤ Landsat:
>   www.facebook.com/NASA.Landsat
> ➤ Planet Labs:
>   www.facebook.com/PlanetLabs
> ➤ Terra Bella:
>   https://www.facebook.com/terrabellatech
> ➤ Tomnod:
>   www.facebook.com/Tomnod
> ➤ UN-SPIDER:
>   www.facebook.com/UNSPIDER

Become a Twitter follower of:

> ➤ **Airbus Defence and Space:** @AirbusDS_GEO
> ➤ **DigitalGlobe:** @digitalglobe
> ➤ **European Space Agency—Earth Observation:** @ESA_EO
> ➤ **International Charter:** @DisastersChart
> ➤ **NASA updates on Landsat:** @NASA_Landsat
> ➤ **Planet Labs:** @planetlabs
> ➤ **Remote Pixel:** @RemotePixel
> ➤ **Surrey Satellites:** @SurreySat
> ➤ **Terra Bella:** @terra_bella
> ➤ **Tomnod:** @tomnod
> ➤ **UN-SPIDER:** @UN_SPIDER
> ➤ **USGS updates on Landsat:** @USGSLandsat
> ➤ **WorldView-4:** @WorldView04

You Tube On YouTube, watch the following videos:

> ➤ **A Planetary Perspective: With Landsat and Google Earth Engine** (a Google video featuring Landsat and its many applications): **www.youtube.com/watch?v=Ezn1ne2Fj6Y**
> ➤ **Amazon Deforestation: Timelapse** (a Google video showing how Landsat imagery is used to track deforestation in the Amazon over time): **www.youtube.com/watch?v=oBIA0lqfcN4**
> ➤ **DigitalGlobe** (DigitalGlobe's YouTube channel showing several different videos about their various satellites): **www.youtube.com/user/digitalglobeinc**

➤ **GeoEye-1 Imagery Collection Animation** (a video showing a simulation of how Geo-Eye-1 collects imagery): **www.youtube.com/ watch?v=yLZR5pg8Amw**

➤ **NASA · Landsat's Orbit** (a NASA video showing the orbital path and 16-day revisit cycle of Landsat): **www.youtube.com/watch?v= P-lbujsVa2M**

➤ **NASA · Landsat 8 Celebrates First Year in Orbit** (a NASA video about Landsat 8 and its imagery): **www.youtube.com/ watch?v=0XVnQEJ6w_Y**

➤ **NASA · TIRS: The Thermal InfraRed Sensor on LDCM** (a NASA video showing the applications of the new thermal instrument onboard LDCM): **www.youtube.com/watch?v=c6OmfYEgzA0**

➤ **NASA · What Doesn't Stay in Vegas? Sprawl** (a NASA video showing the growth of Las Vegas from 1972 to 2010 using Landsat imagery): **www.youtube.com/watch?v=xFzdyxwx50M**

➤ **Planet** (a YouTube channel featuring imagery and examples from Planet Labs series of Dove satellites): **www.youtube.com/channel/ UCiyS6coRsDQ17b3Kj9bl8jw**

➤ **Terra Bella** (a YouTube channel showing off videos and imagery captured by the SkySat satellites): **www.youtube.com/channel/ UC86JFANJdJEbApS6jS0iPhg**

➤ **The View from Space: Landsat's Role in Tracking Forty Years of Global Changes** (a USGS video featuring many Landsat applications): **www.youtube.com/watch?v=IJ2VWKlraSw**

Key Terms

across-track scanning (p. 397)
along-track scanning (p. 397)
Corona (p. 394)
CubeSat (p. 414)
Dove (p. 414)
EROS (p. 398)
ETM+ (p. 402)
Flock-1 (p. 414)
GeoEye-1 (p. 412)
geostationary orbit (p. 395)
hyperspectral sensor (p. 399)

IKONOS (p. 412)
International Charter (p. 417)
Landsat (p. 401)
Landsat 7 (p. 402)
Landsat 8 (p. 402)
Landsat scene (p. 405)
MSI (p. 410)
MSS (p. 402)
multispectral sensor (p. 399)
nanosatellite (p. 414)
NASA (p. 394)

Landsat 8 Imagery

This chapter's lab application builds on the remote sensing basics of the previous chapter and returns to using the MultiSpec program. In this exercise, you'll be starting with a Landsat 8 scene and creating a subset of it with which to work. During the lab, you'll examine the uses for several Landsat 8 band combinations in remote sensing analysis.

Objectives

The goals for you to take away from this exercise are:

➤ To familiarize yourself further and work with satellite imagery in MultiSpec

➤ To create a subset image of a Landsat 8 scene

➤ To examine different Landsat 8 bands in composites and compare the results

➤ To examine various landscape features in multiple Landsat 8 bands and compare them

➤ To apply visual image interpretation techniques to Landsat 8 imagery

Using Geospatial Technologies

The concepts you'll be working with in this lab are used in a variety of real-world applications, including:

➤ Research for the World Wildlife Federation, which utilizes Landsat imagery to study deforestation patterns and their impact on climate and species

➤ Environmental sciences, which have utilized the multispectral characteristics of Landsat imagery to map the locations of coral reefs around the world

[Max Milligan/AWL Images RM/Getty Images]

Obtaining Software

The current version of MultiSpec (3.6) is available for free download at **https://engineering.purdue.edu/~biehl/MultiSpec**.

Important note: Software and online resources can change fast. This lab was designed with the most recently available version of the software at the time of writing. However, if the software or Websites have changed significantly between then and now, an updated version of this lab (using the newest versions) will be available online at **www.macmillanlearning.com/catalog/shellitoigt4e**.

Lab Data

Copy the folder **Chapter 11**—it contains a Landsat 8 OLI/TIRS satellite image (called "landsat8sept272014.img") from 9/27/2014 of northern Ohio, which was constructed from data supplied via LandsatLook. The Landsat 8 image bands (bands 1–9 are sensed by OLI, while bands 10 and 11 are sensed by TIRS) refer to the following portions of the electromagnetic (EM) spectrum in micrometers (µm):

➤ Band 1: Coastal (0.43 to 0.45 µm)

➤ Band 2: Blue (0.45 to 0.51 µm)

➤ Band 3: Green (0.53 to 0.59 µm)

➤ Band 4: Red (0.64 to 0.67 µm)

➤ Band 5: Near infrared (0.85 to 0.88 µm)

➤ Band 6: Shortwave infrared 1 (1.57 to 1.65 μm)

➤ Band 7: Shortwave infrared 2 (2.11 to 2.29 μm)

➤ Band 8: Panchromatic (0.50 to 0.68 μm)

➤ Band 9: Cirrus (1.36 to 1.38 μm)

➤ Band 10: Thermal infrared 1 (10.60 to 11.19 μm)

➤ Band 11: Thermal infrared 2 (11.50 to 12.51 μm)

Keep in mind that Landsat 8 imagery has a 30-meter spatial resolution (except for the panchromatic band, which is 15 meters).

Localizing This Lab

This lab focuses on a section of a Landsat scene from northeast Ohio. Landsat imagery is available for free download via LandsatLook at **http://landsatlook.usgs.gov**. This site will provide raw data that will have to be imported into MultiSpec and processed to use in the program. Information is available online for using this free Landsat data in MultiSpec at **https://engineering.purdue.edu/~biehl/MultiSpec/tutorials/MultiSpec_Tutorial_5.pdf**.

11.1 Opening Landsat Scenes in MultiSpec

1. Start **MultiSpec**.

2. MultiSpec will start with an empty text box, which will give you updates and reports of processes in the program. You can minimize the text box for now.

3. To get started with the Landsat image, select **Open Image** from the **File** pull-down menu. Alternatively, you can select the **Open** icon from the toolbar:

4. Navigate to the **Chapter 11** folder and select **landsat8sept272014** as the file to open, then click **Open**.

5. A new dialog box will appear to let you set the display specification for the landsat8sept272014 image.

6. Under **Channels**, you will see the three color guns available to you (red, green, and blue). Each color gun can hold one band (see the *Lab Data* section for information on which bands correspond to which parts of the electromagnetic spectrum). The number listed next to each color gun represents the band being displayed with that gun.

7. Display **band 5** in the red color gun, **band 4** in the green color gun, and **band 3** in the blue color gun.

8. Accept the other defaults for now and click **OK**.

9. A new window will appear, and the landsat8sept272014 image will load in it.

11.2 Using Landsat 8 Imagery Bands

The Landsat OLI/TIRS image has several different bands (see Table 11.1 for which band represents which wavelengths), each with their own use. For instance, looking at the entire Landsat scene now (the 5-4-3) combination, you've got a broad overview of a large slice of northern Ohio using the near-infrared, red, and green bands.

1. Vegetated areas (such as grass or trees) are reflecting a lot of near-infrared light in the red color gun, causing those areas to appear in shades of red. However, there are a lot of other things in the image as

well. Examine the Landsat scene, zooming in on some of the cyan areas on the lakeshore of Lake Erie, then answer Question 11.1.

> **Question 11.1** The features on the image in cyan are largely urbanized and developed areas. Why are they displayed in cyan on this image with the 5-4-3 band combination?

2. Open the **landsat8sept272014** image again—but this time use **band 10** in the red color gun, **band 10** again in the green color gun, and also **band 10** once more in the blue color gun. This will show band 10 in all three guns, so you will only see this band in grayscale. This version of the landsat8sept272014 image will load in a separate window.

3. Arrange the two windows (landsat8sept272014 in the 5-4-3 combination and landsat8sept272014 in the 10-10-10 combination) side by side to see both of them together. Keep in mind that band 10 in the Landsat 8 imagery is one of the thermal bands sensed by TIRS. Examine both of the Landsat scenes and answer Question 11.2.

> **Question 11.2** What do the brighter places on the 10-10-10 image correspond with? Why do these places mostly appear brighter than their surroundings in the 10-10-10 image?

4. Close the 10-10-10 version of **landsat8sept272014**.

5. Open the **landsat8sept272014** image again, this time using a 9-9-9 combination (i.e., load the image with band 9 in the red gun, band 9 in the green gun, and band 9 in the blue gun). A new window will open with this image, which will be band 9 in grayscale. Place this 9-9-9 image side by side with your original 5-4-3 image.

6. Band 9 in the Landsat 8 imagery is designed for detecting cirrus clouds in imagery. Answer Question 11.3.

> **Question 11.3** Where are the cirrus clouds in this section of northern Ohio in the image? Why are they so hard to see in the regular 5-4-3 image?

7. Close the 9-9-9 image when you're done so you're only working with the 5-4-3 image.

11.3 Subsetting Images and Examining Landsat Imagery

Right now, you're working with an entire Landsat 8 scene, which is an area roughly 170 kilometers long by 183 kilometers wide (as shown in the graphic below—you are using the scene encompassed by path 19, row 31). For this lab, we want to focus only on the area surrounding downtown Cleveland. You will have to create a subset—in essence, "clipping" out the area that you're interested in and creating a new image from that.

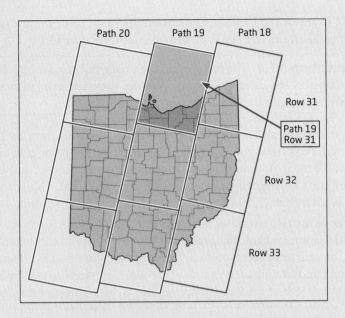

1. Zoom to the part of the **landsat8sept272014** image that shows Cleveland (as in the following graphic):

2. In the image, you should be able to see many features that make up downtown Cleveland—the waterfront area, a lot of urban development, major roads, and water features.

3. In order to create a new image that shows only Cleveland (a suggested region is shown in the graphic above), select **Reformat** from the **Processor** pull-down menu, then select **Change Image File Format**.

4. You can draw a box around the area you want to subset using the cursor, and the new image that's created will have the boundaries of the box you've drawn on the screen. However, for the sake of consistency in this exercise, use the following values for the Area to Reformat:

   a. Lines:

      - Start 8977

      - End 10729

      - Interval 1

   b. Columns:

      - Start 9193

      - End 11161

      - Interval 1

   Leave the other defaults alone and click **OK**.

5. In the **Save As** dialog box that appears, save this new image in the **Chapter 11** folder (call it **clevsub.img**). Choose **Multispectral** for the **Save as Type** option (from the pull-down menu options next to **Save as Type**). When you're ready, click **Save**.

6. Back in MultiSpec, minimize the window containing the landsat8sept272014 image.

7. Open the **clevsub** image you just created in a new window (in the **Open** dialog box, you may have to change the **Files of Type** that it's asking about to "All Files" to be able to select the **clevsub** image option).

8. Open the **clevsub** image with a 5-4-3 combination (band 5 in the red gun, band 4 in the green gun, and band 3 in the blue gun).

9. Use the other defaults for the **Enhancement** options (stretch is Linear and range is Clip 2% of tails).

10. In the **Set Histogram Specifications** dialog box that opens, select the **Compute new histogram** method, and use the default **Area to Histogram** settings.

11. Click **OK** when you're ready. The new subset image shows that the Cleveland area is ready to use.

12. Zoom in on the downtown Cleveland area, especially the areas along the waterfront. Answer Questions 11.4 and 11.5. (You will want to also open Google Earth and compare the Landsat image to its very crisp resolution imagery when you answer these questions.)

> **Question 11.4** What kinds of features on the Cleveland waterfront cannot be distinguished at the 30-meter resolution you're examining?

> **Question 11.5** Conversely, what specific features on the Cleveland waterfront are apparent at the 30-meter resolution you're examining?

11.4 Examining Landsat Bands and Band Combinations

1. Zoom in on the waterfront area, looking at FirstEnergy Stadium, home to the Cleveland Browns, and its immediate surrounding area.

2. Open another version of the clevsub image using the 4-3-2 combination.

3. Arrange the two windows on the screen so that you can examine them together, expanding and zooming in as needed to be able to view the stadium in both three-band combinations.

> **Question 11.6** Which one of the two-band combinations best brought the stadium and its field to prominence?

> **Question 11.7** Why did this band combination best help in viewing the stadium and the field? (*Hint:* You may want to do some brief online research into the nature of the new stadium and its field.)

4. Return to the view of Cleveland's waterfront area. You'll now be examining the water features (particularly Lake Erie and the river). Paying careful attention to the water features, open four display windows, then expand and arrange them side by side to look at

differences between them. Create the following image composites (one for each window):

a. 4-3-2

b. 7-6-5

c. 5-3-2

d. 2-2-2

> **Question 11.8** Which band combination(s) is best for letting you separate water bodies from land? Why?

5. Return to the view of Cleveland's waterfront area. Focus on the urban features and vegetated features (zoom and pan where necessary to get a good look at urbanization). Open new windows with some new band combinations, and note how things change with each one as follows:

a. 7-5-3

b. 2-3-4

c. 5-4-3

> **Question 11.9** Which band combination(s) is best for separating urban areas from other forms of land cover (that is, vegetation, trees, etc.)? Why?

Closing Time

This lab wraps up working with Landsat information as well as working with imagery in MultiSpec. Chapter 12 focuses on a whole different set of remote sensing satellites used for climate and environment monitoring and its lab will return to Google Earth Pro to examine this imagery.

Exit MultiSpec by selecting **Exit** from the **File** pull-down menu. There's no need to save any work in this exercise.

Studying Earth's Climate and Environment from Space

NASA and NOAA Satellites, MODIS, VIIRS, LST, SST, CERES, AIRS, OMI, OMPS, and Jason-3

Climate change and global warming have become hot-button issues in today's world, affecting everything from a country's energy policies, to its economy, to its citizens' choices of homes and vehicles. The study and monitoring of the types of environmental processes that affect the land, oceans, and atmosphere occur on a global scale, and a regular flow of information is needed for analysis. For instance, monitoring changes in global sea surface temperature requires information captured from the world's oceans on an ongoing basis. The same goes for such things as measuring carbon dioxide levels in the atmosphere. In order to acquire information about the whole planet in a timely fashion, remote sensing technology can be utilized. After all, satellites are continuously orbiting the planet and collecting data—they provide the ideal technology for monitoring Earth's climate conditions and changes.

There are a lot of other remote sensing satellites up in orbit besides Landsat and the high-resolution satellites we discussed in Chapter 11. In fact, there's an entire set of science observatories orbiting hundreds of miles over our heads that are dedicated to continuously monitoring things that may be critical to the planet's health, like Earth's temperature and heat radiation, ozone content of the atmosphere, sea levels, and the chlorophyll content of the oceans.

These satellites dedicated to climate and environment monitoring often carry multiple sensor instruments onboard, each having a specific function. Each instrument's capabilities are dedicated to an aspect of environmental monitoring or data collection related to Earth's surface, oceans, atmosphere, or biosphere. These sensors aren't intended for fine details and often have coarse spatial resolution (for instance, some instruments have a spatial resolution of 1000 meters), but they consequently have a much larger swath width—and thus the ability to "see" an area up to 2300 kilometers across.

FIGURE 12.1

The health of land vegetation and the temperature of the sea surface measured through Earth-observing remote sensing instruments. *[MODIS Instrument Team, NASA Goddard Space Flight Center]*

Their instruments are also set up to monitor very specific wavelengths tied to the purpose of their particular sensors. Their revisit times are usually relatively short, imaging the whole Earth in a couple of days to provide regular data streams of observation data about whatever environmental system they're examining.

There are a lot of different uses for these types of remotely sensed data, and many different products are derived from the information the satellites provide. See **Figure 12.1** for an example of the kind of data and imagery that can be collected by one such satellite from a sensor called MODIS (see page 440 for more about MODIS)—the image was collected by a sensor onboard an EOS satellite over 8 days in April 2000. It shows the health of vegetation (the "greening" of springtime) in North America and the sea surface temperature of its surrounding waters. The overall scale of the image allowed most of the continent to be observed with a single remotely sensed image.

 ## What Satellites and Sensors Are Used for Studying Climate and the Environment?

NOAA the National Oceanic and Atmospheric Administration, a U.S. federal agency focused on weather, oceans, and the atmosphere

A wide variety of satellites and sensors are used for acquiring data related to climate and the environment. In this chapter, we'll be focusing on the missions from NASA's Earth Science Division, in conjunction with other satellite missions operated by **NOAA** (National Oceanic and Atmospheric

Administration), **ESA** (European Space Agency), and **JAXA** (Japan Aerospace Exploration Agency). The overall mission of these Earth-observing satellites is to examine global environmental phenomena to advance knowledge and understanding of Earth's systems as a whole.

The NASA missions were begun during the 1990s and there are several ongoing missions through NASA and partnerships with NOAA and other organizations (with more planned for the future). With all of these satellites and sensors, this chapter can't hope to cover all of them; we will just demonstrate some climate and environmental applications from their various sensors. See **Table 12.1** for a list of the earth observation satellites and their sensor instruments that we will be discussing in this chapter. Also see *Hands-On Application 12.1: NASA's Earth Observatory* for a NASA Website that highlights many different applications and examples of these remote sensing satellites.

Terra serves as the "flagship" satellite of the NASA Earth Science missions. It was launched in 1999 as a collaboration among the United States (NASA's Goddard Space Flight Center, NASA's Jet Propulsion Laboratory, and the NASA Langley Research Center) and agencies in Canada and Japan, and it continues its remote sensing mission today. "Terra" means "Earth" in Latin. It's an appropriate name for the satellite, as its instruments provide numerous measures of (among many other functions) the processes involved with Earth's land and climate. There are five instruments onboard Terra, each with its own distinct climate and environmental monitoring capabilities. Terra's orbit is sun synchronous, and it flies in formation as

ESA the European Space Agency; an organization made up of multiple European countries, dedicated to space exploration

JAXA the Japanese Aerospace Exploration Agency; Japan's national space exploration organization

Terra the flagship satellite of the NASA Earth Science program

---

TABLE 12.1 A list of the instruments discussed in this chapter and satellites they are on.

Instrument	Name	Satellite
AIRS	Advanced Infrared Sounder	Aqua
AMSR2	Advanced Microwave Scanning Radiometer 2	GCOM-W1 ("SHIZUKU")
AVHRR	Advanced Very High Resolution Radiometer	NOAA series, MetOp series
CERES	Clouds and the Earth's Radiant Energy System	Terra, Aqua, Suomi-NPP, JPSS-1
MISR	Multi-angle Imaging SpectroRadiometer	Terra
MODIS	Moderate Resolution Imaging Spectroradiometer	Terra, Aqua
MOPITT	Measurement of Pollution in the Troposphere	Terra
OCO-2	Orbiting Carbon Observatory	OCO-2
OMI	Ozone Monitoring Instrument	Aura
OMPS	Ozone Mapping and Profiler Suite	Suomi-NPP, JPSS-1
Posiedon-3B	Posiedon-3B Altimeter	Jason-3
SMAP	Soil Moisture Active Passive	SMAP
VIIRS	Visible Infrared Imaging Radiometer Suite	Suomi-NPP, JPSS-1

Hands-On Application 12.1

NASA's Earth Observatory

Earth Observatory is a NASA Website devoted to global environmental and climate issues and how satellite remote sensing contributes to our understanding of those issues (online at **http://earthobservatory.nasa.gov**). Think of it as a sort of online "magazine" with articles and imagery from 1998 to the present. Earth Observatory features the "Image of the Day" section (accessible under the Images tab) which highlights different satellite images concerning a climate or environmental topic each day.

Also, under the News and Notes tab, NASA will post information about climate and environment monitoring based on current events, while options under the Features tab will allow you to browse imagery by topic (such as Atmosphere, Water, or Human Presence), as well as view topics in the archive going back to 1998. Also under the Features tab is the option for World of Change which will allow you to view satellite imagery over time of areas of interest. Select one or more of the World of Change topics of interest to you to see how long-term environmental remote sensing can monitor land use and land-cover changes.

Expansion Questions

➤ To what kinds of climate and environment phenomena are the most recent "Images of the Day" dedicated? Which satellites and sensors are being used to observe and monitor them?

➤ What kinds of environmental changes have been monitored by your selected World of Change topic of interest and how has satellite imagery been applied to this study and analysis?

Morning Constellation a set of satellites (including Terra, Landsat 7, and two others), each one passing the Equator in the morning during its own orbit

Aqua an Earth-observing satellite whose mission is to monitor Earth's water cycle

Aura a satellite dedicated to monitoring Earth's atmospheric chemistry

A-Train another term for the Afternoon Constellation

Afternoon Constellation a set of satellites that pass the Equator in the afternoon during their orbits

one of four satellites (one of which is Landsat 7—see page 402—called the **Morning Constellation**. This series of satellites gets this name because Terra crosses the Equator at 10:30 a.m. each day.

Launched in 2002 as a joint mission among NASA and agencies in Brazil and Japan, **Aqua** is another key Earth observation satellite. In Latin, "aqua" means "water," indicating the main purpose of the satellite: to examine multiple facets of Earth's water cycle. Analysis of water in all its forms—solid, liquid, and gaseous—is the key element of all the instruments onboard Aqua. Terra and Aqua are designed to work in concert with one another—their sun-synchronous orbits are set up similarly so that while Terra is on a descending path, Aqua is ascending (and vice versa).

Aura (launched in 2004) was designed as a collaboration among NASA and agencies in Finland, the Netherlands, and the United Kingdom. In Latin, "aura" means "breeze," which helps describe the satellite's mission—examination of elements in the air, especially the chemistry of Earth's atmosphere.

Aura orbits in formation with Aqua and other satellites to form what is referred to as the **A-Train** of satellites. This organization of satellites is called the A-Train for two reasons: because it serves as the **Afternoon Constellation** of satellites (as they are timed to pass the Equator in the afternoon on their orbits in response to the Morning Constellation) and also because two of the key satellites (Aqua and Aura) begin with the letter "A."

The lead member of the A-Train is **OCO-2** (Orbital Carbon Observatory), a satellite whose sensors are designed to measure global carbon dioxide. Flying second is the Japanese **GCOM-W1** (the Global Change Observation Mission—Water, also referred to as "SHIZUKU") satellite, dedicated to water monitoring. Aqua comes next, the CALIPSO satellite follows closely after, then the CloudSat satellite follows less than two minutes later, and finally Aura brings up the rear (**Figure 12.2**). Other satellites are currently planned to become part of the A-Train after their launches. The combined data from the A-Train of satellites give scientists a rich data set for analysis of climate change questions.

Suomi NPP, launched in 2011, was designed as a joint partnership among NASA, the U.S. Department of Defense, and NOAA. NOAA also operates and maintains a series of weather satellites and the "NPP" part of the name stands for the National Polar-orbiting Partnership, a reference to Suomi NPP's origins as a bridge between older and newer model NOAA weather satellites. Suomi NPP was named after the late Dr. Verner E. Suomi, a scientist at the University of Wisconsin–Madison who was considered "the father of satellite meteorology."

JPSS-1, scheduled to launch in 2017, is an example of one of the next generation of Earth-observing satellites, a joint mission between NASA and NOAA. JPSS-1 (Joint Polar Satellite System) is planned as the first in a series of missions to study and monitor Earth's climate and environment, carrying many of the same instruments described in this chapter as well as new technologies.

OCO-2 the Orbital Carbon Observatory; a satellite dedicated to measuring global carbon dioxide

GCOM-W1 the Global Change Observation Mission—Water; a JAXA satellite mission dedicated to studying Earth's water cycle

Suomi NPP a joint satellite mission of NASA, NOAA, and the U.S. Department of Defense

JPSS-1 a joint satellite mission of NASA and NOAA, intended as the first in a series of Earth-observing missions

FIGURE 12.2

The configuration of satellites that make up the A-Train. *[NASA]*

FIGURE 12.3

The various satellite missions of the NASA Earth Science Division
[NASA]

There are a lot of other ongoing Earth-observing satellite missions beyond the ones mentioned in this chapter, each dedicated to different aspects of Earth's environment and climate. For instance, data from the CloudSat mission is used in studying questions related to clouds and climate change, while the SORCE mission studies energy from the sun. See **Figure 12.3** for all of these current missions, and also *Hands-On Application 12.2: NASA Eyes on the Earth* for a very cool NASA Website filled with interactive information and data about all of the current satellite missions. The types of data these satellites and sensors acquire relates to a variety of climate change studies including land-cover change, taking Earth's temperature, atmospheric measurements, and water quality issues.

 ## How Is Environmental Monitoring Performed Using Remote Sensing?

Satellite systems such as Landsat and Sentinel (see Chapter 11) are often used for monitoring environmental phenomena—with their multispectral capabilities and moderate swath width, their sensors provide scientists with a wide range of remotely sensed imagery and data about the environment. When studying localized environmental phenomena (such as the aftermath of a forest fire in a national park), systems such as Landsat and Sentinel (are

Hands-On Application 12.2

NASA Eyes on the Earth

NASA has set up a Web application that allows you to view the real-time positions of Terra, Aqua, Aura, and the other satellite missions, and to examine 3D data maps of the products and imagery they capture. Open your Web browser and go to **http://eyes.jpl.nasa.gov/eyes-on-the-earth.html**. If you haven't used Eyes on the Earth before, click on the Download App button, allow your computer to download the file, and then run the executable program that downloads to install the app on your computer. When the app is installed on your computer, click Launch to open the program.

When Eyes on the Earth opens, a separate window will open that allows you to view the orbital paths of several satellite missions (including nearly all of the satellites discussed in this chapter: Terra, Aqua, Aura, Suomi NPP, OCO-2, SMAP, and Jason-3, along with several other missions such as Landsat 7, SORCE, and CloudSat). Pressing the Real Time button will show you the current positions and paths of the satellites. For instance, you can see the paths of five of the A-Train satellites (minus GCOM-W1 which is not a NASA satellite) flying in formation in their current orbital locations. Clicking on a particular satellite in orbit allows you to access information about that mission, view its ground

path, and compare the size of the satellite with other objects.

By selecting one of the icons at the top of the screen, you can view recent satellite imagery and investigate environmental applications based on topics such as global temperature, sea level, and ozone concentrations. After selecting a category, you can choose a range of dates of imagery to view as animation or as a series of images, or you can just view the most recently available data map.

Expansion Questions

➤ View the latest data map of sea level variations (available via the Sea Level icon). What areas of the globe are seeing the greatest rise in sea level (based on the OSTM imagery that loads)?

➤ View the latest data map of carbon monoxide levels (available via the Carbon Monoxide icon). What areas of the globe are experiencing the highest levels of carbon monoxide (based on the AIRS imagery that loads)?

➤ View the latest data map of daytime air temperatures (available via the Global Temperature icon). What areas of the globe are seeing the warmest temperatures during the day (based on the AIRS imagery that loads)?

often utilized due to the size of the study area, their available spatial resolution, and their moderate revisit times. However, when studying environmental effects at a larger national or global scale, their reduced swath wide and higher revisit times aren't ideal.

For instance, Landsat 8 can image a single scene at 30-meter resolution at a swath of 185 kilometers across, and after 16 days will have acquired thousands of individual images that will provide one snapshot of Earth over that time. To really look at a global phenomenon (such as Earth's land surface temperature or the amount of carbon monoxide in the atmosphere) we would need a sensor that could "see" a much larger area of Earth faster. The satellites we'll be discussing in this chapter have these types of capabilities for large-scale environmental monitoring, but their trade-off is often a significantly coarser spatial resolution.

Thinking Critically with Geospatial Technology 12.1

What If We Lose the Capability for Large-Scale Study and Monitoring of Earth's Climate and Environment?

Many of the satellite systems discussed in this chapter are funded and operated by U.S. government agencies, similar to the Landsat program described in Chapter 11. There has been a long chain of NASA and NOAA satellites collecting data and imagery used for monitoring and studying Earth's climate and environment. However, government funding to continue long-running satellite programs like those described in this chapter is not always guaranteed. Different administrations have different budget priorities, and often government budget cuts affect various organizations with each new budget that is passed by Congress. For instance, at the time of writing, the President's preliminary 2018 budget reduced funding for NASA with proposed elimination of four Earth-observing satellites, including OCO-3 and the CLARREO satellite, whose data is intended for climate modeling and prediction.

As priorities for budget resources change, the question becomes "What happens if the funding for continuance of the Earth-observing satellite missions and programs for studying and monitoring climate and environment is discontinued?" Would other space agencies around the world, such as ESA or JAXA, become the foremost sources of remotely sensed environmental and climate data? Will these types of data collection programs be moved away from government-funded venues to the private sector? As we saw in the last chapter, there has been huge growth in commercial development of high-resolution sensors. Could a shift away from government-funded remote sensing fuel a similar growth in private sector climate and environmental monitoring?

MODIS the Moderate Resolution Imaging Spectroradiometer instrument onboard Terra and Aqua

One of the most versatile instruments in this type of rapid large-scale image acquisition is the **MODIS** (the Moderate Resolution Imaging Spectroradiometer) sensor onboard both Terra and Aqua, which produces over 40 separate data products related to environmental monitoring. MODIS's swath is 2300 kilometers, which enables it to image about half of the North American continent in one pass, and allows it to view nearly the entire planet in a single day (see **Figure 12.4**). The trade-off is that most of the MODIS bands have a spatial resolution of 1 kilometer, which makes MODIS useful only for larger scale environmental applications.

Being able to measure 36 spectral bands (including portions of the visible, infrared, and thermal) allows for multiple applications of MODIS imagery—for instance, Figure 12.1 on page 434 shows vegetation health and water temperatures. MODIS products are used to regularly monitor numerous types of environmental features, including (among many others) snow cover, sea surface temperature, sea ice coverage, ocean color, and volcanic activity. See *Hands-On Application 12.3: NASA Worldview* for an online tool for viewing multiple types of MODIS environmental images.

In addition, MODIS's thermal capabilities can track the heat emitted by wildfires, allowing for regular monitoring and mapping of active fires. With MODIS's rapid revisit cycle, it's able to collect near-daily images of wildfires from around the globe. Keep in mind that with MODIS's 1-kilometer

FIGURE 12.4

One day's worth of Earth imagery collected from MODIS (collected on MODIS's first full day of imaging on April 19, 2000). Note the 2300-kilometer width of each MODIS swath.
[NASA]

Hands-On Application 12.3

NASA Worldview

NASA Worldview is a Web application that allows you to view recently acquired imagery from several different Earth-observing satellites and their sensors, including MODIS. Open your Web browser and go to **https://worldview.earthdata.nasa.gov**.

From the Layers panel on the left-hand side of the screen, you can select the type of imagery you want to examine—for instance, one of the default options is Corrected True Color MODIS imagery. By clicking on the Events tab, you'll be able to examine specific occurrences (such as volcanoes, fires, or tropical storms) that were imaged by sensors such as MODIS.

After you've selected a type of imagery or a specific event, press the Camera icon at the bottom of the screen (to the left of the timeline) to set up and then play an animation of imagery. This will show you a different remotely sensed image for each date that is chosen, but set up as a time-lapse

animation. You can also use this technique to go to a specific location on the map (by panning the map) and viewing imagery for a user-defined set of dates.

Expansion Questions

➤ Under Events, select one of the options for Wildfires then examine the animation of imagery (via stepping through the imagery instead of letting it play). How are the capabilities of the MODIS sensor demonstrated here as being extremely useful for monitoring phenomena such as large-scale wildfires?

➤ Under Events, select one of the options for Volcanoes then examine the imagery of that date and the next couple of dates as well (via stepping through the imagery instead of letting it play). How is the large-scale volcanic eruption seen by the MODIS sensor?

spatial resolution you won't be able to see the dimensions of an individual fire—instead, a MODIS image will show the number of fires within that area. In addition to viewing the area during the fire, MODIS's other bands will be able to image the large scale of the area before and after the fires for monitoring of the environmental effects and burn scars. See **Figure 12.5** for an example of MODIS imagery of fires and their aftermath recorded at Wilsons Promontory National Park in Australia.

MODIS is often the go-to source for a variety of environment monitoring applications, but its technology is nearly 20 years old. A newer system in orbit has similar yet updated capabilities. The **VIIRS** (Visible Infrared Imaging Radiometer Suite) instrument onboard Suomi NPP (and also the upcoming JPSS-1) improves upon several aspects of MODIS with a swath width of 3040 kilometers that allows it to image the entire Earth twice each day at 750-meters resolution. With its ability to sense 22 different bands

VIIRS the Visible Infrared Imaging Radiometer Suite instrument onboard Suomi NPP

FIGURE 12.5

MODIS imagery examining the dimensions of a forest fire and resulting smoke plume at Wilsons Promontory National Park. The two smaller MODIS images show the park before and after the fire. *[NASA/GSFC, MODIS Rapid Response]*

April 2, 2005 - Fire in Wilsons Promontory National Park, Victoria, Australia

March 31, 2005

April 9, 2005

Burned area

at once, VIIRS has a wide variety of applications for global monitoring of (among other things) cloud properties, ocean color, vegetation health, snow cover, and chlorophyll content of the oceans. VIIRS imagery is also used for wildfire monitoring like MODIS but capable of producing products at 375-meters resolution.

One new technology that VIIRS brings to the table in terms of environmental monitoring is what's referred to as the day-night band, which allows VIIRS to accurately capture imagery of Earth at night. Satellite imagery data usually is not taken at night, but the VIIRS day-night band is specifically designed to do just that. The day-night band captures the low light of reflected moonlight from clouds, terrain, and water, as well as the light cast from urban features, producing impressive imagery of Earth at night.

An example of using the VIIRS day-night band for environmental monitoring is shown in **Figure 12.6**. In the nighttime image, you can see the glow from urban areas in North Dakota, such as Minot and Bismarck. However, eastern North Dakota is a largely rural or non-urban landscape but the nighttime image shows plenty of lights at night, making the whole region look like a thriving set of towns and small cities. However, those lights in the VIIRS image are actually shale gas drilling operations with the associated housing near the sites. Visible or false color imagery at this scale and resolution would be difficult to use to identify drilling sites, but analyzing VIIRS imagery over time allows us

FIGURE 12.6

Monitoring shale gas drilling in North Dakota at night, as seen by VIIRS.
[NASA]

Hands-On Application 12.4

The VIIRS View Spinning Marble

The Raytheon Corporation has set up a free online tool that lets you examine a virtual globe of different applications of VIIRS imagery called the VIIRS View Spinning Marble. To use this utility, go to **www.raytheon .com/media/spinningmarble/web.html** (you will also have to install and run the Unity player for the Spinning Marble to work properly). When the Spinning Marble opens, you'll have your choice of three different VIIRS global images, accessible by icons at the bottom of the screen: regular visible light imagery, ocean chlorophyll content, and nighttime imagery captured by the day-night band. These three layers can be made more or less prominent by spinning their Globe icons and thus adjusting the lines showing their display intensities at the bottom of the screen. You can pause and zoom in

on the globe; after a brief time it will begin to spin again.

Expansion Questions

➤ With adjusting the nighttime imagery up to its maximum and the daytime visible imagery to its minimum levels, examine some areas around the globe (such as North America and Europe) in comparison to others (such as Africa and South America). How does the nighttime imagery help in determining urban population centers?

➤ By examining the ocean chlorophyll layer at its maximum level, where are the highest concentrations of chlorophyll found around the globe?

to monitor the growth and patterns of shale drilling. For more about working with VIIRS imagery, see *Hands-On Application 12.4: The VIIRS View Spinning Marble*.

MODIS and VIIRS are multipurpose sensors but other satellite missions have much more specific purposes for environmental monitoring. An example of this is the **SMAP** (Soil Moisture Active Passive) satellite, whose sensors are used for measuring the moisture content of Earth's soils. SMAP uses a radar instrument to measure the level of water content in the top layer of soil across the Earth every two to three days. By knowing if soil conditions are wetter or dryer than anticipated, SMAP can be used for monitoring potential drought or flooding conditions. In turn, this type of regular monitoring can be beneficial for assessing crop productivity.

As SMAP can measure moisture content, its data regarding the wetness of soils can provide a link to measuring Earth's temperature. As NASA notes, in places with wet soils (i.e., where their moisture content is higher), the sun's energy will evaporate the moisture, leaving those areas cooler. On the other hand, locations with lower moisture content will be much warmer as the air near the ground will be heated by the sun, without that cooling evaporation from the wetter soils. SMAP data and imagery is just one aspect of a larger study of measuring the temperature of Earth in an ongoing way.

SMAP Soil Moisture Active Passive; a satellite mission dedicated to monitoring soil moisture levels

How Do Satellites Take Earth's Temperature?

A key question in climate studies revolves around Earth's temperature and the rate of a global rise in temperatures. When you think of Earth's temperature, the first thing that usually comes to mind is something like "How warm is it outside?" That type of temperature is measured with a thermometer, which will let you know that the air temperature at that location at that time is 85 degrees. When scientific air temperature readings are taken over time, the thermometers need to be placed at a certain height, out of the sun, and under special conditions. However, that's only the air temperature reading—the actual temperature of Earth is going to vary considerably. For instance, shaded grass and sunlit grass will have differing temperatures, the same with a temperature reading of sunlit versus shaded concrete. Again, that's only going to be one measurement—if we're going to be gathering data related to Earth's temperature and climate change, we're going to need an accurate system that can measure temperatures of the entire planet at regular intervals.

The thermal sensors in the MODIS and VIIRS instruments give us this kind of capability for measuring emitted heat energy from the Earth every one to two days and then translating those measurements into temperature readings. What their sensors are measuring (at very coarse spatial resolutions) are not air temperatures, but instead they measure **LST** (or land surface temperature). This refers to the heat temperature of the top millimeter of the surface of the land, which takes into account the sun, the atmosphere, and the land cover itself. By regularly measuring LST (and having a deep archive of thermal imagery), we can gauge how the Earth's surface heat level fluctuates over time as well as the effect of land cover (such as forests, fields, or developed areas) on Earth's temperature. See **Figure 12.7** for an example of a global LST measurement captured by MODIS.

LST land surface temperature; a measurement of the temperature of Earth's land-cover surface

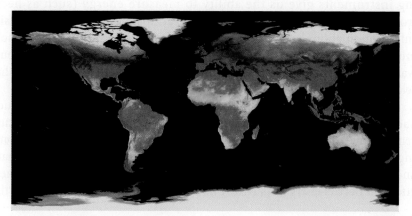

FIGURE 12.7

Global land surface temperature as measured by MODIS.
[NASA]

AVHRR Advanced Very High Resolution Radiometer; the six-band sensor onboard NOAA satellites

SST sea surface temperature; a measurement of the temperature of the surface of Earth's oceans

AMSR2 the Advanced Microwave Scanning Radiometer 2 instrument onboard GCOM-W1

CERES the Clouds and Earth's Radiant Energy System instruments onboard Terra, Aqua, Suomi NPP, and JPSS-1

Similarly, satellites can also measure the temperature of the surface of the oceans. The longest-running record of satellite-based sea surface measurements comes from the **AVHRR** (Advanced Very High Resolution Radiometer) sensors carried on a variety of satellites. AVHRR sensors have been onboard NOAA satellites since 1978, allowing for long-term thermal measurement of Earth's sea surface temperature (**SST**). Today, satellites in the NOAA series as well as the European MetOp series carry AVHRR instruments and collect regular SST data. In addition, MODIS and VIIRS also have the capability to measure the thermal emission from the top millimeter of ocean surface and this can be converted to an SST measurement.

In addition to AVHRR, MODIS, and VIIRS, other instruments collect information used for the computation of SST in a different fashion. For instance, the **AMSR2** (Advanced Microwave Scanning Radiometer 2) instrument onboard the GCOM-W1 satellite (see page 437) measures microwave emissions from the top of the oceans for use in calculating not only SST but also sea ice extents (for frozen portions of the seas). As AMSR2 utilizes the microwave portion of the electromagnetic spectrum and as microwaves can pass through clouds, it can be used to collect SST data in clouded areas that sensors like AVHRR, MODIS, and VIIRS would not be able to image.

Combined with on-site measurements of sea surface temperature from ships and buoys, satellite imagery gives us a good accounting over time of changes and fluctuations in the oceans' temperature. As NASA notes, being able to model the movement and transfer of heat across Earth's oceans is a critical factor in climate studies. Weather forecasting also relies on accurate SST measurements. For instance, hurricanes require water temperature of greater than 82 degrees Fahrenheit in order to form. During hurricane season in the tropics (i.e., late summer to early fall), being able to obtain daily readings of SST provides an immense help in forecasting whether conditions are right for the formation of hurricanes and where to expect this to occur.

Beyond the surface temperature of the land and the sea, remote sensing instruments give us the ability to measure energy (such as heat) from Earth's surface up to the top of the atmosphere on a regular basis. Information about how Earth's energy budget operates (in regard to how much heat energy remains trapped on the planet versus being released outward to space from Earth) is key to understanding long-term processes that relate to climate change.

An instrument used for collecting this type of data is **CERES** (Clouds and Earth's Radiant Energy System), a sensor carried aboard several currently active satellites, including Terra, Aqua, Suomi NPP, and the upcoming JPSS-1. CERES provides critical information about Earth's climate and temperature and one of its primary uses is to study clouds and their effect on Earth's energy. As NASA notes, clouds play a defining role in the

FIGURE 12.8

A global composite of CERES imagery from September 2017, showing the radiated energy from Earth (blue areas are colder, red areas are warmer). *[NASA]*

heat budget of Earth, but different types of clouds affect Earth in different ways—for instance, low clouds reflect more sunlight (and thus, help cool the planet), while high clouds reflect less but effectively keep more heat on Earth. By combining CERES data with data from other satellite instruments, scientists are able to examine cloud properties and gather information about Earth's heat and radiation (see **Figure 12.8** for an example of using CERES to examine Earth's radiated energy).

Climate studies rely on accurate measurements of a variety of different factors as well as a long-term series of observations in order to chart patterns and changes. Sensors such as MODIS and CERES have been in orbit since 1999, collecting imagery for an array of climate and environmental processes. See *Hands-On Application 12.5: NASA Global Maps* for the application of this type of imagery for multiyear maps of energy radiation and temperature changes across the planet. However, there's a lot more to climate studies than just taking Earth's temperature on the land and seas—remote sensing allows us to monitor Earth's atmosphere and address issues related to air quality as well.

How Do Satellites Monitor Earth's Atmosphere and Air Quality?

Earth's atmosphere is home to many different things, including water vapor, clouds, aerosols, pollutants, and gases such as ozone, carbon monoxide, and carbon dioxide, all of which have various impacts on Earth's climate. Remote sensing gives us a means to study, analyze, and monitor all of these types of atmospheric effects to add to our database of climate-related information. A basic type of climate and weather information that satellite sensors can acquire is used to build a profile of atmospheric conditions, such

Hands-On Application 12.5

NASA Global Maps

Part of NASA's Earth Observatory Website includes a section dedicated to global remotely sensed imagery of a variety of different climate and environment phenomena. Open your Web browser and go to **http://earthobservatory.nasa.gov/GlobalMaps**. You'll see several different options for climate and environment imagery, including land surface temperature, sea surface temperature, and net radiation.

Select one of the options (for instance, Land Surface Temperature) and it will open in a new Web page. You'll be able to examine global land surface temperature (as recorded by MODIS) from the year 2000 to the recent past, along with a description of the sensor process and an explanation of what the various colors on the map indicate. By playing the animation (or downloading it to view later) you'll see

global temperature changes take place over a year time period of over 16 years.

Expansion Questions

➤ Choose the option for Net Radiation—this is the data captured by CERES measuring the balance between Earth's incoming and outgoing radiation over time. Play the associate animation for the Net Radiation imagery. How does Earth's energy budget change over time?

➤ Choose the option for Sea Surface Temperature—this is the imagery from MODIS measuring the temperature of the top millimeter of Earth's oceans. Play the associate animation for the Sea Surface Temperature imagery. How does the temperature of the oceans change over time?

sounder the term given to a remote sensing instrument that measures atmospheric phenomena as a function of vertical distribution

as the atmospheric temperature or amount of water vapor in the air. Several satellite sensors that measure atmospheric conditions are referred to as **sounders**, because they examine a three-dimensional atmosphere, in a similar fashion as how "soundings" were used by ships to determine water depths. A satellite sensor can't measure the amount of water vapor directly, so instead a sounder instrument measures things such as water vapor as a function of its height (or vertical distribution) in the atmosphere.

There are many sounding instruments onboard Earth-observing satellites. An example of this type of atmospheric measurement is performed by the **AIRS** (Advanced Infrared Sounder) instrument onboard Aqua, whose uses include measuring temperatures and humidity levels in the atmosphere as well as information about clouds. AIRS senses 2378 bands with different bands allocated to measuring different atmospheric features. AIRS provides a wealth of climate-related data, from daily water vapor and carbon monoxide maps of the globe to daily measurements of the daytime air temperature of the entire planet (see **Figure 12.9** and *Hands-On Application 12.6: Tracking Earth's Climate and Temperature with AIRS*).

AIRS the Advanced Infrared Sounder instrument onboard Aqua

Ozone in Earth's atmosphere is something that's monitored by several different satellite instruments (including AIRS). In Chapter 10 we discussed how the atmosphere absorbs a lot of electromagnetic radiation and also how ozone is the primary greenhouse gas that absorbs harmful ultraviolet

NASA/AIRS 850mb Tair(K): 2014.01.03

FIGURE 12.9

A global temperature map measured by AIRS showing the polar vortex of 2014. *[NASA/JPL]*

light. By monitoring how much ultraviolet energy penetrates Earth's atmosphere to the ground below (and thus is not being readily absorbed by atmospheric ozone) scientists can model the extent of thinning in Earth's atmosphere. By having a long-term record of satellite ozone measurements, we can determine the rate of depletion (or thinning) of the ozone layer (and where "holes" in the ozone layer have occurred).

Two examples of the instruments used in monitoring ozone are **OMI** (Ozone Monitoring Instrument) aboard Aura and **OMPS** (Ozone Monitoring Profile Suite) aboard Suomi NPP and the upcoming JPSS-1. They are both sounders, which (as their names imply) are dedicated to keeping an

OMI the Ozone Monitoring Instrument onboard Aura

OMPS the Ozone Monitoring Profile Suite onboard Suomi NPP and JPSS-1

Hands-On Application 12.6

Tracking Earth's Climate and Temperature with AIRS

AIRS is capable of monitoring a wide range of climate-related conditions for Earth. To see some recent examples, go to **http://airs.jpl.nasa.gov/index.html**. From there, choose the option for Resources and then for Today's Earth Maps. A new set of maps will appear, showcasing the kind of data available through AIRS, such as air temperature, water vapor, carbon monoxide, and ozone levels. Maps of the most recently available data (likely of today's date and a recent time period)

will be displayed. Examine each of them to see how AIRS is used for climate and environmental monitoring.

Expansion Questions

➤ Currently, what geographic areas have the warmest and coldest daytime air temperatures?

➤ Currently, what geographic areas are seeing the highest levels of carbon monoxide in the atmosphere?

eye on changes in ozone levels in the troposphere on a daily basis. OMI and OMPS view sections of the visible-light spectrum as well as the ultraviolet portion of the electromagnetic spectrum. Their data aids scientists in measuring the amount of ultraviolet radiation penetrating to Earth by examining clouds and ozone levels. **Figure 12.10** shows an example of imagery from OMI, showing the thinning (or hole) in the ozone layer above Antarctica.

MOPITT (Measurements of Pollution in the Troposphere) is a different type of sounding instrument onboard Terra, which is used for measuring airborne pollutant levels in the lower atmosphere. MOPITT's sensors collect information to measure the amount of air pollution concentrations across the planet, creating an entire global image about every three days. Specifically, MOPITT monitors carbon monoxide concentrations in Earth's troposphere. Carbon monoxide, a global air pollutant, can be produced naturally as a result of volcanic eruptions and wildfires, or by human causes (such as vehicle emissions or industrial output).

Being able to track long-term carbon monoxide sources can aid in determining areas generating large amounts of pollutants and then monitoring how those pollutants are being circulated through the atmosphere. **Figure 12.11** shows an example of this kind of tracking in two MOPITT images of South America. The first, from March 2000, shows relatively low amounts of carbon monoxide. The second image, from September 2000, shows the dimensions of a large mass of carbon monoxide that was caused

MOPITT the Measurements of Pollution in the Troposphere instrument onboard Terra

FIGURE 12.10

OMI imagery showing a hole in the ozone layer above Antarctica. *[NASA/ Goddard Space Flight Center Scientific Visualization Studio]*

MOPITT Carbon Monoxide at 700 mb

March 2000   September 2000

20   160   300

FIGURE 12.11

MOPITT imagery, captured six months apart, showing growing carbon monoxide concentration levels in South America. *[David Edwards and John Gille, MOPITT Science Team, NCAR/NASA]*

by biomass burning in the Amazon and the movement of carbon monoxide across the ocean to South America from fires in Africa.

While MOPITT tracks carbon monoxide, the sensors onboard the OCO-2 (see page 437) satellite monitor carbon dioxide levels in the atmosphere. OCO-2's sensors are not sounders but instead are regular radiometers that measure in the infrared portion of the spectrum. Carbon dioxide is one of the greenhouse gases that absorbs emitted radiation from Earth and thus keeps radiated heat on Earth, leading scientists to conclude that higher concentrations of carbon dioxide levels are contributing to changing temperatures. OCO-2 gives us a means of studying these global carbon dioxide levels.

Earth's atmosphere also contains aerosols, which are tiny particles caused by human-produced or naturally produced means (such as automobile or factory emissions, or by smoke and dust from wildfires, volcanic eruptions, and other natural phenomena). Air pollution in the form of these tiny particulates can lead to increased health risks. Detailed information on concentrations of aerosols and atmospheric particles can help scientists understand the level of human exposure to these types of pollutants.

The **MISR** (Multi-angle Imaging SpectroRadiometer) onboard Terra is an example of instrumentation used to show how sunlight is scattered in different directions, particularly by aerosols and other atmospheric particles. Gathering such information aids in climate studies. A NASA article by Rosemary Sullivant describes how MISR is extremely sensitive to measuring these types of aerosols, and its data can help scientists tell the difference

MISR the Multi-angle Imaging SpectroRadiometer instrument aboard Terra between what's on the ground and what's in the atmosphere. By using MISR (in conjunction with other instruments) to study aerosols, scientists can help determine areas of high aerosol concentrations and provide data to aid in cleaning up air pollution.

How Do Satellites Monitor the Oceans and Water Quality?

Aside from regularly taking Earth's sea surface temperature and analyzing the water vapor in the atmosphere, remote sensing gives us a variety of means of assessing and monitoring Earth's oceans and water bodies in conjunction with climate-related studies. For instance, as temperatures rise, water bodies (such as oceans) absorb more heat. As water bodies heat up, they expand, and thus the water levels will rise. As this sea level changes, places that were at sea level (such as beachfront areas) can find themselves underwater due to rising water levels. Satellite sensors have the ability to monitor sea levels regularly to see how conditions are changing over time.

Jason-3 a satellite mission dedicated to measuring ocean topography

Poseidon-3B a radar altimeter instrument onboard Jason-3

An example of this is the **Jason-3** satellite mission, whose primary instrument is the **Poseidon-3B** altimeter, a radar device that fires a radar wave at the ocean's surface and measures the distance of that wave's bounce-back to the sensor to determine the height level of the oceans. Jason-3 is the latest in a line of missions stretching back to the early 1990s, which are dedicated to measuring the topography of the oceans. Through the use of its radar altimeter instrument, Jason-3 can calculate the height of the sea surface with an accuracy of about 1 inch. These long-term sea level measurements are of critical aid in determining how the ocean levels are rising.

In addition to sea level measurements, satellite instruments also regularly monitor ocean color. In general, when you think about a lake or the ocean, you'd think about its color being some shade of blue. However, lakes and oceans are home to plenty of marine life and organisms, which can change how the surface color of a body of water appears. In particular, single-cell organisms called phytoplankton live in the water and contain chlorophyll. As mentioned previously (see page 370 in Chapter 10), chlorophyll reflects green light, therefore when a huge concentration of phytoplankton is present, the surface of the water will appear green. As most phytoplankton contain algae, this huge green surface of an area is referred to as an algal bloom. Sensors such as MODIS and VIIRS are used to derive ocean color products to view the surface of water bodies, and their large swath enables us to view the extent of freshwater algal blooms (see **Figure 12.12**).

FIGURE 12.12

An algal bloom in Lake Erie as imaged by MODIS.

[NASA]

Algal blooms can become dangerous (what's referred to as a harmful algal bloom or HAB). Certain species of phytoplankton (such as cyanobacteria or dinoflagellates) produce toxins that cause numerous damaging effects ranging from poisoning shellfish to causing respiratory problems for people near the water to causing severe illness if the algae water is ingested. Harmful algal blooms are sometimes referred to as "red tide" due to the discoloration of an algal bloom. The decay of the harmful algal bloom can also use up the oxygen in the water effectively killing off fish in the bloom area. NOAA, NASA, and other U.S. government agencies have joined resources to utilize satellite imagery and data as an early warning system for detecting harmful algal blooms.

Satellite sensors allow for several means of trying to monitor for harmful algal blooms. MODIS and VIIRS have the capability of monitoring global chlorophyll content of Earth's oceans—see **Figure 12.13**). In addition, an application of MODIS lets researchers study the health of phytoplankton in the oceans. As NASA notes, when phytoplankton are unhealthy or under heavy stress, they take the sunlight they have absorbed and reemit it as fluorescence, which can then be observed by MODIS. Thus, by monitoring fluorescence levels, scientists can gauge phytoplankton health.

FIGURE 12.13

Global ocean chlorophyll content as measured by VIIRS. *[NASA]*

Chapter Wrap-Up

The Earth-observing satellites described in this chapter provide a constant source of remotely sensed data that can be utilized in numerous studies related to climate change and the environment of our planet. Whether they're used in monitoring wildfires, ozone concentrations, algal blooms, or temperature conditions for hurricanes, the applications described in this chapter just scratch the surface of the uses of these satellites, and they give only a modest outline of how these geospatial tools affect our lives.

This chapter's lab will have you start working with imagery from these satellite sensors and Google Earth to get a better feel for some of the applicability of the data. Also, check out *Earth-Observing Mission Apps* for some apps for your mobile device that use data and imagery from various satellites and *Earth-Observing Missions in Social Media* for some Facebook, Twitter, and Instagram accounts, as well as some blogs and YouTube videos showcasing the many satellites of the EOS and their applications.

In the next chapter, we'll dig into a new aspect of geospatial technology: modeling and analyzing landscapes and terrain surfaces.

Important note: The references for this chapter are part of the online companion for this book and can be found at **www.macmillanlearning .com/catalog/shellitoigt4e**.

Earth-Observing Mission Apps

Here's a sampling of available representative Earth-observing satellite apps for your smartphone or tablet. Note that some apps are for Android, some are for Apple iOS, and some may be available for both.

- ➤ **Earth-Now:** An app from NASA's Jet Propulsion Lab featuring global climate data and information from NASA missions
- ➤ **RealEarth:** An app that allows you to view recent imagery from sensors such as MODIS and VIIRS for many different conditions (such as fires or optical aerosol depth)
- ➤ **SatCam:** An app that allows you to capture the conditions of the ground and sky when Terra, Aqua, or Suomi NPP pass overhead
- ➤ **VIIRS View Spinning Marble:** An app from Raytheon that allows you to view global Suomi NPP's VIIRS imagery on your mobile device
- ➤ **WxSat:** An app that displays weather imagery obtained from a geostationary satellite

Earth-Observing Missions in Social Media

Here's a sampling of some representative Facebook, Twitter, and Instagram accounts, along with some YouTube videos and blogs related to this chapter's concepts.

 On Facebook, check out the following:

- ➤ **NASA Climate Change:**
  https://www.facebook.com/NASAClimateChange
- ➤ **NASA Earth:**
  https://www.facebook.com/nasaearth
- ➤ **NOAA Satellite and Information Service:**
  www.facebook.com/NOAANESDIS
- ➤ **NPP—Suomi National Polar-orbiting Partnership:**
  www.facebook.com/NASA.NPP
- ➤ **SMAP:**
  https://www.facebook.com/NASA.SMAP

Become a Twitter follower of:

- ➤ **NASA Climate:** @NASAClimate
- ➤ **NASA Earth:** @NASAEarth

➤ **NASA Earth Data:** @NASAEarthData

➤ **NASA OCO-2 Mission:** @IamOCO2

➤ **NASA SMAP Mission:** @NASASMAP

➤ **NOAA Satellites:** @NOAASatellites

➤ **Suomi NPP Satellite:** @NASANPP

➤ **USGS Climate:** @USGSClimate

Become an Instagram follower of:

➤ **NASA Climate Change:** @nasaclimatechange

➤ **NASA Goddard** (includes remotely sensed imagery from NASA satellites): @nasagoddard

➤ **NOAA Satellites:** @noaasatellites

On YouTube, watch the following videos:

➤ **Airsnews** (a YouTube channel featuring climate science and the AIRS instrument aboard Aqua): **www.youtube.com/user/airsnews**

➤ **NASA | Aqua MODIS: Science and Beauty** (a NASA video that shows several applications of the MODIS instrument on Terra and Aqua): **www.youtube.com/watch?v=1jqFxZI_2XY**

➤ **NASA Aura** (a NASA YouTube channel with several videos showcasing the applications of Aura's instruments): **www.youtube.com/user/NasaAura**

➤ **NASA Climate Change** (a YouTube channel with videos related to monitoring Earth's climate): **https://www.youtube.com/user/NASAClimate**

➤ **NASA Earth Observatory** (a YouTube channel with videos from the Earth Observatory): **www.youtube.com/user/NASAEarthObservatory**

➤ **NASA's SMAP: Mapping the Water Under Our Feet** (a video about SMAP's capabilities): **https://www.youtube.com/watch?v=GCBBEbKVwm4**

➤ **NASA | NPP and the Earth System** (a video about Suomi NPP and what it can do): **www.youtube.com/watch?v=F-ejWjVRoIM**

➤ **OCO-2 to Shed Light on Global Carbon Cycle** (a video about OCO-2 and its capabilities): **www.youtube.com/watch?v=X5VWayNrtWk**

For further up-to-date info, read up on these blogs:

➤ **Earth Matters** (a NASA Earth Observatory blog showcasing applications of Earth-observing satellites): **http://earthobservatory. nasa.gov/blogs/earthmatters**

➤ **Earth Right Now** (a NASA blog about climate change): **https:// climate.nasa.gov/blog/**

Key Terms

A-Train (p. 436)
Afternoon Constellation (p. 436)
AIRS (p. 448)
AMSR2 (p. 446)
Aqua (p. 436)
Aura (p. 436)
AVHRR (p. 446)
CERES (p. 446)
ESA (p. 435)
GCOM-W1 (p. 437)
Jason-3 (p. 452)
JAXA (p. 435)
JPSS-1 (p. 437)
LST (p. 445)
MISR (p. 451)

MODIS (p. 440)
MOPITT (p. 450)
Morning Constellation (p. 436)
NOAA (p. 434)
OCO-2 (p. 437)
OMI (p. 449)
OMPS (p. 449)
Poseidon 3B (p. 452)
SMAP (p. 444)
sounder (p. 448)
SST (p. 446)
Suomi NPP (p. 437)
Terra (p. 435)
VIIRS (p. 442)

Earth-Observing Missions Imagery

This chapter's lab will introduce you to some of the basics of examining imagery from three different Earth-observing satellite missions: Terra, Aqua, and Suomi NPP. You will be examining data from MOPITT as well as many types of imagery from MODIS and VIIRS. You will also be using online resources from NASA and others in conjunction with Google Earth.

Objectives

The goals for you to take away from this lab are:

➤ To utilize the browser version of Google Earth as a tool for examining Earth-observing satellite imagery as an overlay

➤ To examine the usage and functions of the day-night band imagery from VIIRS

➤ To examine the usage and functions of MODIS imagery for various applications

➤ To examine the output from MOPITT imagery of global carbon monoxide

➤ To examine environmental and climate applications of satellite imagery from MODIS for land surface temperature, sea surface temperature, and snow cover

Using Geospatial Technologies

The concepts you'll be working with in this lab are used in a variety of real-world applications, including:

➤ Forestry, which uses MODIS imagery on a near-daily basis to aid in active wildfire and forest fire monitoring

➤ Climatology, which uses MODIS imagery to track and monitor hurricanes and tropical storms and aid in emergency preparedness programs

[© Karen Wattenmaker/www.forestphoto.com]

Obtaining Software

The current version of Google Earth Pro (7.3) is available for free download at **https://www.google.com/earth/download/gep/agree.html**.

The browser version of Google Earth will run in the Google Chrome browser (available for free download at: **https://www.google.com/chrome/browser/desktop/index.html**).

Important note: Software and online resources can change fast. This lab was designed with the most recently available version of the software at the time of writing. However, if the software or Websites have changed significantly between then and now, an updated version of this lab (using the newest versions) is available online at **www.macmillanlearning.com/catalog/shellitoigt4e**.

Lab Data

Copy the folder **Chapter12**—it contains the following KML datasets (for use in Google Earth) from NASA's Earth Observatory:

➤ Fires_russia_tmo_2012170.kml, a MODIS image showing fires in Siberia

➤ MODIS_irene_amo_2011238.kml, a MODIS image showing Hurricane Irene

➤ Lights_bakken_2012317.kml, a VIIRS image showing a section of northwestern North Dakota at night

➤ Lights_samerica_vir_2012202.kml, a VIIRS image showing a section of the South American eastern coastline at night

➤ Phyto_ge_08381.kml, a MODIS image showing a phytoplankton bloom off the coast of New Zealand

The folder also contains the following KML dataset from the University of Wisconsin's Space Science and Engineering Center:

➤ Daily_MODIS_May3.kml, a series of MODIS images covering the United States from May 3, 2017

Localizing This Lab

This lab uses Earth-observing satellite imagery data from a variety of locations across the globe, but there are several ways to examine some satellite imagery at a more local or regional level.

In Section 12.1 take a look through the NASA Earth Observatory Website and archived information, and look for some related phenomena (such as other fires or storms) that are either relatively close to your location or whose MODIS imagery overlaps into your area, download their KML files, then examine those places in Google Earth.

In Sections 12.3 and 12.4, examine some of the MODIS and MOPITT imagery products to determine the climatic interactions in your region, rather than on a global scale. Open and examine images from multiple dates in Google Earth Pro for your area, rather than one month's worth of data.

12.1 Viewing Satellite Imagery Overlays with Google Earth

1. We'll start with using the free browser version of Google Earth to examine several examples of Earth-observing satellite imagery for environmental applications. Go to **https://earth.google.com** and launch **Google Earth**.

2. When Google Earth opens, you'll see a set of controls running down the left-hand side of the screen. Their functions are as follows:

Menu

Search

Voyager

I'm Feeling Lucky

My Places

Share

- The three lines icon is the **Menu** that allows you access to various settings for Google Earth.
- The magnifying glass is the **Search** feature that allows you to query Google Earth to move to various locations.
- The ship's wheel is the **Voyager** feature that allows you to access pre-made tours or guides for Google Earth.
- The dice represents the **I'm Feeling Lucky** feature that will send you to a random location on Google Earth with each click.
- The standard icon is the **My Places** feature that will allow you to add your own data to Google Earth (like we'll do in this chapter).
- The two diagonal lines icon is the **Share** feature that will allow you to post a link to Google Earth imagery using social media.

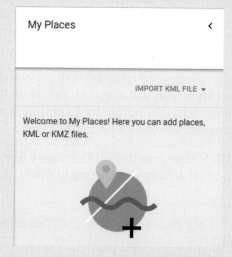

3. To start, click on the **My Places** icon. At the bottom of the My Places box click on the option for **Settings** - this will bring up a dialog that will allow you to **Enable KML File Import**. Do so, then click **SAVE** to close the dialog box. Back in the My Places box, click on **IMPORT KML FILE**.

4. Next, select **Open File** and navigate to the **Chapter12** folder and choose to open the **Lights_bakken_2012317.kml**. A new layer called Gas Drilling, North Dakota, will be added to Google Earth, and the globe will zoom in to a section of northwestern North Dakota. Although North Dakota is sparsely populated, there are a lot of lights that can be seen by VIIRS at night. These represent gas and oil drilling sites of the Bakken shale play in the region. More information about this is available through NASA's Earth Observatory at **http://earthobservatory.nasa.gov/IOTD/view.php?id=79810**. You may have to zoom out and pan around the imagery to see the whole North Dakota region.

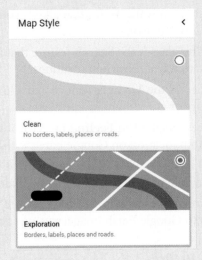

5. To put this imagery into geographic context, click on the **Menu** icon, then select **Map Style**. This will allow you to turn on features such as borders and labels. Select the option for **Exploration** and you'll see labels for locations appear on Google Earth.

> **Question 12.1** How does the VIIRS imagery help determine where shale and oil drilling is occurring in this section of North Dakota?

6. Next, click on the **My Places** icon again and then click on **DISCARD** to remove the KML file and clear the display.

7. Open the next KML file to Google Earth: **Lights_samerica_vir_2012202.kml**. This is a VIIRS image captured during the night in the eastern coast of South America in July 2012. More information about this is available through NASA's Earth Observatory at **https://earthobservatory.nasa.gov/IOTD/view.php?id=79822**. You may have to zoom out and pan around to see the whole region covered by the imagery.

> **Question 12.2** How does the VIIRS imagery help determine the spread of cities and populations across this section of South America? How can this imagery be used to aid in measuring demographics such as population density?

8. Discard the KML file and open the next one to examine: **Fires_russia_tmo_2012170.kml**. This is a MODIS image of wildfires burning in Siberia. More information about this is available through NASA's Earth Observatory at **http://earthobservatory.nasa.gov/IOTD/view.**

php?id=78305. You may have to zoom out and pan around to see the whole region covered by the imagery.

> **Question 12.3** How are the fires being shown in this MODIS imagery? How is the extent of the fires being tracked via MODIS? (*Hint*: What else is visible in the MODIS scene aside from the fires themselves?)

9. Discard the KML file and open the next one to examine: **Phyto_ge_08381.kml**. This is a MODIS image of a phytoplankton bloom off the coast of New Zealand. More information about this is available through NASA's Earth Observatory at **https:// earthobservatory.nasa.gov/IOTD//view.php?id=8381**. Again, you may first have to zoom out to see the extent of the MODIS image and then zoom in to examine some of the details.

> **Question 12.4** How is the phytoplankton bloom shown in this MODIS image (i.e., what distinguishes the bloom from the surrounding ocean)? Beyond the large bloom, a second bloom is also present—how is it displayed on the MODIS imagery?

10. Once more, discard the KML file and open the next one to examine: **MODIS_irene_amo_2011238.kml**. This is a MODIS image of Hurricane Irene from 2011, a massive destructive storm formation. More information about monitoring this storm is available at **http:// earthobservatory.nasa.gov/NaturalHazards/view.php?id=51931**. As with the other KML images, you'll have to zoom out to see the extent of the imagery.

> **Question 12.5** Why is MODIS used for monitoring weather and storm formations such as Hurricane Irene rather than other satellite systems we've discussed, such as WorldView-3 or Landsat 7 or 8?

11. Discard the Hurricane Irene KML file. We won't be using the browser version of Google Earth any longer in this lab, but keep your browser open as we'll be doing more with various Websites.

12.2 Viewing Recent Earth-Observing Imagery with Google Earth Pro

1. A system such as MODIS can image almost the entire Earth in a single day, so we will have access to very recent imagery from it. To examine some examples of this, first start **Google Earth Pro** (GEP). When GEP opens, from its **File** pull-down menu select **Open**, then navigate to the **Chapter12** folder and open a KML file called **Daily_MODIS_May3**.

When the KML opens it will begin zooming very closely to Earth—you may want to stop it in mid-zoom and then zoom out to see the entire United States in Google Earth.

2. This KML consists of a series of MODIS images from May 3, 2017, that cover nearly all of the United States—pan and zoom across the imagery to see the entire area covered by these MODIS images.

> **Question 12.6** By examining the imagery you'll see where the overlapping images were taken from and where small pieces of imagery are missing. Approximately how much area of the United States is covered in one MODIS swath? Why do you think there are pieces missing from the imagery?

[Source: University of Wisconsin SSEC]

3. These MODIS images were downloaded from the University of Wisconsin's Space Science and Engineering Center's MODIS Today online tool. To see what's happening with current MODIS imagery, point your Web browser to **http://ge.ssec.wisc.edu/modis-today**. When the Website opens, you'll see the currently available real-time MODIS imagery available. Click on the radio buttons for **Terra** and **Aqua** to see what type of coverage of MODIS imagery is currently available for each one (as both satellites carry a MODIS instrument).

4. Choose whichever satellite (Terra or Aqua) gives you the better overall coverage. If neither is acceptable, click on the **Previous Day** button and try again with either satellite. When you have a good MODIS

image of the United States to work with, click on **Open in Google Earth**. If you are prompted with a dialog box, choose to open the file in Google Earth Pro. If not, save the KML to the **Chapter12** folder and then manually open it in Google Earth Pro.

5. You'll see in Google Earth Pro's **Places** box, under Temporary Places, the name of the initial KML you added (t1.17122) placed at the top of the list and then the second KML you added directly from the Website (shown in the graphic above as aqua_today.kml) placed second at the bottom. When you add KML/KMZ files to Google Earth Pro, their order of priority for displaying them goes from top to bottom in the Places box. So, whatever is at the bottom of the stack gets displayed first, then the next up from the bottom displayed on top of it, until whatever is at the top of the stack gets displayed over everything (this will also become important in the next section of the lab). You can turn layers on and off by placing a checkmark in their boxes. For now, only display the new KML file you downloaded from the Website.

> **Question 12.7** From examining today's MODIS imagery you selected, are there any notable weather patterns you can see forming (large storm, etc.)? How much of the United States is visible and not clouded in?

6. Turn off both of the MODIS images.

12.3 Using the NASA Earth Observations (NEO) Web Resources

1. In your Web browser go to **http://neo.sci.gsfc.nasa.gov**. This is the Website for NEO (NASA Earth Observations), an online source of downloadable Earth observation satellite imagery. In this portion of the lab, you'll be using NEO's imagery in conjunction with Google Earth Pro in a KMZ format. Rather than providing you with only a view of a flat satellite image, NEO gives you the option of examining the imagery draped across a global view in Google Earth Pro.

[NASA NEO]

2. Click on the **Atmosphere** option at the top of the page.

3. Select the option for **Carbon Monoxide**.

4. This is an image of global carbon monoxide concentration for one month, collected using the MOPITT instrument aboard Terra (see the section on **About this dataset** for more detailed information).

5. Select **2016** for the year and then choose the option for **June 2016**. Select **Google Earth** from the **Downloads File Type** pull-down menu, then click on the option for **1440 × 720** to begin the download of the KMZ file. If prompted, open the file in Google Earth Pro. If you had to download the KMZ file, locate it on your computer and manually open the file in Google Earth Pro.

[NASA NEO]

6. Rotate Google Earth Pro to examine the MOPITT imagery draped over the globe. You'll want to make sure there's a checkmark in Google Earth Pro's **Borders and Labels** box.

> **Question 12.8** Geographically, where were the highest concentrations of carbon monoxide for this month?

7. Turn off the MOPITT Carbon Monoxide image in Google Earth Pro.
8. Back on the NEO Website, select the **Energy** tab, then select **Land/ Surface Temperature [Day]**.
9. Choose the option for **2017**, then select **February**.
10. Use **Google Earth** for the **Downloads File Type** option and click on the **1440 × 720** option. If prompted, open the file in Google Earth Pro. If you had to download the KMZ file, locate it on your computer and manually open the file in Google Earth Pro.
11. Rotate Google Earth Pro to examine the new MODIS image.

> **Question 12.9** Geographically, where were the lowest daytime land surface temperatures for this month?

> **Question 12.10** Geographically, where specifically in South America are the lowest daytime land temperatures for this month? Why is this when the rest of the continent has higher daytime land surface temperatures? (*Hint:* You may want to zoom in on some of these areas then examine them with and without the MODIS imagery turned on.)

12. Turn off the MODIS **Land/Surface** Temperature image in Google Earth Pro.
13. Back on the NEO Website, select the **Ocean** tab, and then select **Sea Surface Temperature 2002 + (MODIS)**. Choose **2017** for the year, then select **March**. Use **Google Earth** for the **Downloads File Type** option and click on the **1440 × 720** option. If prompted, open the file in Google Earth Pro. If you had to download the KMZ file, locate it on your computer and manually open the file in Google Earth Pro.
14. Rotate Google Earth Pro and zoom in to examine the new MODIS composite image.

> **Question 12.11** Geographically, where were the areas with the warmest sea surface temperatures on the planet for this month? (*Hint:* You may want to turn on the Grid functions in Google Earth Pro—under the View pull-down menu—to add further geographic context to help answer this question.)

> **Question 12.12** How do the sea surface temperatures of the
> Atlantic Ocean off the coasts of Lisbon, Portugal; Saint Pierre,
> Newfoundland; London, England; and New York City, New York
> all compare to one another (i.e., rank them from warmest to coolest
> temperatures) for this month?

15. In Google Earth Pro, turn off this last MODIS image.

12.4 Earth-Observing Imagery from the National Snow and Ice Data Center

1. NASA NEO is not the only source of Earth-observing satellite data
   for viewing in a virtual globe format. Another example is the National
   Snow and Ice Data Center (NSIDC). Go to their Website at **http://
   nsidc.org/data/google_earth**.

2. Go to the **View More Snow, Ice, Glaciers, Permafrost, and Sea
   Ice on the Globe** section of the Website. Look for the section called
   Global Snow Cover from MODIS.

Global Snow Cover from MODIS

View global, monthly average snow cover by month for 10 years: 2000-2009.
Download:

2009 snow cover (KML, 670 KB)	2004 snow cover (KML, 670 KB)
2008 snow cover (KML, 670 KB)	2003 snow cover (KML, 670 KB)
2007 snow cover (KML, 670 KB)	2002 snow cover (KML, 670 KB)
2006 snow cover (KML, 670 KB)	2001 snow cover (KML, 670 KB)
2005 snow cover (KML, 670 KB)	2000 snow cover (KML, 670 KB)

Data Source: MODIS/Terra Snow Cover Monthly L3 Global 0.05Deg CMG, Version 5

Note: This time series runs from October 2000 to May 2009. The following months are missing: March 2002 and December 2003

3. Read about what the imagery is showing, and then click on the **2009
   snow cover** KML option. Like with NEO, if prompted, open the file
   in Google Earth Pro. If you had to download the KML file, locate it on
   your computer and manually open the file in Google Earth Pro.

4. In Google Earth Pro's Temporary Places, there will be a new
   expandable item for **NSIDC**. By expanding this item and looking
   through the options, you'll see another expandable item called **Snow
   Cover**. This shows that snow data for each month of 2009 is available
   as an option that can be clicked on and off. To examine just one month
   of data, turn off all of the other months. You can also use the slider bar
   that appears at the top of the Google Earth Pro view to adjust the snow
   data in Google Earth Pro from month to month.

5. Rotate Google Earth and zoom in to examine the new MODIS imagery.

> **Question 12.13** Geographically, where were the greatest concentrations of snow cover in the southern hemisphere in June 2009?

> **Question 12.14** Geographically, where were the greatest concentrations of snow cover in the northern hemisphere in December 2009?

6. At this point, you can exit Google Earth Pro by selecting **Exit** from the **File** pull-down menu.

Closing Time

This exercise showed off a number of different ways of viewing and visually analyzing climate and environmental data available from Earth-observing satellites. The next chapter's going to change gears and we'll start looking at the ground that those satellites are viewing.

Digital Landscaping

Topographic Maps, US Topos, Contours, Digital Terrain Modeling, Digital Elevation Models (DEMs), Lidar, 3DEP, and Applications of Terrain Data

There's one big element in geospatial technology that we haven't yet dealt with: the terrain and surface of Earth. All of the maps, imagery, and coordinates discussed have dealt with two-dimensional (2D) images of either the surface or the developed or natural features on the land. We haven't yet described how the actual landscape can be modeled with geospatial technology. This chapter delves into how terrain features can be described, modeled, and analyzed using several of the geospatial tools (such as GIS and remote sensing) that we're already familiar with. Whether you are modeling topographic features for construction or recreational opportunities, or you're just interested in seeing what the view from above looks like, the landforms on Earth's surface are a principal subject for geospatial analysis (see **Figure 13.1** for an example of visualizing Mount Everest with Google Earth).

First, when we're examining landforms on Earth's surface, we will have to assign an elevation to each location. In terms of coordinates, each *x/y* pair will now have a *z*-**value** that will indicate that location's elevation. These elevations have to be measured relative to something—when a point has an elevation of 900 feet, this indicates that it is 900 feet above something. The "something" represents a baseline, or **vertical datum**, that is the zero point for elevation measurements. Some maps of landforms indicate that the vertical datum is taken as mean sea level (represented by the National Vertical Datum of 1929). When this vertical datum is used and an elevation value on a map indicates 1000 feet, that number can be read as 1000 feet above mean sea level. Much geospatial data in North America utilizes the North American Vertical Datum of 1988 (**NAVD88**) as the origin point for elevations. The U.S. National Geodetic Survey has announced

z-value the elevation assigned to an *x/y* coordinate

vertical datum a baseline used as a starting point in measuring elevation values (which are either above or below this value)

NAVD88 the North American Vertical Datum of 1988, the vertical datum used for much geospatial data in North America

Mount Everest modeled as a digital terrain landscape and shown in Google Earth.

that it is currently developing a new vertical datum to take the place of NAVD88, but it will not be ready until 2022. Coastal terrain models may also use mean high water as their vertical datum.

How Can Terrain Be Represented on Topographic Maps?

topographic map a map created by the USGS to show landscape and terrain as well as the location of features on the land

A common (and widely available) method of representing terrain features is the **topographic map**. As its name implies, a topographic map is a printed paper map designed to show the topography of the land and the features on it. A U.S. topographic mapping program that ran from 1945 to 1992 produced topographic maps that were published by the USGS (United States Geological Survey) at a variety of map scales. A common topographic map was a 1:24000 quadrangle, also referred to as a 7.5 minute topographic map, insofar as it displays an area covering 7.5 minutes of latitude by 7.5 minutes of longitude. Topographic maps were also available in smaller map scales, such as 1:100000 and 1:2000000.

DRG Digital Raster Graphic; a scanned version of a USGS topographic map

Topographic maps are available in a digital format as a **DRG** (Digital Raster Graphic). DRGs are scanned versions of topographic maps that have been georeferenced (see Chapter 3) so that they'll match up with other geospatial data sources when used in GIS. Usually, DRGs are available in **GeoTIFF** file format (see Chapter 7 for more information about TIFFs), which allows for high-resolution images while also providing spatial reference.

GeoTIFF a graphical file format that can also carry spatial referencing

However, DRGs are just scanned images. To use the specific features on a topographic map (such as the streets or rivers) as separate

GIS datasets, those features would have to be digitized (see Chapter 5) into their own geospatial layers. This is what a Digital Line Graph (**DLG**) is—the digitized features from a topographic map. DLGs are vector datasets representing transportation features (such as streets, highways, and railroads), hydrography features (such as rivers or streams), or boundaries (such as state, county, city, or national forest borders (see **Figure 13.2** for an example of several DLGs and their corresponding DRG).

DLG Digital Line Graph; the features (such as roads, rivers, or boundaries) digitized from USGS maps

Topographic maps (and thus, DRGs) model the landscape through the use of **contour lines**—imaginary lines drawn on the map that join points of equal elevation (see **Figure 13.3**). These contour lines are available in DLG format as a hypsography layer. A particular elevation on the surface is represented by a contour line drawn on the map. However, because elevation is really a continuously variable phenomenon, it's often difficult to represent every change in elevation (such as from 8 feet of elevation to 10 feet of elevation) without overloading the map to the point of uselessness with contour lines. Thus, contour lines are drawn a certain elevation distance apart (say, a new contour line for every 50 feet of elevation).

contour line an imaginary line drawn on a map to connect points of common elevation

FIGURE 13.2

1:24000 DLG data showing roads, railroads, boundaries, and hydrologic features of Youngstown, Ohio, along with the corresponding 1:24000 DRG of Youngstown.

Youngstown DLGs
— Roads DLG
— Rails DLG
— Hydro DLG
— Boundaries DLG

FIGURE 13.3

Contour lines as shown on a DRG (at a 20-foot contour interval). *[USGS]*

contour interval the vertical difference between two adjacent contour lines drawn on a map

This elevation or vertical difference between contour lines is called the **contour interval** and is set up according to the constraints of the map and the area being measured. In general, a wider contour interval is selected when mapping more mountainous terrain (since there are numerous higher elevations), and a narrower contour interval is used when mapping flatter terrain (because there are fewer changes in elevation and more details can be mapped). Small-scale maps (see Chapter 7) tend to use a wider contour interval because they cover a larger geographic area and present more basic information about the terrain, whereas large-scale maps generally utilize a narrow contour interval for the opposite reason (they show a smaller geographic area and thus can present more detailed terrain information).

A big thing to keep in mind when dealing with topographic maps is that they are no longer being produced by the USGS. The original USGS topographic mapping project lasted until 1992, with sporadic revisions made to a small percentage of the maps for a few years after that. The topographic map program was replaced by The National Map (see Chapter 1) to deliver digital geospatial data (including contour line data) for the United States. Since these topographic maps are no longer being produced, neither are their corresponding DRGs or DLGs. However, see *Hands-On Application 13.1: The Historical Topographic Map Explorer* for an online tool for examining series of historic USGS topographic maps.

US Topo a digital topographic map series created by the USGS to allow multiple layers of data to be used on a map in GeoPDF file format

Printed or scanned topographic maps haven't been produced for several years, however the next generation of digital topographic mapping is available through the **US Topo** series. Like the old topographic maps, the US Topos are delivered in 7.5 minute quad sections, but in GeoPDF file format (see Chapter 7). A US Topo has multiple layers of data stored on the same

Hands-On Application 13.1

The Historical Topographic Map Explorer

The Historical Topographic Map Explorer is a great USGS online tool for examining historic topographic maps over time. To get started, go to **http://historicalmaps.arcgis.com/usgs**. When the Website opens, you'll need to type in a location. Start with Orlando, FL. When the basemap of Orlando appears, click on it to open the historical timeline of topographic maps at the bottom of the screen. In the timeline, the dates of available topographic maps (and their map scales) are shown. By clicking on one of the dates, the corresponding topographic map will be added to the view in its georeferenced position. By clicking on a second date, that map will be added as well.

In the table of contents on the left side of the screen, maps can be reordered (the map at the bottom of the stack is drawn first and maps at the top of the stack are drawn over it), removed (by clicking the X for the map) or made semi-transparent

(with the slider bar under each map). First add the 1:24000 scale topographic map for 1956, then add the map for 1970. Use the slider bar to make the 1970 map semi-transparent to see the changes from one time period to the other. Next add the 1980 map and compare it to the 1970 map, then add and compare the 1995 map to the 1980 map.

Expansion Questions

➤ How has Orlando changed and where has it expanded as a city over time from 1956 to 1970? From 1970 to 1980? From 1980 to 1995? And also from 1956 to 1995?

➤ Look for your own local area and see what historical topographic maps are online for your location. What map dates are available to view? How has your area changed significantly on the maps from the earliest time period available to the latest available time period?

map, which can be turned on and off according to what features you want to display on the map. US Topo maps include contours, recent transportation features (roads, railroads, airports), transportation names, contours, boundaries, and orthophoto images, all as separately accessible layers. A US Topo also contains a lot of data (including the scale, the dates of map creation, and similar information) in the white border that surrounds the map. This information-filled border is referred to as the map's **collar**, and like the other layers in the US Topo it can be displayed or hidden. See **Figure 13.4** for an example of a US Topo.

collar the white, information-filled border around a US Topo

The data contained in the US Topos (such as the contours or transportation layers) is derived from the national GIS data. The USGS produces over 100 US Topos per day and each US Topo is on a three-year revision cycle. Also, historic topographic maps are being converted over to the GeoPDF format as part of the Historic Topographic Map Collection (HTMC) and freely distributed online.

As a GeoPDF, each US Topo is a stand-alone product that can be used on a computer or on a mobile device. The GeoPDF can also store coordinate information for each location on the map, and a set of tools

FIGURE 13.4

A 2013 US Topo of Youngstown, Ohio, showing the available layers for the map.
[USGS]

available from a special free plug-in allows you to measure distances, calculate the size of areas, and also connect the US Topo to a GPS device for obtaining your current location information (see *Hands-On Application 13.2: US Topos as GeoPDFs* for how to obtain and work with US Topos).

Hands-On Application 13.2

US Topos as GeoPDFs

Before working with US Topos, you'll need to install the TerraGo Toolbar in order to take advantage of all of the GeoPDF features available with the US Topo. The TerraGo Toolbar can be downloaded directly from **http://www.terragotech.com/products/ terrago-toolbar**. With the GeoPDF downloaded and the toolbar installed, now it's time to get some US Topos to work with.

US Topos can be obtained easily through The National Map (see Chapter 1) in the same way you would get other data layers. An alternative way to browse current and historic US Topos is through the USGS Store at **http://store.usgs.gov**. Click on the link for Map Locator & Downloader and you'll be taken to a separate Website to search for available maps. At this point, you can first type the name of a place to search for available US Topos. For instance, search for East Lansing, Michigan, and the display map will shift to show that. Next, click on the blue marker in the center of the East Lansing view and a new pop-up box will appear to show you all of the available maps for that location. For each of the maps labeled as 7.5 × 7.5 GRID, click on the circular icon with the blue-and-white cross on it to add that map to the download cart.

As you add the 7.5 maps (these will be US Topos or Historic Topographic Map Collection maps) to the download cart, you'll see your cart begin to fill. Once you have the 7.5 × 7.5 GRID maps for all available dates, click on Download All Cart Items. A zip file will download to your computer containing all of the GeoPDFs that you have added to the cart. When the zip file finishes downloading, unzip the US Topos and open the one with the most recent date. For further information on accessing and using US Topos, see the Quickstart guide from the USGS, online at **http://nationalmap.gov/usto-po/quickstart.pdf**.

Expansion Questions

➤ What dates of US Topos and historical topographic maps are available for East Lansing, Michigan?

➤ What is the approximate elevation of Spartan Stadium on the Michigan State University campus? (*Hint:* You can use the information on the Orthoimage and Geographic Names layers to locate the campus and stadium and the Contours layer for elevation.)

➤ What is the contour interval of the US Topo? What is the highest elevation shown on the map?

Thinking Critically with Geospatial Technology 13.1

If Everything's Digital, Do We Still Need Printed Topographic Maps?

A similar question was raised back in Chapter 2: If all of the historic USGS topographic map products are available digitally and in easily accessible formats that can be used on a desktop, laptop, or tablet, is there still a need to have printed copies (at a variety of map scales) on hand? Keeping in mind that the topographic mapping program ran through 1992 and only a small percentage of the maps were updated after that (and the estimated median date for the currentness of a map was 1979), are

the available printed maps up-to-date enough to be of use? Do surveyors, geologists, botanists, archeologists—or any other professionals who require topographic information for fieldwork—need to carry printed topo quads with them? Is there a need for geographers to have printed versions of several quads when seamless digital copies (which can be examined side by side) are so readily available? Are there situations in which a printed topo map is necessary?

How Can Geospatial Technology Represent Terrain?

DTM a representation of a terrain surface calculated by measuring elevation values at a series of locations

In geospatial technology, terrain and landscape features can be represented by more than just contour lines on a map. A digital terrain model (**DTM**) is the name given to a model of the landscape that is used in conjunction with GIS or remotely sensed imagery. The function of a DTM is both to represent the features of the landscape accurately and to be a useful tool for analysis of the terrain itself. The key to a DTM is to represent a z-value for x and y locations properly. With a z-value, the model can be shown in a perspective view to demonstrate the appearance of the terrain, but this doesn't necessarily make it a three-dimensional model. In fact, a DTM is usually best described as a two-and-a-half–dimensional model. In a 2D model, all coordinates are measured with x and y values, without a number for z at these coordinates. In a **two-and-a-half–dimensional (2.5D) model**, a single z-value can be assigned to each x/y coordinate as a measure of elevation at that location. In a full **three-dimensional (3D) model**, multiple z-values can be assigned to each x/y coordinate. Most DTMs have one elevation value measured for the terrain height at each location, making them 2.5D models (**Figure 13.5**).

two-and-a-half–dimensional (2.5D) model a model of the terrain that allows for a single z-value to be assigned to each x/y coordinate location

three-dimensional (3D) model a model of the terrain that allows for multiple z-values to be assigned to each x/y coordinate location

An example of a type of DTM that's used in terrain modeling is a **TIN** (Triangulated Irregular Network), in which selected elevation points of the terrain, those that the system deems the "most important," are used in constructing the model. Points are joined together to form nonoverlapping triangles, representing the terrain surfaces (see **Figure 13.6**). While TINs are often used in terrain modeling in GIS, another very widely used type of digital terrain model is the DEM.

TIN Triangulated Irregular Network; a terrain model that allows for non-equally spaced elevation points to be used in the creation of the surface

DEM digital elevation model; a representation of the terrain surface, created by measuring a set of equally spaced elevation values

What Is a DEM?

A **DEM** is a digital elevation model, a specific type of model of the terrain and landscape, produced by the USGS and others. A DEM is based on regularly spaced point data of elevations, but it can be converted to a raster

(a) 2D

(b) 2.5D

(c) 3D

FIGURE 13.6

A TIN representation of the terrain around Gate City, Virginia.

grid representation (see Chapter 5) for use with other geospatial data, such as satellite imagery or GIS layers. In grid format, the elevation values are represented by the grid cell value, while the grid resolution represents the size of the cell being measured. Thus, DEM resolution is measured in much the same way that we measure the resolution of a remotely sensed image (see Chapter 10). If a DEM has 30-meter resolution, then each of the DEM's raster grid cells is set up at 30 meters in size.

DEMs have been created through a variety of methods, including using DLG contour information, photogrammetry, and remotely sensed data and stereo imagery (see Chapter 14). The ASTER sensor onboard the Terra satellite (see Chapter 12) was used to create a global DEM at 30-meter resolution. Another source of terrain data is the Shuttle Radar Topography Mission (**SRTM**), which originated as a part of a mission of the space shuttle *Endeavor* in February 2000. For 11 days, *Endeavor* used a special radar system to map Earth's terrain and topographic features from orbit, and the result was a highly accurate digital elevation model. At the mission's close, roughly 80% of Earth was examined and modeled as 90-meter DEMs (30-meter DEM data is also available for the United States). Updates and edits to the initial SRTM dataset are available as well.

An additional remote sensing method of terrain mapping is called **lidar** (a mashup of the words "light" and "radar" but also an acronym for Light Detection and Ranging). Rather than firing a microwave pulse at a ground target as a radar system would, lidar uses a laser beam to measure the terrain. In lidar, a plane is equipped with a system that fires a series of laser beams (between 30,000 and 250,000 pulses per second) at the ground. The laser beams are reflected from the ground back to the plane, and based on the distance from the plane to targets on the ground, the elevation of the landscape (as well as objects on the surface of the terrain) can be determined. GPS (see Chapter 4) is used in part to determine where the beams are striking the ground. See **Figure 13.7** for an example of how lidar data is collected.

SRTM the Shuttle Radar Topography Mission, flown in February 2000, which mapped Earth's surface from orbit for the purpose of constructing digital elevation models of the planet

lidar a process in which a series of laser beams fired at the ground from an aircraft is used both to create highly accurate elevation models and also to measure the elevation of objects from the ground

FIGURE 13.7

Gathering point cloud data: the process behind measuring terrain heights using lidar.

point cloud the name given to the massive number of elevation data measurements collected by lidar

DSM digital surface model; a measurement of the heights of ground elevations as well as the objects on top of the ground as captured by lidar

LAS the industry standard data format used for lidar data

The data collected by lidar is often referred to as **point cloud** data, owing to the high volume of point locations measured (you could potentially measure millions of locations in a single flight). After the data is collected and processed, the end result of a lidar mission is a highly accurate set of x/y locations with a z-value (see **Figure 13.8** for an example of a lidar-derived elevation model). These "bare earth" elevation models of open terrain have a high degree of accuracy for their vertical measurements (between 15 and 30 centimeters), and today lidar equipment is also attached to UAS drones and can achieve 3- to 5-centimeter vertical accuracy.

However, Earth is covered by more than just open terrain. Lidar points will also measure the heights of features such as tree canopies or roofs of buildings and can be used to create realistic-looking 3D models of these objects on Earth's surface. We'll get to 3D visualization in Chapter 14, but these digital surface models (**DSMs**) can be developed using lidar to measure the heights of all types of things on Earth's surface, not just the ground. The heights of objects like trees or buildings are removed to create an elevation model of the actual Earth's surface. As such, lidar has become a very versatile remote sensing tool that is used extensively in both government and private sector fields. A **LAS** file is the industry standard for storing lidar data, and many software packages (such as ArcGIS) are able to generate elevation models and calculate object heights using this data.

FIGURE 13.8

Mount St. Helens as viewed from lidar data. *[USGS]*

Digital elevation data is available from the USGS through the **3DEP** (3D Elevation Program) initiative. 3DEP data is available as a free download from The National Map (see *Hands-On Application 1.3: The National Map* on page 11 of Chapter 1 for instructions on how to obtain various geospatial datasets). The purpose of 3DEP is twofold: to have a digital elevation product that covers the entire United States at the best possible resolution, and to eliminate any gaps in the data. 3DEP is designed to be "seamless", insofar as its data uses the same datum, projection, and elevation unit. With 3DEP, users select which sections of the national elevation dataset they need and then download those areas. 3DEP data is available at 1 arc second (about 30-meter), 1/3 arc second (about 10-meter), and 1/9 arc second (about 3-meter) resolutions, depending on the region. At the time of writing, some areas around the United States also had 1-meter resolution 3DEP data available. To see what's available for 3DEP data for your own area, see *Hands-On Application 13.3: U.S. Elevation Data and The National Map.* 3DEP draws from a variety of different sources for its elevation data. Lidar is a key method used to derive the DEMs that comprise 3DEP data, as well as radar remote sensing in areas such as Alaska. Point cloud data for areas as well as derived DSMs are also part of the 3DEP data sets.

3DEP the 3D Elevation Program, a U.S. government program that provides digital elevation data for the entire United States

Hands-On Application 13.3

U.S. Elevation Data and The National Map

The USGS distributes 3DEP data through The National Map. To see what types of elevation data are available, go to the TNM download viewer platform at **http://viewer.nationalmap.gov/basic**. Under the options for Data on the left side, put a checkmark in the box next to Elevation Products 3DEP. This will allow to you see all of the available elevation resolutions for 3DEP. Under this, click on the option for Product Availability and a series of radio buttons will appear, allowing you to see what 3DEP elevation data resolutions are available at areas across the United States. Then click on Show Availability to see what DEM data is available. For instance, clicking the radio button for DEM 1-meter will show you which areas of the country have available 1-meter 3DEP elevation data.

Next, you can go to a particular location to see what elevation data is available. For instance, in the Search location box, type in Topeka, Kansas. The Viewer will center on Topeka, but you may also need to zoom out a little bit from the initial search area to see the entire area. Select each of the radio buttons to see the kinds of 3DEP data available for download for Topeka.

Expansion Questions

➤ What resolutions of 3DEP elevation data are available for Topeka, Kansas?

➤ For your own local area, what types of elevation data and which resolutions of data are available?

How Can Digital Terrain Models Be Utilized?

viewshed a data layer that indicates what an observer can and cannot see from a particular location due to terrain

With terrain data, many different types of terrain analysis can be performed. DTMs are used in geospatial technology for creating **viewsheds**—maps that show what can be seen (or not seen) from a particular vantage point. A viewshed is used for determining how far a person's visibility is (that is, what they can see) from a location before his or her view is blocked by the terrain. A viewshed can be created by selecting a vantage point and then computing how far a person's line of sight extends in all directions until something blocks the view. DTMs are also used for a variety of hydrologic applications, for example calculating the accumulation of water in an area or delineation of stream channels or watersheds.

slope a measurement of the rate of elevation change at a location, found by dividing the vertical height (the rise) by the horizontal length (the run)

slope aspect the direction that a slope is facing

DTMs can be used to derive a new dataset of **slope** information—rather than elevation values, slope represents the change of elevations (and the rate of change) at a location by calculating the rise (vertical distance) over the run (horizontal distance). When a slope surface is created from a DTM, information can be derived not just about heights, but also about the steepness or flatness of areas. Similarly, a surface of **slope aspect** can be computed, which will show the direction that the slope is facing for each location. Now, you not only have information about where the

FIGURE 13.9

A DEM of a portion of the Laurelville/Hocking region of Ohio and the slope and slope aspect grids derived from it.

DEM Terrain Elevations

hockdem

Value

High : 1172

Low : 650

Slope of DEM

	0 - 7.604005223
	7.604005224 - 15.79293392
	15.79293393 - 23.98186263
	23.98186264 - 31.58586785
	31.58586786 - 38.31248785
	38.31248786 - 44.16172264
	44.16172265 - 49.71849569
	49.7184957 - 55.27526873
	55.27526874 - 74.57774353

Slope Aspect of DEM

	Flat (-1)
	North (0-22.5)
	Northeast (22.5-67.5)
	East (67.5-112.5)
	Southeast (112.5-157.5)
	South (157.5-202.5)
	Southwest (202.5-247.5)
	West (247.5-292.5)
	Northwest (292.5-337.5)
	North (337.5-360)

steepest slopes are located, but also about whether they are north-facing, east-facing, and so on. **Figure 13.9** shows a comparison of a DEM with the slope and slope aspect maps derived from it.

Digital terrain models aren't just limited to flat images on a computer screen or a paper map—remember, constructs like DTMs are 2.5D models, and because they have a z-value, that z-value can be visualized in a 3D view. It's more accurate to say these types of visualizations are really **pseudo-3D** views because they're really 2.5D, but they are not really full 3D models. Setting up a pseudo-3D view of a digital terrain model involves examining it at a **perspective view** (or an oblique view). Points or raster cells are elevated to the height of their z-value, and the resultant model is shown from an angular view. In this way, mountain peaks can be seen jutting up from the surface, and a meteor crater looks like a depression in the ground.

In many geospatial technology applications, the model becomes interactive, allowing the user to move or "fly" over the terrain, skimming over the surface and banking past mountains. Beyond viewing the terrain in perspective view, there are numerous ways of artificially altering the appearance of the terrain to make it more realistic looking. One way is to drape imagery over the surface (see **Figure 13.10** for an example of a terrain surface shown in perspective view with a digital topographic

pseudo-3D a term often used to describe the perspective view of a terrain model because it is often a 2.5D model, not a full 3D model

perspective view the oblique angle view of a digital terrain model from which the model takes on a three-dimensional appearance

FIGURE 13.10

A pseudo-3D view of a digital terrain model enhanced with a digital topographic map displayed over the terrain.

hillshade a shaded relief map of the terrain created by modeling the position of the Sun in the sky relative to the landscape

Sun altitude the value between 0 and 90 used in constructing a hillshade to model the Sun's elevation above the terrain

Sun azimuth the value between 0 and 360 used in constructing a hillshade to model the Sun's position in the sky to show the direction of the Sun's rays striking the surface

map overlaid to show off the contour lines in relation to the pseudo-3D model).

Remember that a DTM is just a map or grid of elevation surfaces. Like any other map, it can be displayed with various color ramps to show the differences in elevations, but there are several ways to make the terrain model look more realistic (and start to resemble that image of Mount Everest at the beginning of this chapter). The first of these techniques is called a **hillshade**, which models how the terrain would look under different lighting conditions (based on the location of the Sun in the sky). When you're constructing a hillshade, you can calculate where the light source (that is, the Sun) will be located at different times with respect to the terrain and simulate the effects of shadows cast on the landscape by the various terrain features.

By setting two parameters for the Sun, you can determine different appearances of the terrain at different times of day. The first parameter is the **Sun altitude**, a value between 0 and 90 representing the angle of the Sun in the sky between 0 degrees (directly at the surface) and 90 degrees (directly overhead). The second parameter is the **Sun azimuth**, a value between 0 and 360 that shows where its rays are coming from. The values represent a circle around the landscape, with 0 being due north, 90 being due east, 180 being due south, and 270 being due west (the values are measured clockwise from due north). By setting values for Sun altitude and Sun azimuth, the terrain can take on various appearances to simulate how the landscape looks at different times and in different conditions during the day. A hillshade using a Sun altitude of 45 degrees and a Sun azimuth of 315 degrees is shown in **Figure 13.11**.

Hillshading provides a good shaded map of what the terrain will look like under various lighting conditions, but there are plenty of features on the landscape (for instance roads and land cover) that aren't shown with a hillshade. In order to see these types of features, we can use a process called **draping**, which essentially shows the terrain model with a remotely sensed image (or another dataset) on top of it. **Figure 13.12** shows an example of a Landsat TM image draped over a DEM. Draping is achieved by first aligning the image with its corresponding places on the terrain model, then

draping a process in which an image is given z-values to match the heights in a digital terrain model

FIGURE 13.11

A DEM of Columbiana County, Ohio, and a hillshade of the DEM made using a Sun altitude of 45 degrees and a Sun azimuth of 315 degrees. *[Source: USGS/Esri]*

FIGURE 13.12

A DEM of the Laurelville/ Hocking region of Ohio and a Landsat TM image draped over the DEM.

base heights the *z*-values of a digital terrain model that can then be applied to an image in the process of draping

assigning the *z*-values from those locations on the terrain (in Esri terminology, these are referred to as **base heights**) to those locations on the image. In essence, locations on the image are assigned a *z*-value that corresponds with the terrain model.

Draping is a common technique to show remotely sensed imagery on terrain features (the terrain model of Mount Everest in Figure 13.1 on page 472 appears in this way). Programs such as Google Earth and Google Earth Pro can show pseudo-3D landscapes by draping imagery over the underlying terrain models (see **Figure 13.13**). By creating a new draped image

FIGURE 13.13

Examining imagery on a digital terrain model of Glacier National Park in Google Earth.

Hands-On Application 13.4

Terrain and Imagery Examples in Google Earth

To see some examples of three-dimensional terrain, go to **earth.google.com** and start using Google Earth. Use the search function to look for Glacier National Park, MT. When Google Earth arrives at Glacier, click on the 3D button to change to a perspective view of the national park to see the mountains and terrain. By double-clicking on the red and white arrow button, you'll gain access to the navigation controls for Google Earth. You can use the controls to tilt the view up or down, then navigate and spin the view. Use the mouse and/or the arrow keys on the keyboard to fly around the landscape. Remember that what you're examining here is imagery that has been draped over a digital terrain model representing the landscape of this section of the country. Fly around the Glacier area (see Figure 13.13 for an example) and get a

feel for navigation over draped imagery. You'll be doing more of this (among many other things) in Geospatial Lab Application 13.1 but using Google Earth Pro.

Expansion Questions

➤ Try using Google Earth to fly to other areas and view their terrain, such as Mount Everest, Death Valley, and Machu Picchu. How does the addition of draped remotely sensed imagery on the terrain add to your sense of visualizing the features of the terrain?

➤ From examining Glacier as well as the other three options above, you'll see that in some locations some features on the landscape (such as trees or buildings) may just be shown flat and not in 3D. Why is this?

in perspective view, we can get new visual information about the appearance of the landscape that's not directly obtainable through an examination of contours or non-perspective DTMs (see *Hands-On Application 13.4: Terrain and Imagery Examples in Google Earth* for more information). For example, draping a dataset of contours over a DTM and looking at it in perspective can demonstrate visually how contour lines match up with the elevations that they represent.

Even with hillshading or draping to improve the terrain's appearance, there's no getting around the fact that some sections of terrain don't have much variety in terms of changing elevations and features. In these cases, differences in landscape elevations or slopes may be difficult to see when viewing or interacting with a digital terrain model. The solution is to enhance the differences between elevations artificially so that the landscape can be better visualized. **Vertical exaggeration** is the process of artificially altering the terrain model for visualization purposes so that the vertical scale of the digital terrain model is larger than the horizontal scale. If the model's vertical exaggeration is "5×," then the vertical scale (the z-values) will be five times that of the horizontal scale. These types of artificial changes can really enhance certain vertical features (for example, valleys appear deeper and peaks appear higher) for visual purposes. The downside to vertical exaggeration is that it alters the scale of the data and should be used only for visualization. For a comparison of different vertical exaggerations applied to a DTM, see **Figure 13.14**.

vertical exaggeration
a process whereby the z-values are artificially enhanced for purposes of terrain visualization

FIGURE 13.14

Different levels of vertical exaggeration for a DEM (values of none, 2, 5, and 10).

VE = None

VE = 2

VE = 5

VE = 10

Chapter Wrap-Up

This chapter has explored both methods of modeling and ways of visualizing the landforms on Earth's surface. In the next chapter, we'll do a lot more with the 3D aspect of visualization. After all, there's more than just the terrain that can be viewed in 3D—there are plenty of structures, buildings, and natural growth that can be added to a pseudo-3D terrain model to improve the realism of the scene. Chapter 14, which picks up where this one leaves off, continues pursuing new avenues of 3D visualization.

Before then, check out *Terrain and Topography Apps* for some downloadable apps that will allow you to use topographic maps on your mobile device. Also take a look at *Digital Terrain in Social Media* for information from Twitter and Instagram accounts as well as some YouTube videos related to this chapter's concepts. This chapter's lab will allow you to implement the chapter topics using Google Earth Pro.

Important note: The references for this chapter are part of the online companion for this book and can be found at **www.macmillanlearning.com/catalog/shellitoigt4e.**

Terrain and Topography Apps

Here's a sampling of available representative terrain apps for your smartphone or tablet. Note that some apps are for Android, some are for Apple iOS, and some may be available for both.

➤ **Avenza Maps:** An app that allows you to purchase maps, and also to import GeoPDF maps for use on a mobile device

➤ **NP Maps:** An app that allows you to download and view topographic maps of the U.S. National Parks

Digital Terrain in Social Media

Here's a sampling of some representative Twitter and Instagram accounts, along with some YouTube videos related to this chapter's concepts.

🐦 **Become a Twitter follower of:**

➤ **OpenTopography:** @OpenTopography

➤ **USGS (United States Geological Survey):** @USGS

➤ **USGS News about Mapping:** @USGSNewsMapping

Become an Instagram follower of:

➤ **OpenTopography:** @OpenTopography

➤ **USGS:** @usgs

You Tube On YouTube, watch the following videos:

➤ **How does LiDAR remote sensing work?** (a basic introduction to lidar): **www.youtube.com/watch?v=EYbhNSUnIdU**

➤ **How to use a GeoPDF** (a video from TerraGo about GeoPDFs): **https://www.youtube.com/watch?v=XUI_FFiv1kg**

➤ **Using US Topo and Historic Topo Maps on your Mobile Device** (a USGS video about using US Topos): **https://www.youtube.com/watch?v=nbCGoYJJwZs**

➤ **US Topo** (a USGS video concerning the development of US Topos): **www.youtube.com/watch?v=hv0jxsW3qgY**

➤ **What is the USGS 3DEP Program?** (a video interview concerning the 3DEP program): **https://www.youtube.com/watch?v=GBdGWWMvO-Y**

Key Terms

Digital Terrain Analysis

This chapter's lab will introduce you to some of the basics of digital terrain modeling—working with DTMs, slope, viewsheds, and imagery draped over the terrain model. You'll be using the free Google Earth Pro for this lab.

Objectives

The goals for you to take away from this lab are:

➤ To examine pseudo-3D terrain and navigate across it in Google Earth Pro

➤ To examine the effects of different levels of vertical exaggeration on the terrain

➤ To create a viewshed and analyze the result

➤ To create an animation that simulates flight over 3D terrain in Google Earth Pro

➤ To create an elevation profile for use in examining a DTM and slope information

Using Geospatial Technologies

The concepts you'll be working with in this lab are used in a variety of real-world applications, including:

➤ Civil engineering, in which slope calculations from digital elevation models are utilized to aid in determining the direction of overland water flow, a process that contributes to calculating the boundaries of a watershed

➤ Military intelligence, which makes use of terrain maps and digital elevation models in order to have the best possible layout of real-world areas (as opposed to two-dimensional maps) when planning operations involving troop deployments, artillery placements, or drone strikes

[PATRICK HERTZOG/AFP/Getty Images]

Obtaining Software

The current version of Google Earth Pro (7.3) is available for free download at **https://www.google.com/earth/download/gep/agree.html**.

Important note: Software and online resources can change fast. This lab was designed with the most recently available version of the software at the time of writing. However, if the software or Websites have changed significantly between then and now, an updated version of this lab (using the newest versions) will be available online at **www.macmillanlearning.com/catalog/shellitoigt4e**.

Lab Data

There is no data to copy in this lab. All data comes as part of the Google Earth Pro data that is installed with the software or is streamed across the Internet when using Google Earth Pro.

Localizing This Lab

The lab can be performed using areas close to your location, as Google Earth Pro's terrain and imagery covers the globe.

13.1 Examining Landscapes and Terrain with Google Earth Pro

1. Start **Google Earth Pro (GEP)**. By default, GEP's Terrain option will be turned on for you. Check in the **Layers** box to be sure that there is a checkmark next to **Terrain** (and don't uncheck that for the remainder of the lab). Having this layer turned on activates the DTM that underlies the imagery within GEP.

2. Next, as this lab is focused on digital terrain modeling, we'll want to use the most detailed DTM possible. From the **Tools** pull-down menu, select **Options**. In the Google Earth **Options** dialog box that appears, choose the **3D View** tab and place a checkmark in the box next to **Use high quality terrain (disable for quicker resolution and faster rendering)**. Next, click **Apply** and then click **OK** to close the dialog box and return to GEP.

3. For this lab, we'll be using GEP to examine a terrain model of Zion National Park in Utah—in the **Search** box, type **Zion National Park, UT**. GEP will zoom down to the area of Zion National Park. In the **Layers** box, expand the **More** option and place a checkmark next to **Parks/Recreation Areas**. This will show you the outline of Zion in green. Zoom out so that you can see the entire layout of Zion in the view. For more information about Zion National Park, check out **www.nps.gov/zion/index.htm**.

4. You will see a question mark symbol labeled as "Visitor Center" next to a label for "Park Headquarters." Center the view on this area of Zion.

5. From the **Tools** pull-down menu, choose **Options**. In the dialog box that appears, click on the **Navigation** tab and make sure that the box that says **Automatically tilt when zooming** has its radio button filled

in. This will ensure that GEP will zoom in to areas while tilting to a perspective view. Next, click **Apply** and then click **OK** to close the dialog box and return to GEP.

6. Use the **Zoom Slider** to tilt the view down, so that you have a good perspective view on Zion (see *Geospatial Lab Application 1.1* for more info on using this tool). Basically, push down the **plus** button on the Zoom Slider, and your view will zoom in and tilt down to a perspective where you can see the sides of the mountains and canyons in Zion in pseudo-3D. You can also hold down the **Ctrl** key on the keyboard while moving the mouse forward or backward to tilt the view.

7. Use your mouse wheel to move the view in and out (do this instead of using the Zoom Slider, because otherwise you'll keep tilting backward as well) as well as holding the **Ctrl** key until you can position the view as if you were at the park headquarters/visitor center and facing south (see below for an example).

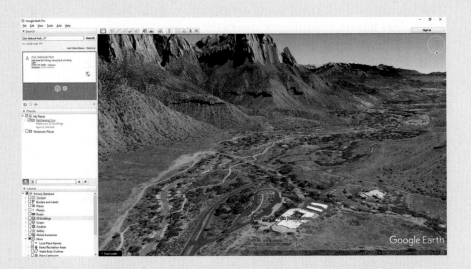

8. There are two references to heights or elevation values in the bottom portion of the view:

   • The heading marked "elev" shows the height of the terrain model (that is, the height above the vertical datum) where the cursor is placed.

   • The heading marked "Eye alt" shows GEP's measurement for how high above the terrain (or the vertical datum) your vantage point is.

   Maneuver your view so your Eye alt is a good level above the terrain (perhaps about 4100–4200 feet), and you'll be able to see the mountain area that surrounds the park headquarters in Zion.

> **Question 13.1** How does the pseudo-3D view from this position and altitude aid in bringing out the terrain features of Zion (compared to what you originally saw in the overhead view earlier)?

13.2 Vertical Exaggeration and Measuring Elevation Height Values in Google Earth Pro

Google Earth Pro also allows you to alter the vertical exaggeration of the terrain layer. As we discussed on page 487, vertical exaggeration changes the vertical scale but keeps the horizontal scale the same, therefore it should be used for visualization purposes only.

1. To look at different levels of vertical exaggeration, select **Options** from the **Tools** pull-down menu. Select the **3D View** tab.

2. In the box marked Elevation Exaggeration, you can type a value (between 0.01 and 3) for vertical exaggeration of GEP's terrain. Type a value of **2**, click **Apply** and **OK**, and then reexamine the area surrounding the park headquarters in Zion.

> **Question 13.2** How did the vertical exaggeration value of 2 affect the view of Zion?

> **Question 13.3** Try the following values for vertical exaggeration: 0.5, 1, and 3. How did each value affect the visualization of the terrain? In addition to the value of 2 you examined in Question 13.2, which value of vertical exaggeration was the most useful for a visual representation of Zion and why?

3. Reset the Elevation Exaggeration to a value of 1 when you're done.

4. From here, we'll examine the elevation values of the terrain surface itself. Although Google Earth Pro will always show the imagery over the terrain model, elevation information of each location is available. Wherever you move the cursor on the screen, a new value for elevation is computed in the elev option at the bottom of the view. By zooming and placing the cursor on its symbol on a spot on the screen, you can determine the elevation value for that location.

> **Question 13.4** At what elevation is the height of the park headquarters/visitors center?

13.3 Working with Viewsheds in Google Earth Pro

Now that you have a pretty good idea of how the landscape of Zion looks in the areas near the park headquarters, we'll now create a viewshed that will allow you to see what is visible and what is blocked from sight at a certain location.

1. To begin, we'll see what areas of Zion can be seen and cannot be seen from the park headquarters. Zoom in closely on the question mark symbol that GEP uses to mark the park headquarters. This symbol is what we'll use as the location for creating a viewshed.

2. Select the **Add Placemark** icon from the toolbar and put the placemark right on the park headquarters question mark symbol (see *Geospatial Lab Application 1.1* for how to do this).

3. In the **New Placemark** dialog box, name this new placemark **Park HQ**. Click **OK** in the New Placemark dialog box to accept the new name and close the dialog box.

Google Earth – New Placemark

Name: Park HQ

Latitude: 37°12′33.32″N

Longitude: 112°58′49.10″W

4. You'll see the new **Park HQ** placemark in the Places box. Right-click on it and select **Show Viewshed**. When prompted about the placemark being too low, click on **Adjust automatically**.

5. GEP will compute the viewshed. Zoom out a bit to see the extent of the viewshed—all of the areas covered in green are the places that can be seen from the location of the Park HQ placemark and the areas not in green cannot be seen from there. After zooming out, look at the areas immediately south and southeast of the Park HQ placemark.

> **Question 13.5** Can the ranger station, the campground, or the two picnic areas just south and southeast of the park headquarters be seen from the Park HQ vantage point?

6. Click on **Exit viewshed** to remove the viewshed layer.

7. About 2 miles to the northeast of the park headquarters is a scenic overlook. Its symbol in GEP is a small green arrow pointing to a green star. Move over to that location so that you can see it in the view and zoom in closely on the overlook symbol.

8. Put a new placemark directly on the green star and name this placemark **Zion Overlook**.

9. Create a viewshed for the Zion Overlook point and answer Questions 13.6 and 13.7. Exit the viewshed once you're done.

> **Question 13.6** Can the area labelled as parking to the immediate north of your Zion Overlook point be seen from the overlook?

> **Question 13.7** What is blocking your view of Sand Beach trail from the overlook? Be specific in your answer. *Hint:* You may want to zoom in closely to your Zion Overlook placemark and position the view as if you were standing at that spot and looking in the direction of Sand Beach trail.

13.4 Flying and Recording Animations in Google Earth

Google Earth Pro allows you to record high-definition videos of the areas you view in GEP. In this section, you'll be capturing a video of the high-resolution imagery draped over the terrain to record a virtual tour of a section of Zion. Before recording, use the mouse and the Move tools to practice flying around the Zion area. You can fly over the terrain, dip in and out of valleys, and skim over the mountaintops. Don't forget you can also hold down the Ctrl key and move the mouse to tilt your view. It's important to have a good feel of the controls, as any movements within the view will be recorded to the video. When you feel confident in your skills for flying over 3D terrain in Google Earth Pro, move on to the next step.

1. The tour you will be recording will start at the park headquarters (your Park HQ placemark), move to the scenic overlook (your Zion Overlook placemark), and then finish at the lodging/ parking area about a mile to the north of the overlook. We'll do a short "dry run" of this before you record. To begin, double-click on the **Park HQ** placemark in the Places box and the view will shift there.

2. Next, double-click on the **Zion Overlook** placemark in the Places box and the view will jump to there.

3. Lastly, use the mouse and **Move** tools to fly manually over the terrain a mile north of the overlook to the parking and lodging area and end the tour there.

4. If you need to, try this dry run of maneuvering among the three points so that you feel comfortable prior to recording. When you're ready to make your tour video, double-click on the **Park HQ** placemark in the Places box to return to the starting point of the tour. Also, take away the checkmarks next to the **Park HQ** and **Zion Overlook** placemarks in the Places box so that the two placemarks will not appear in the view (and thus not appear in the video) and that all you will see is the GEP imagery and terrain.

5. On Google Earth's toolbar, select the **Record a Tour** button.

6. A new set of controls will appear at the bottom of the view:

7. To start recording the video, press the red record button.

   Important note: If you have a microphone hooked up to your computer, you can get really creative and narrate your tour—your narration or sounds will be saved along with your video.

8. After showing the park headquarters in the view for a couple seconds, double-click on the **Zion Overlook** placemark in the Places box to jump to there. Show the overlook for a few seconds, then use the mouse and move commands to fly to the lodging/parking area to the north.

9. When you're done, press the **red** record button again to end your recording of the tour.

10. A new set of controls will appear at the bottom of the screen, and Google **Earth Pro** will begin to play your video.

11. Use the **rewind** and **fast-forward** buttons to skip around in the video, and also the **play/pause** button to start or stop. The button with the two arrows will repeat the tour or put it on a loop to keep playing.

12. If the tour is how you want it, save the tour by pressing the **disk icon** (Save Tour) button. Call it **Ziontour** in the dialog box that opens. The saved tour will be added to your Places box (just like all other GEP layers). If the tour is not looking how you want it, return to Step 4 and remake the tour.

13. Right now, the tour can only be played in GEP. You'll want to export your tour to a high-definition video that can be watched by others or shared over the Web. To start this process, first close the DVR control box in the view by clicking on the x in the upper right-hand corner of the controls.

14. Next, from the **Tools** pull-down menu, select **Movie Maker**.

15. In the **Movie Maker** dialog box, choose **1280 × 720 (HD)** for **Video parameters**. This will create a high-definition (HD) video of your tour.

16. Next, under **Record from…** choose the radio button for **A saved tour** and choose **Ziontour (My Places)**. This will be the source of your video.

17. From the pull-down menu next to File type, choose **.MJPEG (.mp4)**. If you are using Windows, you may want to instead choose **Windows Media (.asf)**.

18. Under **Save to…** use the **browse** button to navigate to the drive to which you want to save your video and call it **Ziontourmovie**.

19. Leave the other defaults alone and click **Create Movie**. A dialog box will appear showing the conversion process to create your video.

20. Return to the folder where you saved your movie (either as .mp4 or .asf) and view it using your computer's media player.

> **Question 13.8** Show (or submit) your final video file of your video tour of Zion to your instructor, who will check it over for its quality and completeness for you to get credit for this question.

13.5 Measuring Profiles and Slopes in Google Earth Pro

1. In the last part of this lab, you will be examining profiles of the digital terrain model and examining the slope information. To begin, return the view to the **Park HQ** placemark.

2. Zoom out so that you can see both the Park HQ and the Zion Overlook placemarks clearly in the view. What we'll do next is construct a profile (a two-dimensional view) of the digital terrain between the two placemarks and then examine the elevations and slope of the terrain.

3. To examine the terrain profile, you must first draw a path between the two points. To begin, choose the ruler tool from the GEP toolbar.

4. In the **Ruler** dialog box that opens, select the **Path** tab.

5. In the view, you'll see that your cursor has changed to a crosshairs symbol. Left-click once on the **Park HQ** placemark location and then click the left mouse button once on the **Zion Overlook** placemark position. You'll see a yellow line drawn between them and the Length of this line will be shown in the Ruler dialog box.

6. In the Ruler dialog box, click **Save**.

7. A new dialog will open and allow you to give this path you've drawn a name. Call it **ParkHQOverlookPath** and click **OK** in the **naming** dialog.

8. You'll see a new item added to the Places box called ParkHQOverlookPath. Right-click on it and select **Show Elevation Profile**.

9. A new window will open below the view, showing the elevation profile between the two points. This profile will show you, in two dimensions, the outline and elevations of the terrain in between the two points.

10. As you move your cursor along the outline of the terrain in the profile, information will appear about the elevation and the slope at the location of your cursor. You'll see the corresponding location highlighted in the view as well. The slope information is in a positive percentage as you move uphill from the park headquarters to the overlook and a negative percentage as you move downhill. Carefully examine the profile (and the digital terrain) and answer Questions 13.9, 13.10, and 13.11.

> **Question 13.9** What is the average slope for the 2.2-mile path between the park headquarters and overlook (both uphill and downhill)?

> **Question 13.10** What is the steepest uphill slope between the two points and where is this located?

> **Question 13.11** What is the steepest downhill slope between the two points and where is this located?

11. You can also select a small section of the path and only examine the profile of that small subset. To do so, click the mouse at a place in the profile you want to begin. Hold down the left mouse button and move

the mouse to the place in the profile where you want to end. Release the mouse button and you will see that small section of the profile you chose in a darker red. Statistics for the elevation and slope of that subset will be shown in a red box at the top of the profile.

12. We're going to examine one small part of the profile. On the far right side of the profile (the terrain nearest the overlook), you'll see that a section of the landscape drops off sharply (this should be around the 1.6-mile mark between the Park HQ and the overlook). Using the method described above, highlight the subset of the profile from the location at the bottom of the drop-off to the overlook itself, then answer Question 13.12.

> **Question 13.12** For this section of the terrain, what is the maximum uphill and downhill slope? What is the average uphill and downhill slope?

Closing Time

Google Earth Pro is a very powerful program with a lot of functionality and options for working with DTMs and digital terrain analysis. Google Earth Pro's detailed terrain model under the draped imagery can be used as the base for many different types of analysis. When you're finished, exit GEP by selecting the **File** pull-down menu and choosing **Exit**. There's no need to save any of your temporary places data for this lab.

14 See the World in 3D

3D Geovisualization, 3D Modeling and Design, Prism Maps, SketchUp, and Stereo Imagery in 3D

Up until the previous chapter, all the topics in this book dealt with geospatial technology in a two-dimensional format, whether as geospatial data, measurements, maps, or imagery. Chapter 13 began to go beyond two dimensions and start on a third, presenting terrain in a perspective and pseudo-3D view. This chapter looks entirely at presenting geospatial information in three dimensions, as well as designing and visualizing 3D data and concepts.

There's no doubt that 3D visualization of data is impressive—as technology has improved and computers have become faster and more powerful, 3D rendering has become more familiar to us. Video games and simulators are extremely impressive to watch, and computer-animated movies have made huge strides in recent years. Although geospatial technologies haven't (yet) reached the technical level of some of the state-of-the-art CGI we see in some big-budget movies, many 3D modeling and visualization techniques are capable of creating perspective views, 3D maps, and realistic-looking 3D objects. For instance, Google Earth and Google Earth Pro both support viewing and interacting with 3D geospatial data. See **Figure 14.1** for an example of realistic-looking 3D geospatial visualization in Google Earth.

Before proceeding, keep in mind the discussion from Chapter 13 concerning two-and-a-half–dimensional (2.5D) data versus three-dimensional (3D) data. If only one height value can be assigned to an *x*/*y* location, then the data is considered 2.5D; if multiple height values can be assigned to an *x*/*y* location, then the data is fully 3D (3D data also has volume). For ease of reading and usage, this chapter uses the term "3D" throughout to refer to all geospatial phenomena that incorporate a third dimension into their design or visualization, although technically, some examples will be 2.5D or pseudo-3D.

FIGURE 14.1

Lower Manhattan
(New York City), rendered
in 3D.

 ## What Is 3D Modeling?

3D modeling (in geospatial terms) refers to the design of spatial data in a three-dimensional (or pseudo-3D) form. Designing a 3D version of the buildings in a city's downtown district or a 3D version of the Eiffel Tower would be examples of 3D modeling of geospatial data. Like all of our spatial data, the 3D models should be georeferenced to take advantage of real-world coordinate systems and measurements. If you're designing a 3D model of your house, you'll want to start with a georeferenced base so that you can make accurate measurements of your house's footprint. An orthophoto, high-resolution satellite image, or architectural diagram with spatial reference are all good starting points for a georeferenced base. Digitizing the house's footprint will give you a polygon (with two dimensions, x and y), but a 3D-style object has to have a third dimension—a z-dimension. This z-**value** will be the height of the object above whatever surface it's standing on—in the case of your house, the ground on which it's built.

In order to create a 3D version of the polygon, it will have to be extruded to reach the height specified by the z-value. **Extrusion** is the process of giving height to an object. If your house is 15 feet high, the footprint polygon can be extruded to a height of 15 feet. Extrusion will change the objects—extruding a polygon will transform it into a **block**. In this case, the building footprint will change into a block 15 feet high (see **Figure 14.2** for the comparison between the two-dimensional polygon footprint and the extruded block).

Any object can be extruded to a particular height by assigning it a z-value. Keep in mind that these items extrude from ground level up and some objects can't be designed this way. For example, suppose you want to design a 3D model of a 10-foot-high pedestrian walkway that's 15 feet over a busy road. If you digitize the bridge's footprint and extrude it to 10 feet, it will look like a wall blocking the road rather than a bridge over it. To get

3D modeling designing and visualizing data that contains a third dimension (a z-value) in addition to x- and y-values

z-value the numerical value representing the height of an object

extrusion the extending of a flat object to have a z-value

block a flat polygon that has been extruded to transform it into an object with a z-value

offset a value applied to an object that is not attached to the ground, but is entirely above ground level

around this, you'll have to apply an **offset**, or a *z*-value, where the initial object is placed, before you do the extrusion. In this case, the digitized polygon representing the bridge will have to be offset to a height of 15 feet (its height from the ground as it crosses over the road), then extruded to a height of 10 feet (to capture the dimensions of the walkway). **Figure 14.3** illustrates the difference between a regular extruded block and an extruded block that has been offset from the surface.

A building being modeled will likely consist of multiple blocks (or other shapes), each representing a different portion of the building. Multiple levels of a house or different-sized parts of a building can each be represented by blocks. Just as a GIS dataset can consist of multiple polygons, a 3D representation can consist of multiple extruded shapes and blocks (see **Figure 14.4**). Each of these blocks can then be treated as its own object for further modeling.

FIGURE 14.4

Multiple extruded blocks making up a single building (a very early version of Weller House, on Youngstown State University's campus).

In 3D modeling it's not enough just to place an object in its correct georeferenced location—it must also conform correctly to the real-world terrain and to real-world elevations. If your house is located 900 feet above sea level, you don't want to base your house at a ground level of 0 feet. If you do, your house will be shown not on Earth's surface, but 900 feet underground when you merge your 3D model with other geospatial data. Terrain modeling was discussed back in Chapter 13, but the key element to remember with 3D modeling is the concept of **base heights** (to use Esri terminology). These values represent the elevation of the terrain above a vertical datum.

base heights the elevation values assigned to the terrain upon which the objects will be placed

When you're designing a 3D object (for instance a model of your house) that you intend to place on an area of the terrain, you should make sure that the terrain's base heights are applied to the object you're designing. In this way, the software will understand that the base of your house begins at an elevation of 900 feet, and extruding your house's footprint to a height of 15 feet means its rooftop will be 915 feet above sea level, rather than 15 feet above ground level (0 feet). Many geospatial software packages enable the user to combine 3D modeling techniques with terrain modeling to help with the design of more realistic 3D visualizations (some specific programs that do this are described later in this chapter).

Once the footprints and elevations are correctly set in place, it's time to start making those gray blocks look like the thing you're trying to model. Every block has several **faces**, each face representing one side of the object. For example, an extruded rectangular block will have six faces: top, bottom, and four sides. Each face can be "painted" to give it a more realistic appearance. This painting can take a variety of forms—doing something as simple as changing the color of one face to, say, yellow (thus repainting one side of your house to make it yellow) can make a considerable difference in its appearance.

face one side of a block

textures graphics applied to 3D objects to give them a more realistic appearance

More realistic appearances can be achieved by applying a **texture** to a face. A texture is a graphic designed to simulate the appearance of materials (such as brick, wood, stone, glass, etc.). With textures, you don't have to draw individual bricks on the side of a building—you just apply a red-brick texture and the face's appearance is changed to red bricks. Similarly, windows can be designed by applying a translucent texture to a window shape drawn onto a face, and various roofing materials can be applied to the faces that make up the roof of your house (see **Figure 14.5** for an example of a 3D building with various textures applied to the block faces).

Because a texture is just a graphic, it's possible to apply other graphics to a face. For instance, you could take a digital photo of your home's particular brick and apply it to a face of your model. If it's accuracy you want, you might even take a picture of an entire side of your house, resize it, and "paste" that image onto the entire corresponding side of your model. However, it's not just buildings and objects that can be visualized using 3D modeling techniques: 3D maps can also quickly communicate spatial information to the viewer.

FIGURE 14.5

A version of Youngstown State University's Weller House with features added and colors, textures, and additional details applied to the block faces (compare with Figure 14.4).

Thinking Critically with Geospatial Technology 14.1

What Is the Advantage of Using 3D Design?

The use of 3D visualization has become increasingly widespread as a means of viewing and using geospatial data. As processors have gotten faster and graphics and animation have improved, it's become easier than ever to render large amounts of geospatial data in realistic 3D form. Why is visualization of geospatial data so critical to conveying a message to the user or viewer of the data? Beyond the cool factor, what makes 3D visualization a better conveyor of data than a simple 2D map? What advantages does 3D visualization have over regular 2D for the presentation of data and communication of geospatial information?

How Are 3D Maps Made?

A 3D-style map can be made by applying the kind of 3D-modeling concepts that we've just discussed. However, this time you won't be working with extruding polygons that represent building footprints on the ground, but with the shapes and objects on a map and what they can represent. A regular choropleth or thematic map (see Chapter 7) can be transformed into a 3D-style map by extruding the polygons that are displayed on the map. The extruded polygons are called raised **prisms**, and a map that features these extrusions is referred to as a **prism map**. In a prism map, the shapes on the map are extruded to a height that reflects the data value assigned to them. For instance, if a thematic map of the United States is intended to show population by state, then each state's shape will be extruded to a "height" of the number of units equal to that state's population. Thus, the polygon shape of California will be a raised prism extruded to a value of 36.7 million, while neighboring Nevada's shape will only be extruded to a value of 2.6 million.

Figure 14.6 shows an example of a prism map dealing with world population. Each country's polygon has a value of the population statistics for that country. Those countries with the highest raised prisms have the largest number of persons (in this example, the map shows China and India's huge populations relative to those of neighboring countries). The polygon shape of each country is extruded relative to the number of units being measured.

3D design is becoming more and more prevalent as technology improves. In Chapter 13, you used Google Earth Pro to look at pseudo-3D versions of landscapes, but Google Earth Pro can be used for viewing the

prisms the extruded shapes on a prism map

prism map a 3D thematic map from whose surface shapes (such as the polygons that define states or nations) are extruded to different heights, which reflect different values for the same type of data

FIGURE 14.6

A prism map showing the 2010 populations of every country in the world.

results of 3D design as well, as done back in Chapter 1 when you looked at 3D versions of the Corn Palace or Mount Rushmore. There are numerous types of 3D modeling and design software and visualization tools out there; a commonly used 3D design software tool (that also works with Google Earth Pro) is SketchUp.

 ## What Is SketchUp and How Is It Used for 3D Modeling and Visualization?

SketchUp a 3D modeling and design software program distributed online by Trimble

It's easy to create 3D objects, structures, buildings, landforms, and more through the use of the **SketchUp** software program and then view them in Google Earth Pro. The regular version of SketchUp (called "SketchUp Make") is made freely available for download by Trimble, while a SketchUp Pro version with additional features is available for purchase (you'll be downloading and using the free version of SketchUp Make in this chapter's *Geospatial Lab Application*). SketchUp is a very versatile and intuitive program that allows for a full range of 3D design capabilities. Structures created with SketchUp can be as realistic and detailed—or as basic—as you want them to be. See **Figure 14.7** for an example of a football stadium designed in SketchUp.

Shapes can be molded, extruded, bent, and moved to build any type of structure you can envision (see *Hands-On Application 14.1: My.SketchUp* for a quick way to get your hands dirty with basic usage of SketchUp). Realistic textures can be applied to faces for a better appearance, and digital photos can also be resized and applied to a face for a more photorealistic appearance. SketchUp also features a process called Match Photo, which enables you to design the polygon dimensions of a building straight

FIGURE 14.7

A version of Youngstown State University's football stadium complex, displayed in SketchUp Make.

Hands-On Application 14.1

My.SketchUp

To jump right in and start using some of the 3D design features of SketchUp, a version—My.SketchUp—is available that runs in your Web browser without having to download any additional software. To get started, go to **https://my.sketch up.com/app**. When My.SketchUp opens, a figure of a person will be there to help establish a scale and reference for your 3D models. By clicking on a control on the left-hand side, more drawing options will appear for you. For instance, clicking on the Rectangle option will expand to give you choices such as "rotated rectangle," "circle," and "polygon." For a simple way to start, draw a rectangle or polygon on the "ground" (using the figure of the person as a guide), which would represent the footprint of a structure. Next, click on the Push/Pull tool and grab hold of your footprint and drag it upward to extrude that shape. Click on the Line tool to be able to draw things such as a door or windows on the faces of the extruded shape. You can use the Orbit tool and

its options to rotate around the 3D model you've made. Lastly, click on the Paint tool to access various options for applying textures to the faces and features you drew on them. When you're done, there are options in the upper-left corner for saving or downloading your model (to work within the regular SketchUp).

Expansion Questions

➤ What type of 3D structures or objects can you design quickly just using the My.SketchUp tools available in the browser? How do the Line and Paint tools help to bring these models to life?

➤ By clicking on the Components option on the right-hand side, you'll have access to Trimble's 3D Warehouse to add pre-made 3D models and objects to your own model. What types of objects are available in the 3D Warehouse for you to utilize with your My.SketchUp model?

from a digital photo. From there, the photo can be applied directly to the faces for a realistic-looking 3D design.

SketchUp also enables the use of **components** (or pre-made 3D objects) for added realism and detail in designing 3D models. Components give you a quick way to insert trees, cars, furniture, landscaping, and similar features into 3D models (see **Figure 14.8** for examples of components used in SketchUp). Like other 3D objects, components are constructed using shapes and textures, but can be stored for re-use throughout the construction of a model. For instance, if you're designing a 3D version of a car dealership, you can quickly add a couple dozen different car components and customize them, rather than design each 3D car model individually. Components are also useful for elements you'll use multiple times in a model. For example, if you're designing a 3D version of a school building, you could design a window once, store it as a component, and then easily reuse the same model of a window multiple times around the school.

Geospatial Lab Application 14.1 will introduce you to several aspects of using SketchUp for 3D design. You'll begin with a building's footprint, use tools to create a shape, design elements of the building (such as the

components pre-designed (or user-designed) objects in SketchUp

FIGURE 14.8

Sample components, displayed in SketchUp.

FIGURE 14.8

Sample components, displayed in SketchUp.

SKP file the file format and extension used by an object created in SketchUp

KML Keyhole Markup Language—the file format used for Google Earth data

KMZ the compressed version of a KML file

entranceway), and adapt components to fit in with your own ideas. A file created with SketchUp (like the one you'll make in the lab) uses the ".skp" file extension. The **SKP file** consists of information about the polygons that make up the structure, the textures applied, and the geometry of how these objects fit together. However, the SKP format is native to SketchUp. You (or whoever you share your data with) have to be running SketchUp to open and examine an SKP file. In addition, you can open only one SKP file at a time, which causes some difficulties if you've designed multiple buildings (each in its own SKP file) and want to see how they look next to one another.

One way around these issues is to convert the SKP files into another format, called **KML** (short for Keyhole Markup Language), which can be used in Google Earth Pro. KML files are also used as **KMZ** files, which is the compressed version of KML. Through this conversion, each building becomes its own KMZ file, and many KMZ files can be opened at once in Google Earth or Google Earth Pro. By converting an SKP file into a KMZ file, your SketchUp model can be viewed interactively in Google Earth Pro. **Figure 14.9** shows an example of this—Youngstown State University's stadium complex (containing both sides of the stadium, stadium lights, scoreboard, and goal posts) was converted from an SKP file into a KMZ file. The KMZ file can be opened in Google Earth Pro, which overlays the 3D model of the stadium over the Google Earth Pro imagery and allows the user to rotate the stadium, fly across the field, zoom into a bleacher seat, or utilize any of the other usual uses of Google Earth Pro. KML/KMZ has become a standard for creating objects or overlays to view in Google Earth Pro—other file formats (such as satellite imagery formats) can be converted to KML or KMZ, and then be placed or draped across Google Earth Pro (recall how you examined KMZ files of satellite imagery in Google Earth Pro in *Geospatial Lab Application 12.1*).

FIGURE 14.9

Youngstown State University's Stambaugh Stadium complex, converted from an SKP file in SketchUp and displayed as a KMZ file in Google Earth.

The big question that should be asked at this point is: How does Google Earth Pro know where to place that KMZ file? If Google Earth Pro knows just where to place it, the Youngstown State University stadium must have had some sort of spatial reference attached to it, so that it ends up placed over the imagery of the stadium's location and not at some random point on the globe. The best way of going about this is to know the object's spatial location before you start to do any modeling—otherwise, you'll find yourself trying to georeference the 3D model's location after the fact. SketchUp works hand in hand with geospatial technologies to obtain this data. SketchUp has the capability to take a "snapshot" of map or satellite imagery (like the area you can see with Google Earth Pro), and import that snapshot into SketchUp. If you're using SketchUp Make you can obtain a georeferenced map of the area to use as a starting point, but if you're using SketchUp Pro you can obtain satellite imagery complete with a digital terrain model of the landscape being shown along with the relevant spatial reference information.

With SketchUp, you'll have a color overhead image with spatial reference information to use as a starting point for your modeling efforts (see Figure 14.10 for examples of this geolocation). When you've completed your work in SketchUp, you can convert your SKP file (.skp) into a KMZ file (.kmz), which will then contain the necessary information to place the model in its proper spatial location in Google Earth Pro. This is how you'll begin 3D modeling in *Geospatial Lab Application 14.1*—first by obtaining a "snapshot" of the overhead map of the building, and then by creating the building's footprint from this starting point.

A section of Theodore Roosevelt National Park and Medora in North Dakota in Google imagery (above) and in an image version (including terrain) in SketchUp (below).

ArcGlobe a 3D visualization tool used to display data in a global environment, which is part of Esri's ArcGIS Desktop program

ArcScene the 3D visualization and design component of Esri's ArcGIS Desktop program

What Kinds of 3D Design and Visualization of Geospatial Data Are Available?

There are numerous 3D design and modeling programs on the market today, and many commercial geospatial programs offer some sort of capability of visualizing or working with 3D data (there's no way this chapter can get into all of them, but we will discuss a few). For instance, Esri's Arc-GIS Desktop program contains the 3D Analyst extension (which enables numerous 3D features related to terrain and modeling), as well as two other interfaces for working with 3D data—**ArcGlobe** and **ArcScene**. As its name implies, ArcGlobe allows for 3D visualization on a "global" level. If you're examining 3D surfaces and objects in Los Angeles, you can fly north across the globe to view other 3D spatial data in San Francisco, then continue north to look at the landscape in Seattle. ArcGlobe functions like a virtual globe program and works with all types of Esri file formats. ArcScene contains a full set of 3D visualization tools for ArcGIS and allows for the same type of 3D modeling, except on a "local" scale. If you were modeling one section of a 3D Los Angeles, that's the extent of the data with which you'd be working. Similarly, ArcGIS Pro (see page 145 in Chapter 5) uses the 3D Analyst to support taking your 2D GIS data and displaying it in a pseudo-3D view.

3D modeling techniques and other methods of shaping and designing geospatial data are advancing rapidly. Esri's **CityEngine** program uses a rule-based structure to enable users to work rapidly on the construction of entire rendered 3D cities without having to extrude and model individual buildings. See *Hands-On Application 14.2: 3D CityEngine Web Scenes* for some interactive online examples of this. Other companies, such as PLW Modelworks, utilize remotely sensed imagery and aerial photography to create remarkably realistic large-scale 3D models (see **Figure 14.11** for an example). In addition, like many of the other geospatial technology applications we've discussed in this book, 3D modeling and visualization is being done online—see *Hands-On Application 14.3: Cesium and 3D Maps* for some very neat examples of Web-based 3D visualization.

With these types of 3D modeling and design techniques, all manner of 3D features can be created and utilized with other geospatial data. It's one thing to look at a flat map of a university campus, but it's another thing entirely to have it laid out in a realistic 3D environment with which you can interact. Prospective students are likely to get a better sense of place by

CityEngine a powerful rule-based 3D visualization program designed for quickly modeling 3D cities

FIGURE 14.11

A 3D model of the city of Naples, Italy.
[PLW Modelworks]

Hands-On Application 14.2

3D CityEngine Web Scenes

For some interactive examples of using CityEngine, go to **http://www.arcgis.com/home/group .html?id=d731c2294ded45f6a1ad37ef5f87cd6d**. All of the options on the ArcGIS Online Website are various city area Web Scenes generated using CityEngine. By selecting one of them, the CityEngine Web Viewer will open and load the 3D city data. Within the CityEngine Web Viewer, controls for navigating the scene are in the left-hand corner, various layers that can be turned off and on will be on the right-hand side of the screen, and a series of bookmarks will be along the bottom. By clicking on a bookmark, the view will shift to show the camera angle at that point. Check out several of the examples, particularly Desert City, Esri Campus, and Philadelphia Redevelopment.

Expansion Questions

➤ How does the 3D visualization of Philadelphia in its current state and with the proposed redevelopment aid in showing which areas, spatially, will be impacted by the redevelopment plan?

➤ How are various exterior and interior city features shown and visualized in both the Desert City and the Esri Campus Web Scenes?

Hands-On Application 14.3

Cesium and 3D Maps

Cesium is a free Javascript API (see page 549 in Chapter 15 for more about using Javascript for Web-based programming) that can be used for designing 3D Web applications. The open-source Cesium can be used for rendering geospatial data on a 3D virtual globe and accessing it through your Web browser. To see some examples of using Cesium for 3D on the Web, go to **https://cesiumjs .org/demos.html**. Check out some of the links to many Websites that use Cesium for 3D visualization, including the Swiss Federal Geoportal, FodarEarth, and Shake Finder (for viewing earthquake data).

Expansion Questions

➤ How are many of these demos using 3D visualization with Cesium for showcasing their applications?

➤ How does the addition of visualizing these kinds of geospatial data on a 3D virtual globe in your browser aid in interpreting the data or communicating the concept to the viewer?

viewing a 3D representation of their dorms as well as understanding where they are in relation to the urban areas surrounding the campus.

The same idea holds true for designing any area in interactive 3D. Future planning efforts can be implemented better when the planners can actually see the visual impact that various types of building plans would have on the neighborhood (with respect to height, texture, and landscaping). With 3D geospatial data, the locations of items and structures can be seen in relation to each other and to the surrounding terrain. For example, in Google Earth and Google Earth Pro, a wide range of 3D content created with various 3D techniques is tied to spatial locations with which you can interact (see **Figures 14.12** and **14.13** for examples of rendered 3D features in Google Earth, and *Hands-On Application 14.4: 3D Buildings and Structures in Google Earth* for how to interact with this 3D data in Google Earth).

Hands-On Application 14.4

3D Buildings and Structures in Google Earth

Figures 14.12 and 14.13 are both images of 3D structures in Google Earth, but they're far from being the only places that have been designed in 3D for the software. For this application, open your Web browser and go to **https://earth.google.com** and launch Google Earth. In the search box, look for Boston, Massachusetts, and look around Boston Common and the downtown area. By clicking on the button marked 3D, Google Earth will change into a perspective view and the buildings and skyline of Boston should appear in 3D. Also, by double-clicking on the small white and red arrow control, it will exchange places with the large globe control and this new control will allow you to navigate around the 3D scene of Boston. You can also hold down the Ctrl key on the keyboard and move the mouse to tilt the perspective on the view. Use these controls to

fly around the city and see how Boston has been put together in 3D.

When you're done, try looking at the design of 3D buildings in other cities, such as San Francisco, London, and Paris. Check out how 3D visualization of objects in Google Earth extends past just buildings to features such as the Golden Gate Bridge, the London Eye, and the Eiffel Tower.

Expansion Questions

➤ Check out your own local city in Google Earth. Has the city been created in 3D for Google Earth? What prominent local features (beyond buildings) can be viewed in 3D?

➤ Use Google Earth to fly through and examine the Las Vegas Strip in 3D. What areas and features of the Strip really pop when viewed in 3D with Google Earth?

FIGURE 14.12

Tokyo Tower (and its surrounding areas) as shown in 3D using Google Earth.

FIGURE 14.13

A section of the Las Vegas Strip as shown in 3D using Google Earth.

 ## How Can Geospatial Data Be Visualized in 3D?

Even with all of these great 3D modeling and design programs, when the products are viewed they are still only seen in a two-dimensional environment on a computer screen. It's one thing to view a 2D choropleth map (as in Chapter 7), another thing to create a pseudo-3D representation of the mapped data (such as a prism map), and another thing altogether to be able to interact with the data in an immersive environment. Examining a realistic geospatial landscape and flying across it (as in Chapter 13) is very neat, but what if you could simulate the full 3D effect of diving through canyons or banking over ridgelines? As computers get faster and develop the ability to handle geospatial data more quickly and easily, new methods of visualizing data are being developed.

Immersive "virtual reality" technology allows you to become part of the data, move with the landscape, have yourself placed inside a structure, or move and see the surface of the map. This sense of immersion, or interactivity, with geospatial data adds immensely to our understanding of the nature of spatial data, whether we are researchers or students. Virtual reality technology is becoming more available and is utilizing geospatial technology. For instance, Google Earth can be used with VR headsets such as the Oculus Rift or Vive to give you a full 3D immersive experience of flying across Earth's landscape as you would with using Google Earth on your Web browser.

Another way of making geospatial data appear more realistic (and adding a sense of depth and immersion) that's not reliant on virtual reality is to view the data in stereo. Stereo imagery shows two images (for instance, two images of the same landscape), rendered apart from each other. When you look at something, your eyes are seeing the same thing from two slightly different points of view—which is why you see depth and distance (and view the real world in three dimensions). Stereo images use a similar principle: They are two representations of the same thing, taken from two slightly different positions of the camera or sensor. Remote sensing has long been making use of stereo imagery for measuring the heights and depths of areas. Overlapping aerial photography and images acquired from sensors (such as the MISR sensor onboard Terra—see page 451 in Chapter 12) can be used for creating stereo data.

This kind of stereo concept is also utilized when UAS (see pages 334–338 in Chapter 9) are used for collecting imagery with a significant amount of overlap, which can then be processed into 3D renditions of the land-scape that was just imaged. Software such as the popular Pix4D program can take UAS photos and search for thousands of common points between the multiple overlapping photographs. Pix4D will then use that informa-tion (together with features such as the GPS location of the UAS and the focal length of the drone's camera) to create a 3D point cloud of these com-mon locations. Using that, the imagery from the UAS can then be knitted together to form a mosaic and then displayed in 3D (see **Figure 14.14**). In this way, heavily overlapping imagery and modern photogrammetric tech-niques (see page 348 in Chapter 9) can be used for viewing remotely sensed imagery in 3D.

Without relying on software post-processing, a simple way of view-ing imagery or data in stereo (which several geospatial software programs have the capability for) is to create a red-and-blue **anaglyph** of the data—in which the two images are slightly separated, but one is highlighted in red and the other in blue. By wearing a pair of 3D glasses with colored gels for lenses (one red and one blue), the anaglyph images appear to be raised or

anaglyph a new image created by slightly offsetting the features in an existing image from each other and highlighting them in two different colors

FIGURE 14.14

A 3D model of a petrochemical facility, generated from UAS imagery processed using Pix4D. *[Jarlath O'Neill Dunne, University of Vermont]*

FIGURE 14.15

An anaglyph of a terrain model of the Los Angeles area of southern California. *[NASA]*

elevated off the screen in a 3D effect (see **Figure 14.15**). This type of stereo effect (where one eye views the red and the other eye views the blue) creates a type of **stereoscopic 3D** that gives the illusion of depth in parts of the image. Some items may appear in the foreground and others in the background. Red and blue anaglyphs are certainly nothing new, but today, when stereoscopic 3D movies have become all the rage (including the top-grossing film of all time), there are more high-tech ways of using these stereo concepts with geospatial technology.

stereoscopic 3D an effect that creates the illusion of depth or immersion in an image by utilizing stereo imaging techniques

If you go to a 3D movie today, you're not going to be wearing red and blue lenses, but rather a pair of polarized glasses. The same type of 3D glasses can be used with the **GeoWall**, a computer tool that's used for examining imagery in an immersive three-dimensional environment with stereoscopic 3D viewing techniques. First developed by the members of the GeoWall Consortium in 2001, it has spread to more than 400 systems, which have been developed for use in schools and colleges (along with such places as the EROS Data Center and NASA's Jet Propulsion Laboratory). The GeoWall can be used for viewing geospatial data in a stereoscopic 3D format.

GeoWall a powerful computer tool for displaying data and imagery in stereoscopic 3D

The GeoWall consists of several components that can be purchased separately, to keep the cost of the whole setup affordable (around $10,000). First, two high-power Digital Light Processing (DLP) projectors project the two images that are necessary for stereo viewing. These projectors are hooked into a powerful computer with a dual-output video card (so that the computer can send its output into both of the projectors). The images are projected through special filters in front of the projectors' lenses, which polarize the light. The light is projected onto a special screen made of material that preserves the integrity of the polarized light (rather than scattering it). Lastly, the user wears a pair of polarized 3D glasses to view the imagery. See **Figure 14.16** for the components of the GeoWall projectors and their filters.

Wearing the 3D glasses, the viewer perceives the images in a 3D effect as if the items and images are appearing off the screen—similar to the effect

FIGURE 14.16

The basic setup of the GeoWall includes twin DLP projectors connected to the computer and polarizing filters in front of the lenses. *[Courtesy Bradley A. Shellito]*

FIGURE 14.17

Imagery projected onto the screen by the GeoWall system (as viewed without the glasses). *[Courtesy Bradley A. Shellito]*

created by a stereoscopic 3D movie. Viewing and interacting with imagery like this gives the viewer a sense of depth or immersion, as some items will be in the foreground of the image and some will be in the background. **Figure 14.17** shows an image of a GeoWall in operation, from the point of view of an outsider who is not wearing the polarized glasses (without the glasses, the images appear blurry and slightly askew). GeoWall viewers position themselves in front of the screen, and are able to see the images projected in stereoscopic 3D.

Using the GeoWall for visualizing geospatial data carries a lot of benefits. The GeoWall enables the viewer to see maps draped over a surface model or terrain data displayed in stereoscopic 3D. Certainly, being able to dive inside a meteor crater or view volcanic features, coastlines, or river valleys in a stereoscopic 3D environment allows the viewer to get a realistic sense of the terrain features. 3D cities in Google Earth, landscapes, geologic processes, draped satellite imagery, draped topographic maps, stereoscopic photos, and tours of virtual environments are all ways of viewing types of data with a GeoWall setup.

Chapter Wrap-Up

The goal of this chapter was to provide an introduction to some of the basic concepts of 3D visualization and how they can be incorporated into geospatial technology. *Geospatial Lab Application 14.1* will get your hands dirty working with 3D design and visualization using SketchUp and Google Earth. However, 3D modeling goes well beyond SketchUp—designing, programming, and working with 3D modeling and visualization comprise a whole industry on its own: video games, movies, television, and Web animation, and rapidly developing simulators of many kinds are constantly serving up extremely impressive and ever-more realistic and immersive 3D designs. There's no way to cover all the possible applications, software, companies, and changes going on in this rapidly developing field in just one short chapter.

Check out *3D Visualization Apps* for some apps related to this chapter's contents, as well as *3D Visualization in Social Media* for some Facebook pages, Twitter accounts, Instagram accounts, YouTube videos, and blogs dealing with geospatial technology and 3D visualization.

3D visualization is a very appealing and quickly evolving area of geospatial technology, but there are many new developments yet to come as technology advances. Chapter 15 explores some of these new frontiers in geospatial technology.

Important note: The references for this chapter are part of the online companion for this book and can be found at **www.macmillanlearning .com/catalog/shellitoigt4e.**

3D Visualization Apps

Here's a sampling of available representative 3D visualization apps for your smartphone or tablet. Note that some apps are for Android, some are for Apple iOS, and some may be available for both.

➤ **Google Earth:** An app for viewing 3D visualization in Google Earth

➤ **Maps 3D and Navigation:** A mobile app for navigation and routing that also contains pseudo-3D versions of the maps

3D Visualization in Social Media

Here's a sampling of some representative Facebook, Twitter, and Instagram accounts, along with some YouTube videos and blogs related to this chapter's concepts.

On Facebook, check out the following:

➤ **CityEngine:**
www.facebook.com/CityEngine

➤ **SketchUp:**
www.facebook.com/sketchup

Become a Twitter follower of:

➤ **Cesium** (Cesium for 3D visualization): @CesiumJS

➤ **CityEngine:** @CityEngine

➤ **Mapzen** (makers of the open source Tangram, used for 2D and 3D modeling): @mapzen

➤ **SketchUp:** @SketchUp

Become an Instagram follower of:

➤ **SketchUp:** @sketchup_official

On YouTube, watch the following videos:

➤ **CityEngine TV** (a YouTube channel featuring several different applications of CityEngine): **www.youtube.com/user/CityEngineTV**

➤ **Esri CityEngine Trailer** (a short demo of CityEngine's capabilities): **www.youtube.com/watch?v=aFRqSJFp-I0**

➤ **Esri CityEngine Mobile Virtual Reality Solution** (a short video showing how CityEngine 3D models can be used with a VR headset): **https://www.youtube.com/watch?v=lcLEOGO1JPE**

➤ **The SketchUp Channel** (a YouTube channel featuring videos about a variety of SketchUp applications and tutorials): **www.youtube.com/user/SketchUpVideo**

For further up-to-date info, read up on these blogs:

➤ **ArcGIS Blog—3D GIS Category** (blog posts concerning topics related to working with 3D GIS and 3D visualization): **http://blogs.esri.com/ esri/arcgis/category/subject-3d-gis**

➤ **The Cesium Blog** (a blog about applications of Cesium for 3D visualization): **https://cesiumjs.org/blog.html**

➤ **SketchUpdate Blog** (the official blog for SketchUp): **http://blog .sketchup.com**

Key Terms

3D modeling (p. 503)
anaglyph (p. 517)
ArcGlobe (p. 512)
ArcScene (p. 512)
base heights (p. 505)
block (p. 503)
CityEngine (p. 513)
components (p. 509)
extrusion (p. 503)
face (p. 505)
GeoWall (p. 518)

KML (p. 510)
KMZ (p. 510)
offset (p. 504)
prism map (p. 507)
prisms (p. 507)
SketchUp (p. 508)
SKP file (p. 510)
stereoscopic 3D (p. 518)
textures (p. 506)
z-value (p. 503)

3D Modeling and Visualization

This chapter's lab will introduce you to concepts of modeling and design using three-dimensional objects. Working from an aerial image of a building with spatial reference, you will design a building in 3D using SketchUp, incorporating accurate height measurements, textures, and other effects, and place it in its proper location in Google Earth Pro. While there are many methods that can be used for designing a building, this lab will show you how to use a variety of SketchUp tools through the active process of creating a building.

Important note: Even though this lab uses a fixed example, it can be personalized by selecting another structure, one that you're more familiar with—such as your home, workplace, or school—and using the same techniques described in this lab to design that structure instead.

As in *Geospatial Lab Application 7.1,* there are no questions to answer—the product you'll create in this lab will be a 3D representation of a building in SketchUp, and you'll place a version of it in Google Earth Pro. The answer sheet for this chapter will be a checklist of items to help you keep track of your progress in the lab.

Objectives

The goals for you to take away from this lab are:

➤ To obtain a basemap to use in a design environment

➤ To familiarize yourself with the uses of the SketchUp software package, including its operating environment and several of its tools

➤ To apply textures to objects

➤ To utilize components in 3D design

➤ To transform your 3D model into the KMZ file format and place it in Google Earth Pro

Using Geospatial Technologies

The concepts you'll be working with in this lab are used in a variety of real-world applications, including:

➤ Urban planning, where 3D models make it possible to visualize new development projects, assess the visual impact of new buildings on the surrounding area, and develop projects that will be in character with the cityscape

➤ Archaeology, where 3D modeling can be used to re-create ancient cities and the layouts of archeological sites

[photobank.kiev.ua/Shutterstock]

Obtaining Software

SketchUp comes in two versions—the free SketchUp Make as well as SketchUp Pro, which you have to purchase. We'll be using the free SketchUp Make. The current version is available for free download at **www.sketchup.com**. The current version of Google Earth Pro (7.3) is available for free download at **https://www.google.com/earth/download/gep/agree.html**.

Important note: Software and online resources can change fast. This lab was designed with the most recently available version of the software at the time of writing. However, if the software or Websites have changed significantly between then and now, an updated version of this lab (using the newest versions) will be available online at **www.macmillanlearning.com/catalog/shellitoigt4e**.

Lab Data

There is no data to be copied for this lab. You will, however, use the "Bird's Eye" imagery from Bing Maps (see *Hands-On Application 9.5: Oblique Imagery on Bing Maps* on page 344).

Localizing This Lab

In this lab you will design a 3D model of a building on Youngstown State University's (YSU's) campus, and then view it in Google Earth Pro. In Section 14.2, you will start by taking a "snapshot" of the map of the building. Rather than examining a campus building at YSU, you can start at this point with a base image of any local building—such as the school building you're in, a nearby landmark or government building, or your house. You can use the same steps in this lab and the SketchUp Make tools to design any building or structure you so choose in place of the one used in this lab—the same lab techniques of using SketchUp Make will still apply; you'll just have to adapt them to your own 3D design needs.

14.1 Starting SketchUp Make

1. Start **SketchUp Make**. Select the option for **Choose Template**, and then from the available options select **Simple Template—Feet and Inches**. There are several different templates available, depending on what type of 3D design you want to do.

2. Next, select the **Start using SketchUp** button.

3. SketchUp Make will open into a blank working environment, with three intersecting lines at the center. These represent the axes you will use for 3D design—red and green are the horizontal (x and y) axes, while blue is the vertical (z) axis.

4. You'll see the SketchUp Make toolbar at the top of the screen. You'll use many of these tools during the lab—the tools that you'll use most often in this lab (from left to right on the toolbar) function as follows:

- The black arrow is the **Select** tool that lets you choose or select objects.

- The pink eraser is the **Eraser** tool that allows you to remove lines and objects.

- The pencil is the **Line** tool that lets you draw, or construct, lines.

- The box with the red arrow coming up from it is the **Push/Pull** tool that allows you to alter the shape of polygons.

- The four red arrows represent the **Move/Copy** tool that can be used to drag objects around the screen and also to create copies of objects.

- The extended yellow tape measure is the **Tape Measure** tool that allows you to measure the length or width of objects.

- The bucket spilling paint is the **Paint Bucket** tool that allows you to change the color or texture of objects.

- The swirling red and green lines around the pole represent the **Orbit** tool that allows you to change your view and maneuver around the scene.
- The hand is the **Pan** tool that allows you to move around the view.
- The magnifying glass is the **Zoom** tool that allows you to zoom in and out of the scene.

14.2 Obtaining OpenStreetMap Imagery

SketchUp Make can use OpenStreetMap imagery as a starting point for your design work in SketchUp Make, and then you can transfer your 3D structures into Google Earth Pro (GEP).

1. To begin, start **Google Earth Pro**.

   Important note: This lab will have you design a simplified representation of the Lincoln Building on YSU's campus. All of the measurements of such things as heights, window lengths, window measurements, and so on are either just approximations or values generated for this lab and for use only in this simplified model. To use real values for Lincoln (or your own buildings) you would have to make measurements of the structure or consult blueprints for actual heights and sizes.

2. The Lincoln Building is located on YSU's campus at the following latitude/longitude decimal degrees coordinates:

   - Latitude: **41.103978**
   - Longitude: **–80.647777**

3. Zoom GEP into this area and you'll see the top of the building (see the following graphic). Fill the GEP view with the entirety of the building and return to SketchUp Make. This should give you a good look at the overhead view of the Lincoln Building for reference in the next step. You'll use GEP later, so minimize it for now.

4. Back in SketchUp Make, from the **File** pull-down menu, select **Geo-location**, then select **Add Location**. Add Location allows you to take a "snapshot" of OpenStreetMap imagery and import it into SketchUp Make (with information on the coordinates of the view) for use as a digitizing source. The SketchUp Pro version will allow you to take this snapshot using satellite imagery as well as terrain information, but for our purposes the OpenStreetMap imagery available in the free version of SketchUp Make will work just as well.

5. In the **Add Location** box that appears, type the coordinates of Lincoln Building and click **Search**. The map will adjust itself to see the Lincoln Building and its surroundings. Use the Zoom and Pan tools to expand the view so that you can see all of Lincoln Building in the window. *Important note:* If the map does not appear, click the icon in the upper-right corner of the **Add Location** dialog box and select **Trimble Map**.

6. When you've got the view set up how you want it, click the **Select Region** button.

7. A new image will appear—a bounding box with white markers at the corners. Drag the corners of the box so that it stretches all the way across Lincoln Building (as in the following graphic). This area will be the "snapshot" that SketchUp Make will take of map, which you'll use in the modeling process—so be sure to capture the whole area.

8. When everything is ready, click **Grab**.

9. After the importing process is done, the image (in color) should appear in SketchUp Make, centered on the axes.

14.3 Digitizing the Building Footprint

1. You can digitize the dimensions of Lincoln Building directly from the imported map view. The easiest way to do this is to position the view so that it's looking directly down on the image. From the **Camera** pull-down menu, select **Standard Views**, then select **Top**.

2. The view will switch to show the image as seen from directly above. To zoom in or out so that you can see the entire building, select the **Zoom** icon from the toolbar (or if your mouse has a scroll wheel, you can use that to zoom as well). To zoom using the icon, select it, then roll the mouse forward or back to zoom in and out.

3. When you can see the outline of the entire building, it's time to start digitizing. Click on the **Pencil** icon (the Line tool) on the toolbar to start drawing.

4. Your cursor will now switch to a pencil. Start at the lower left-hand corner of the building and click the pencil onto the building's corner. A purple point will appear. Now, drag the pencil over to the lower right-hand corner and hover it there for a second. The words "on face in group" will appear. This is SketchUp Make's way of telling you just what surface you're drawing on (in this case, the face is the surface of the image).

5. Click the **Pencil** on the lower right-hand corner, then at the upper right-hand corner, then at the upper left. The guide may change to "Perpendicular to the edge," indicating that you've drawn a perpendicular line. Finally, drag the Pencil back to where you started. When you reach the beginning point, the cursor should turn green and the word "Endpoint" will appear. This indicates that you've closed the polygon of the building's footprint in on itself. Click the Pencil at the endpoint and the polygon will appear, covering the entire building.

Important note: If you have only lines and not a complete polygon, return to the lower corner and remake the footprint polygon. Also, due to how the building footprints are laid out in OpenStreetMap your drawn lines may be slightly off from the sketch shown on the map. For this 3D model it is more important to have right-angle edges at the best approximation of the basemap possible than to match the sketch drawn on the OpenStreet-Map base layer exactly.

14.4 Extruding the Building Footprint

1. With the building's footprint successfully digitized, the next step is to give it some height.

2. From the **Camera** pull-down menu, select **Standard Views** and select **Iso**. The view will switch to a perspective view.

3. From the toolbar, select the **Push/Pull** tool.

4. Your cursor will change again. Hover it over the footprint polygon and you'll see it change texture slightly. Click on the **footprint** and hold down the mouse button, then use the mouse to extrude the polygon by pushing forward.

5. In the lower right-hand corner of SketchUp Make, you'll see the real-world height to which you're extruding the building. The Lincoln Building is approximately 72 feet high. During the extrusion operation, you can type 72´ and hit the **Enter** key on the keyboard while holding down the mouse button and the polygon being pushed/pulled will automatically extrude to that height.

6. You can double-check the measurement by selecting the **Tape Measure** icon from the toolbar.

7. Click the measuring tape at the base of the Lincoln Building block and drag it to the top. The measurement (in feet and inches) will appear. You can use the push/pull tool to readjust the building height if necessary.

14.5 Maneuvering in SketchUp Make

1. An easy way to maneuver and examine the 3D objects you're creating is to use the Orbit and Pan tools, which are available on the toolbar.

2. Use the **Pan** tool (the hand) to drag the current view in any direction.

3. Use the **Orbit** tool (the red and green arrows curved around a pole) to rotate the view around an object.

4. Practice with the two tools for a couple of minutes to get used to the feeling of maneuvering in the SketchUp Make environment.

5. Maneuver around so that you can view the front of the building (the part that opens onto the street).

14.6 Adding Features to the Building

1. Use Bing Maps' Bird's Eye feature to get a closer look at Lincoln Building. Viewing from the north, you may be able to see a recessed entranceway leading into the front of the Lincoln Building.

2. Set up the entranceway approximately 27 feet (27′) across, and set it to begin approximately 34 feet (34′) from each side of the front of the building. The opening itself is about 9 feet (9′) tall. This is the size of a block you want to draw on the front of the building.

3. In order to find these dimensions, you can use the Tape Measure to create some guide lines (or "construction lines") on the face of the building.

4. Select the **Tape Measure** icon and start from the lower-left corner (as you face the building). Measure upward approximately 9 feet, and then double-click the mouse.

    Important note: A more precise way to use the Tape Measure is to type **9′** (for 9 feet) while dragging the tape in the direction you want to measure. A small mark will appear at the 9′ length, and you can use that as a marker for more measurements.

5. Next, drag the mouse over to the other side of the building and double-click. A dashed line will appear on the face of the building at a height of 9 feet. This is a construction line—it's invisible in the modeling process, but it's useful for drawing measurements.

6. Starting along one side of the building, measure across 34 feet, double-click the mouse, and then measure down to the base of

the building. Do the same on the other side. You should have construction lines set up that block out the entranceway on the face of the building.

7. If you find you're creating too many construction lines, you can always select them with the **Cursor** icon and delete them by using the **Eraser** icon on the toolbar (in fact, anything you create can be removed with the eraser).

8. Now, use the pencil to draw the entranceway on the face of the building, tracing over the construction lines.

9. Finally, select the **Push/Pull** tool and use it to recess the entranceway by approximately 6 feet. Use the mouse to push the block inward, and watch the distance bar in the lower-right corner.

10. Examine the building with Bing Maps' Bird's Eye imagery, and you'll see that Lincoln Building has several sets of windows that run around the top of the building—two sets of 4 windows on the front and back, and 18 windows on either side. Drawing all 52 of these identical windows individually would be a huge chore, so to save time and effort, you'll be designing the window only once and reusing it throughout the model.

11. One way of doing this is to set up multiple construction lines across the front of the building. Orbit around to see the top front of the building. Set up construction lines as follows (these are simplifications of the real measurements for purposes of this lab):

    a. The first window begins 6 feet from the edge of the building.

    b. The first window begins 3 feet from the top of the building.

    c. The window is 3 feet wide and 21 feet high.

    d. There are five divisions on the window, and their measurements are (from the top):

       - 2 feet
       - 5 feet

- 7 feet
- 5 feet
- 2 feet

12. Next, draw the outline of the window and the five divisions by using the **Pencil** tool in conjunction with the construction lines.

13. With the outlines drawn, it's time to make it actually look like a window by adding some textures.

14. Select the **Paint Bucket** icon from the toolbar.

15. A new window of **Materials** will open. These are the various textures that can be "painted" on a face. For the purposes of this simplified model, select **Asphalt and Concrete** from the **Materials** window pull-down menu, then select the option for **Polished Concrete New**. Back in the model of the window, click the **Paint Bucket** on the top and bottom sections of the window. They should change appearance to a "concrete block" finish.

16. Next, select **Glass and Mirrors** from the **Materials** window pull-down menu and choose the option for **Translucent_Glass_Blue**. Click on the second and fourth blocks of the window with this new material.

17. Lastly, choose **Colors-Named** from the **Materials** window pull-down menu, select **0137_Black** (it's a black color at the bottom of the list), and then click on the final (third) section of the window. All five sections of the window should now be colored in. You can now close the **Materials** box.

14.7 Working with Components

As we noted before, you've created one window, but the model requires a total of 52 of them across the four sides of the building. If you're going to re-use something many times, the easiest option is to create a component out of the object.

1. Select the **Cursor** icon and drag a box over all sections of the window. They'll become highlighted in blue, signifying they're now selected. Be very careful not to select any of the building faces themselves—you want only the sections of the window. (To de-select items, hold down the **Shift** key and click on the items with the mouse.)

2. Right-click on the selected window and choose **Make Component**.

3. In the **Create Component** dialog box, type in the name "**Lincwind**" for the new component.

4. Select **Any** for "Glue to" from the drop-down box.

5. Place a checkmark in the **Cut opening** option.

6. Place a checkmark in the **Replace selection with component** box.

7. Click **Create** when you're done.

8. The selected section of the window will now be a SketchUp Make Component and it will be stored for later use in the model.

14.8 Making Multiple Copies

Now that the window is a component, it'll be easy to work again with the same item. In this section of the building, there are four windows, each approximately 2.5 feet apart and 3 feet wide. What you're going to do is make multiple copies of the component window, then have SketchUp Make automatically space out the windows for you.

1. Create a construction line 13.5 feet away (the space of two window lengths and the distances between them) from the window's edge, stretching from the top to the bottom of the building. This is where the copy of the window will go. *Important note:* Be sure you are drawing

this construction line on the face of the building and not out into space. The line will show up in purple when drawn on the face.

2. Next, use the cursor to select the window component. It will be outlined in blue when it's selected.

3. Next, choose the **Move/Copy** tool from the toolbar.

4. The cursor will change to the new icon.

    Important note: Hold down the **Ctrl** key on the keyboard—this tells SketchUp Make that you want to copy the window instead of moving it somewhere. If you don't hold down the Ctrl key, SketchUp Make will try to move the window somewhere else instead of making a copy.

5. Use the mouse to select and drag the copy of the window over to the right of the new 13.5-foot distant construction line. Make sure that the Inference dialog box indicates that you're copying it "On Face" (that is, the copy will be moved to the face of the building, instead of into space somewhere).

6. Once the copy of the window is positioned properly, type /3 on the keyboard. This will tell SketchUp Make to make a total of three copies, evenly spaced between the original and the location of the new copy. You will now have four windows on the front left-hand side of the building.

7. Now it's time to put windows on the other side of the front. Create a construction line 6 feet from the right-hand side of the building.

8. To add the component of the window, first open the Tray on the right-hand side of the screen (like where you chose textures from) by going to the **Window** pull-down menu, choosing **Default Tray**, and then **Show Tray**. From the options that appear in the Tray, expand the **Components** section.

9. In the Components section, select the **House** icon (under the **Select** tab) to utilize only those components that exist in the current model (the only component should be your "Lincwind" window, in addition to the default person added by SketchUp Make).

10. Click on the **Lincwind** component itself in the window and drag it onto the face of the building in the model, positioning it properly on the face in conjunction with the construction line you just drew.

11. Now, repeat the previous steps (draw a construction line 13.5 feet away and copy the windows, creating a total of three copies properly spaced out the correct distance from one another).

12. You should now have two sets of four windows each on the front of the building.

13. Rotate to the back of the building. You'll want to create the same layout (two sets of four windows each), using the same measurements on the back face of the building.

14. Next, rotate to one side of the building. Each side of the building has 18 windows on it, so you'll use the same technique—position the component, create a copy at the farthest point away, then have SketchUp Make automatically fill in the evenly spaced copies.

15. Starting at the left-hand side of the building, create a vertical construction line 6 feet from the edge and a horizontal construction line 3 feet from the top. Create another vertical construction line 93.5 feet from the building's edge—this is where the copy will be placed.

16. Place the window component 6 feet from the building's left-hand edge and 3 feet from the top of the building.

17. Drag a copy to the left of the 93.5-foot construction line.

18. Type **/17** on the keyboard.

19. A total of 18 windows should be created on the side of the building.

20. Repeat this process on the other side of the building to create an additional 18 windows.

21. At this point, all 52 windows should have been created on the four sides of the building.

14.9 Working with Textures

Now, it's time to give the building a textured appearance to make it look like the orange brick in the photos instead of the plain white of the block.

1. Select the **Paint Bucket** icon again, and select **Brick, Cladding, and Siding** from the **Materials** pull-down menu. Next, choose the option for **Brick_ Rough_Tan**. Click on each of the sides of the building and each will be "painted" with the new texture.

2. For the roof and the entranceway, select **Colors** from the **Materials** pull-down menu, and then select **Color_M06**. "Paint" the roof and the entranceway faces with this color.

14.10 Working with 3D Text

The words "LINCOLN BUILDING" appear on the front of the building, near the lower right-hand side. 3D text can easily be added to a model to create names, signs, or other uses of block lettering.

1. From the **Tools** pull-down menu, select **3D Text**.

2. In the text box, type **LINCOLN BUILDING** (all caps—and hit the enter key between the two words so that they'll appear on two different lines).

3. Choose **Arial** for the Font.

4. Type **8"** (for 8 inches) for the Height of the letters.

5. Type **3"** (for 3 inches) for the Extruded letters.

6. Make sure the **Filled** and **Extruded** checkboxes are marked.

7. When ready, click **Place**.

[Source: SketchUp Make/Trimble]

8. The 3D Text will appear as a selected object—use the cursor to place it on the face of the building near the right-hand side of the entrance.

14.11 Adding Other Features

The purpose of this lab is to familiarize you with using SketchUp Make's features and to create an approximation of a building—not to model it right down to its last detail (though SketchUp Make is certainly powerful enough to do just that). However, there are other features on the building that you may want to add to continue utilizing your SketchUp Make skills.

1. If you look at the oblique imagery, you'll see some other 3D features of the building:

    • The black section of the building stretching up from the entranceway that also contains windows

    • The black pillars in the front of the building that stretch up to the roof

    • The units on the roof of the building itself

    • The side entrance on the west side of the building

2. Using the techniques we've discussed in this lab, you can create representations of these other features by using a mixture of extruding items (like drawing a polygon on the roof and extruding it), or by applying textures (you could draw in the blocks for the front windows above the entrance and fill them in).

3. When you have the model completed the way you want it, there is only one more thing to do.

14.12 Placing the Model in Google Earth Pro

1. The final step is to place the finished product into GEP as a 3D model. To do this, go to the **File** pull-down menu, select **Export**, and then select **3D Model**.

2. In the *Export Model* dialog box, select a location to save your model and call it **Lincoln**, then click **Export** then choose **KMZ** from the dropdown menu. SketchUp Make will export the model to a KMZ file and save it in the location you indicated.

3. Return to Google Earth Pro. From GEP's **File** pull-down menu, choose **Open**. Then navigate to where you just saved your Lincoln.KMZ file and open it. The KMZ will appear in GE's Temporary Places (named "Lincoln").

4. In GEP, use the Zoom and Rotate tools to examine your 3D model.

14.13 Saving Your Work (and Working on It Later)

1. You can save your work at any time in SketchUp Make by selecting **Save** from the **File** pull-down menu.

2. You can reopen your SketchUp Make model to work on it later by choosing **Open** from the **File** pull-down menu.

Closing Time

This lab was involved, but it served to show off many of the 3D modeling features available within SketchUp Make and dealt with how to interface SketchUp Make together with GE. Exit SketchUp Make by selecting **Exit** from the **File** pull-down menu. Make sure you saved the final version of your SketchUp Make model as well as exported the KMZ version of it.

Final SketchUp Make Modeling Checklist

_____ Obtain overhead view of the Lincoln Building and place it in SketchUp Make

_____ Digitize building footprint

_____ Extrude building to proper height

_____ Set building properly on the terrain

_____ Create entranceway opening on front of building

_____ Create components for windows

_____ Apply windows to all sides of building

_____ Apply appropriate colors and textures to all faces

_____ Create 3D text on front of building

_____ Add other features (front columns, windows, roof unit)

_____ Place 3D model in appropriate place in Google Earth Pro

Life in the Geospatial Cloud and Other Current Developments

15

Using the Cloud with Geospatial Technology, Web Maps, Story Maps, Who's Involved with Geospatial Technology, Geospatial Technologies in K–12 Education, and College and University Geospatial Educational Programs

It's an understatement to say that geospatial technology is a rapidly changing and advancing field. New Websites, imagery, tools, gadgets, and developments are coming at an incredible rate. GPS, digital maps, and satellite images are part of everyday life. When an app on your smartphone can provide you with the coordinates of your location, a map of what's nearby, a high-resolution remotely sensed image of your surroundings, and directions to wherever you want to go, you know that geospatial technologies have become an essential part of your life—and they're not going away. Or, as the Geospatial Revolution Project at Penn State University put it: "The location of anything is becoming everything." Back in Chapter 1, we discussed the incredibly wide variety of human activities that utilize some form of geospatial technology—if it involves location, it's going to involve geospatial technologies. We've covered a lot of ground over the last 14 chapters of this book, but every day there are new frontiers opening up, and every technological breakthrough brings the promise of new worlds to explore.

This chapter explores these new advances in the geospatial world and takes a look at who's making this progress possible through their participation, support, and sponsorship. How can you get involved with creating your own maps online? What kind of educational opportunities are there in higher education, and how is geospatial technology being used at the K–12 education levels? We'll look into all of these topics and review some examples of what's happening in geospatial technology (and, again, there's no way we're going to be able to cover everything or salute everybody—these pages aim to provide a sampling of what's out there).

FIGURE 15.1

Using the cloud for geospatial technology applications.

The previous chapters have described numerous datasets and how to obtain them (either free or at minimal cost), often through the hands-on applications in the chapters. Online resources are the quickest and easiest way to access geospatial data. Whether you want to acquire archived satellite imagery or get the newest updates for Google Earth, you can do everything via the Internet. The same holds true with distributing your own data and information—setting up a Web interface where your data can be accessed interactively is the way to go.

Geospatial technology is increasingly becoming part of a field referred to as the **cloud**, wherein resources are stored at another location but served to a user across the Internet (see **Figure 15.1**). The cloud is structured to let you store your data on other servers at other locations, but allows you to access it when you need to via the Internet. Similarly, other people would be able to access your data through these servers. However, there's much more to the cloud than simply serving datasets to users, as the cloud structure can be used for a wide variety of geospatial applications.

cloud a computer structure wherein a user's data, resources, or applications are kept at another location and made available to the user over the Internet

 ## How Is the Cloud Used with Geospatial Technology?

Chances are good that you use the cloud on a regular basis. For instance, if you take a photo with your smartphone, you can store your photos in the cloud rather than taking up space on your phone. Whenever you want to show your photos to a friend, your smartphone will retrieve your photos from the cloud to display on your phone. What this means is that your photos are stored somewhere else—on a data server at a remote location—and then you can access them on demand via the Internet.

The same goes for a service such as YouTube—if you film a video and upload it to your YouTube account for people to watch, someone else doesn't have to download your video to see it. Your video is stored on a server elsewhere and another person uses an app or a Web browser to connect to that server via the Internet to view your video. Similarly, when you want to watch a movie from a service such as Netflix, you use the cloud to see it. The digital files for that movie are stored on a server at another location and can be streamed to your television via the Internet whenever you feel like watching *Sharknado 5*. The cloud is really synonymous with the Internet—whatever you want to use (photos, movies, etc.) is stored on a remote server and you can get access to it via the Internet.

If you can use the cloud to store and access movies or photos, you can do the same with data. For instance, when you used The National Map back in Chapter 1, all of that GIS data was stored on servers that you downloaded via the Internet. If you have your own GIS data, you can store it in the cloud and others can access it. Services such as Dropbox, Google Drive, Apple's iCloud, and Amazon Web Services work this way. Posting parcel databases and DEMs for download on a server will be of great aid to anyone with some geospatial knowledge, but those who don't know what GIS is, what a shapefile is, or how to georeference imagery may be baffled about how to use the data. For instance, say you've used GIS to map the floodplain for your local area, and you've used that information to determine what land parcels are on the floodplain as well as the zoning type and property value of each parcel. If you posted the floodplain layer, the parcels layer, and any tabular data that would need to be joined to the parcels on a server for others to download, people with GIS knowledge may find this collection of data very useful. However, casual Internet users might be able to download these files, but probably would not know how to view them, overlay them, or query them to find out if their properties are on the floodplain or not.

It would be more helpful if you could construct an interactive map of this data and post it online. This type of map would allow casual users to see where their property boundaries are in relation to the floodplain and find out whether or not their house is likely to be flooded. They could also click on their parcel to see the property value and zoning information. Perhaps an aerial photo of their housing development might also be available, showing the locations of nearby rivers and other bodies of water. In addition, the users of the map would have easy access to it through their Web browsers. Even though you may have created the map using ArcGIS, the users wouldn't need an expensive piece of software—they would need only a computer with a Web browser or a smartphone or tablet with access to the Web. Suddenly your online map has the potential to become a utility for realtors to use on-site as well as a mine of useful information for prospective homebuyers.

The cloud makes an ideal platform for delivering these types of online maps to end users (and you've already used some examples of these cloud-based online geospatial tools in the *Hands-On Applications* of previous chapters). All of this geospatial information can be assembled into a **Web map** and placed on a server in the cloud so that anyone can use and interact with its resources by simply using the Web.

When you're setting up (or "publishing") a Web map, you'll be using some type of software that will be hosted online in the cloud. There are several of these types of hosting services, but one of the most popular is Software as a Service (**SaaS**), in which the software you're using to run your application or map will be hosted in the cloud. With SaaS, you don't need to install software on your own computer—all of the software, tools, and accessories are stored in the cloud and you access them through a Web browser. For example, Google's Gmail is an example of SaaS. When you want to use your Gmail account, you log on to the Gmail Website using your browser and you have access to all of the Gmail features—you don't actually download a Gmail program to your computer.

In terms of geospatial technology, Esri's **ArcGIS Online** is a good example of SaaS wherein everything you need to create and publish Web maps is available to you through a Web browser. To make a Web map, sign in to the Website of ArcGIS Online and start using all of the available tools to create your Web map. ArcGIS Online allows users to share datasets with others by using the cloud, but it also gives users the tools to create Web maps and applications that can be served to others via the cloud.

ArcGIS Online comes in two versions. The first of these is free to use— all you need is to create a free Esri account to use when you log on to the ArcGIS Online Website and you can begin creating Web maps to share online. You can create maps from the huge amount of data made available by users of ArcGIS Online, or you can add your own data to the map. This is what you'll do in *Geospatial Lab Application 15.1*—use the free resources and data to create and share a Web map.

The second version of ArcGIS Online enables you to use a lot more features, but also involves paying to use them. In this case, your free Esri account is tied to an organizational account that pays for the use of these added features. For instance, a university or a business may have an organizational account to which your free account can be linked. The organization has a certain number of **credits** that are consumed when you do things with ArcGIS Online. Think of credits as cell phone minutes—you get a certain number of minutes to make phone calls with and when they're used up you need to purchase more. ArcGIS Online credits work the same way—whenever you use an ArcGIS Online feature it deducts credits from the account. For example, the organization version of ArcGIS Online allows you to do batch geocoding (see Chapter 8) and geoprocessing and spatial analysis (see Chapter 6), but these things use up the organization's credits.

The current version of ArcGIS Desktop also interfaces directly with ArcGIS Online, allowing you to upload GIS data and maps as **services** to your ArcGIS Online account. You can then use these services to create Web

Web map an interactive online representation of geospatial data, which can be accessed via a Web browser

SaaS Software as a Service; a method by which the software and data for an application are hosted in the cloud and accessed through a Web browser

ArcGIS Online Esri's cloud-based GIS platform where data and Web mapping software can be accessed via the Internet

credits a system used by Esri to control the amount of content that can be served by an organization in ArcGIS Online

service a format for GIS data and maps to be distributed (or "served") to others via the Internet maps or applications in ArcGIS Online (see **Figure 15.2**). For instance, you can add data and perform analysis in ArcGIS for Desktop, then take your final results, publish them as a service in ArcGIS Online, and create a Web map from them to share with the world.

A **Story Map** is an example of a Web application that goes beyond a simple Web map in ArcGIS Online. In a Story Map, you can tailor your Web map to include pictures, videos, and text that help to communicate a certain theme or information (the "story" of the title) to the user. Story Maps can be freely created using the resources of ArcGIS Online. **Figure 15.3** shows an example of a Story Map created to examine the shale gas boom and where hydraulic fracturing (better known as "fracking") is taking place across the United States. The Story Map contains three tabs, each showing a different map and containing different information—the first shows the location of various shale plays, the second shows the production rates per state, and the third shows the actual locations of fracking sites as points with information about who owns the well pad and when the location was fractured. In this case, the Story Map

Story Map an interactive Web map created using ArcGIS Online resources that can add descriptive text, photos, or videos to locations on a map to inform the user about the map's topic

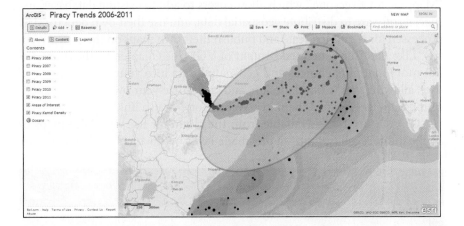

FIGURE 15.2

Somali piracy trends as mapped in ArcGIS Online.

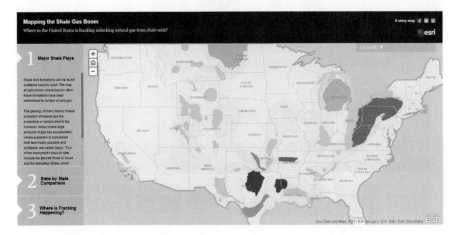

FIGURE 15.3

A Story Map showing the major shale plays across the United States.

provides a lot of information about each development in a series of different interactive Web maps to tell the full "story" of the shale plays and fracking.

Figure 15.4 shows a different type of Story Map, referred to as a "Map Tour." In this type of Story Map, you can select different locations on a map and have descriptive text, photos, or videos linked to those places. In the Map Tour, clicking on a location or the photo thumbnail at the bottom of the screen will bring up the text and media for that place. For instance, in Figure 15.4, selecting a location in San Diego, California, will give you information about the transportation innovations at that place, such as the San Diego Green Line Trolley, the Sprinter light rail, and electric car recharging stations. Text and photos for each location will also appear to help tell the "story" about transportation in San Diego. A Map Tour can also be used to give a traditional tour of areas (such as a walking tour of historic places in a city or a tour of highlights at a popular tourist destination). There are many different types and formats of Story Maps available beyond these two—for more about using Story Maps, see *Hands-On Application 15.1: Esri Story Maps.*

Tools like these are part of a larger initiative that Esri refers to as "Mapping for Everyone," which allows users to add to existing databases, use special software to distribute geospatial data, and use their own data (or data processed for them) to create and publish their own maps on their own Website. Like the user-generated content of VGI (see Chapter 1), these options put more mapping choices and geospatial data into your hands as an end user, allowing you to build, design, and customize your own mapping applications and then embed them in a Website or e-mail a link to other people.

One service that's available from the cloud for use with Web maps or applications is a **basemap**—a pre-made layer over which you can lay other

basemap an image layer that serves as a backdrop for the other layers that are available in ArcGIS Online

FIGURE 15.4

A Map Tour showing various transportation innovations across the city of San Diego, California. *[Used by permission. Copyright © 2015 Esri. All rights reserved.]*

Hands-On Application 15.1

Esri Story Maps

To examine a bunch of different Story Maps, point your Web browser at **http://storymaps.arcgis .com.** This Website is the place to begin with Story Maps—if you want to create a certain type of Story Map, you would choose the Apps option and then select the type of Story Map you want to make. However, in order to view the different types of Story Maps available, choose the option for Gallery. In the Story Map Gallery, you'll see a wide variety of Story Maps created by Esri as well as people from the larger geospatial community. Under the choices for Story Map App you can view different kinds of Story Maps that use the various map types—this includes Map Tours as well as other Story Map types such as Map Journal, Shortlist, Swipe, and Spyglass. Story Maps are also grouped together by theme—by choosing one of the options under

Collections, you can find Story Maps related to topics of interest to you, including those designed for Events and Disasters, Nature and Environment, and Sports and Entertainment. Investigate some of the different types of Story Maps as well as a sampling from various collections.

Expansion Questions

➤ Choose three collections related to your own interests. What topics (or "stories") are being presented with Story Maps in these collections?

➤ Look at Story Map examples from the different types available under Story Map Apps. How does each one present the topic (or "story") in a different fashion?

geospatial data. When you're using a utility such as Google Maps (as in *Hands-On Application 8.3: Online Mapping and Routing Applications and Shortest Paths* on page 309), you can select a view of the shortest path between locations to be displayed over a street map or a satellite image of the area. The street map and the satellite image are examples of basemaps on which other layers can be displayed, and they can be streamed to you via the cloud and displayed in your Web browser. Geospatial data can be obtained through the cloud structure in the form of aerial or satellite imagery, digital topographic maps, and digital street maps, all of which can be used as basemaps with your own applications. Because this data is being served to you over the Internet, you don't have to download it and set it up on your own computer.

For instance, Esri makes a number of such basemaps available through ArcGIS Online. When you're creating a Web map in ArcGIS Online, you can choose the type of basemap you want and display your data overlaid on it (see **Figure 15.5** for examples of the basemap data available through ArcGIS Online). In addition, these same basemaps can be streamed from the cloud into ArcGIS Desktop as well, so that you can have on-demand access to georeferenced high-resolution satellite imagery or street maps to aid in your GIS analyses.

There are many programs available for taking geospatial data and setting it up online in an easy-to-use interactive format as well as creating an interactive map to incorporate into a Website. **ArcGIS Server** is a good example—this Esri software is designed to facilitate the distribution

ArcGIS Server an Esri utility for distribution of geospatial data in an interactive mapping format on the Internet

(a) (b)

(c) (d)

FIGURE 15.5

Examples of basemaps available in ArcGIS Online: (a) Imagery with Labels, (b) Streets, (c) Topographic, and (d) OpenStreetMap.

API Application Programming Interface; the functions and commands used when making requests of software

of interactive maps and geospatial data across the Internet for viewing and analysis. (See **Figure 15.6**, which shows an online mapping and analysis application created for the Pennsylvania Department of Conservation and Natural Resources using ArcGIS Server.)

Setting up interactive online tools for managing and analyzing geospatial data is becoming increasingly common for a variety of utilities. Many other cloud-based geospatial resources are also available for creating Web maps. Products such as CARTO or Mapbox enable users to create quickly and easily powerful maps and geospatial applications that are stored in and served from the cloud.

Web applications are created with the help of an **API** (Application Programming Interface) that can be used for Web development. An API describes the available commands you can use and the format of those commands that

FIGURE 15.6

Online GIS for the Pennsylvania Department of Conservation and Natural Resources, created using ArcGIS Server.

allow two or more types of software or applications to interface or "talk" with one another. APIs are used by programmers when building software or applications. The W3C Geolocation API described on page 17 of Chapter 1 is an example of this—it allows programmers to obtain the location information of your computer or mobile device and plot that on a map.

For example, if you were running a bakery at a particular location, you'd have your shop's address on its Website and when people wanted to get to your shop, they'd have to leave your Website, go to Google Maps, type in the address, and then toggle back and forth between two Websites. What would be more beneficial is if you could embed an interactive version of Google Maps in your shop's Website to show the bakery's location. To do this, you would use a special type of API (in this case, the Google Maps Embed API) to add this to your bakery's Website. This kind of technique was used by the Guggenheim Museum in New York City to add Google Maps interactivity to its Website (see **Figure 15.7**).

APIs are used for creating geospatial Web applications. For example, you would use the Javascript API for developing Web applications for

NATIONAL MUSEUM OF THE AMERICAN INDIAN

The National Museum of the American Indian is the sixteenth museum of the Smithsonian Institution. It is the first national museum dedicated to the preservation, study, and exhibition of the life, ... [more]

ADDRESS

National Museum of the American Indian in Washington, D.C.
Fourth Street & Independence Ave., S.W.
Washington, DC 20560
Phone:301-238-6624
Email:info@si.edu

Additional Information

District of Columbia State   >
Tourism

District of Columbia State   >
Road Conditions

Weather Information

National Weather Service
(NOAA)

7-day forecast

	Today	Tomorrow
		60%
High:	86° F	High: 84° F
Low:	63° F	Low: 69° F
	Partly Sunny	Chance Rain Showers

	High° F	Low° F
Sun	80	69
Mon	79	69
Tue	84	65
Wed	78	58
Thurs	72	·

DIRECTIONS

The National Museum of the American Indian in Washington, D.C., is located on the National Mall between the Smithsonian's National Air & Space Museum and the U.S. Capitol Building.

Metro: L'Enfant Plaza Station (all lines except Red); exit Maryland Avenue/Smithsonian Museums

FIGURE 15.7

The Website of recreation.gov which uses embedded interactive Google Maps.
Recreation.gov

Hands-On Application 15.2

More Than a Map—The Google Maps API

To examine some of the applications of the Google Maps API, point your Web browser to **https://developers.google.com/maps.** Click on Get Started to see the various categories of Google APIs and select one to see a demo of it, including utilizing basemaps, satellite imagery, street views, routing, and data visualization. Within each one of these options, you can view different presentations of data—for instance, on the Satellite link you can examine high-resolution overhead views, as well as 45-degree oblique imagery. Under the View Showcase link, you can read about different Google Maps API applications.

Expansion Questions

➤ What sort of real-world applications have been developed using the Google Maps API, and what have they been used for?

➤ How are the applications for satellite imagery, street views, and data visualization being used to visualize and communicate geospatial information?

Thinking Critically with Geospatial Technology 15.1

Who Owns Geospatial Data?

Here's something for you to think about: You start with a basemap sample (say, a world topographic map), and you obtain data from the USGS about earthquake locations and other natural hazards. You create a Web map using ArcGIS Online, and post it on your Website. Who owns this geospatial resource, and who owns the geospatial dataset you've created? In essence, you've taken other freely available data and created your own unique geospatial resource—but can you really

lay claim to it as something you own? Similarly, if you make changes or updates to a resource such as OpenStreetMap, can you claim ownership of a shared online resource? With more and more geospatial data and customization functions becoming available via cloud resources (such as the high-resolution imagery that's available through ArcGIS Online), who really owns the end products created from them? Does anybody really own them?

ArcGIS. Other companies, for instance Google and Microsoft, provide APIs that allow you to use their mapping products (such as Google Maps or Bing Maps) as functions on Websites. Also commonly used are open source products, such as Leaflet or OpenLayers, which provide libraries of Javascript functions for creating maps and geospatial applications for mobile devices and the Web. By combining several map layers, a user can create a new map **mashup**, which uses different datasets together in one map. Check out *Hands-On Application 15.2: More Than a Map—The Google Maps API* for some examples of the Google Maps API in action as an agent for developing a wide variety of geospatial Web applications.

mashup the combination of two or more map layers thereby creating one new application

Users can create Web maps or applications that can be distributed to everyone via the cloud, thereby circulating geospatial knowledge and making geospatial tools freely available. Users of an interactive Web map of a local metropark may have never before seen the letters G, I, and S together in an acronym, but they can still access the Web map on their smartphone or tablet to locate service stations, places to eat, points of interest, and their location on a hiking trail. GIS analysts can carry out extensive studies of crime hotspots, and law enforcement officers can make use of this data via an updated Web application served to them from the cloud. In the aftermath of a disaster, rescue and relief teams can examine frequently updated maps showing blocked roads, locations of emergency calls, and remotely sensed imagery of the area, all available through Web maps or via applications shared in the cloud. All these new developments in the geospatial world open an increasingly wide variety of professional and educational opportunities to people who are interested or trained in geospatial technology.

 ## Who Is Involved with Geospatial Technology?

There are numerous power players in the geospatial world—we've already discussed the role of companies such as Microsoft, Google, Esri, Digital-Globe, Planet Labs, and a growing number of government agencies, including NASA, NOAA, and the USGS within the United States, or international groups such as the ESA, in generating and providing geospatial data and infrastructure. Other federal government agencies, including the EPA (Environmental Protection Agency) and the NGA (the National Geospatial-Intelligence Agency), play important roles in applications of geospatial technologies. Many state and local agencies utilize GIS, GPS, or remotely sensed data through city planning, county assessors' offices, or county engineers' offices. Also, the Open Geospatial Consortium (**OGC**) is dedicated to developing new standards of geospatial data interoperability.

There are also several prominent professional organizations in the geospatial field. The American Association of Geographers (**AAG**) is the national organization for geographers in all fields, whether physical, regional, cultural, or geospatial. Boasting several thousand members, AAG is a key organization for geographers, hosting a large annual national meeting, multiple regional branches, and publishing key geographic journals. The American Society for Photogrammetry and Remote Sensing (**ASPRS**) is a major professional organization dedicated to remote sensing and geospatial research. ASPRS publishes a prominent academic journal in the field and hosts national and regional meetings that are well attended by geospatial professionals. The Urban and Regional Information Systems Association (**URISA**) sponsors numerous conferences and publications dealing with the academic, professional, and practical aspects of GIS. Finally, but of equal importance, the National Council for Geographic Education (**NCGE**) is a nationally recognized organization of educators dedicated to promoting geographic education through conferences, research, and journal publications.

There are many ongoing research and educational initiatives dedicated to furthering geospatial technology and its applications, as well as

OGC the Open Geospatial Consortium; a group involved with developing new standards of geospatial data interoperability

AAG the American Association of Geographers; the national organization for the field of geography

ASPRS the American Society for Photogrammetry and Remote Sensing; a professional organization in the geospatial and remote sensing field

URISA the Urban and Regional Information Systems Association; an association for GIS professionals

NCGE the National Council for Geographic Education; a professional organization dedicated to geographic education

OhioView a research and education consortium consisting of the state universities in Ohio, along with related partners

to improving and fostering access to geospatial data and tools. For example, **OhioView**, created in 1996, is an initiative dedicated to remote sensing research and educational programs in Ohio. OhioView's efforts in part helped lead to the free distribution of Landsat imagery (previously, a single Landsat scene cost $4000). OhioView serves as a consortium for remote sensing and geospatial research, and is composed of members from each of the state universities in Ohio, along with other partners. With all of these statewide affiliations, OhioView serves as a source of many initiatives within the state, including educational efforts.

AmericaView a national organization dedicated to remote sensing research and education

StateView the term used to describe the programs affiliated with AmericaView in each state

This type of statewide consortium may have originated with OhioView, but it has expanded to a national program called **AmericaView**. Administered in part by the USGS and set up as a 501(c)(3) organization, AmericaView aims to spread remote sensing research and promote educational initiatives all over the United States (see **Figure 15.8**). Many states have developed a **StateView** program similar to OhioView (for instance, there's VirginiaView, WisconsinView, TexasView, and so on), with the eventual goal of having a complete set of 50 StateView programs. Each StateView is composed of state universities and related partners from that state and offers various geospatial services and data (such as Landsat imagery and other remotely sensed data) for that state. For instance, OhioView servers permit the free download of Ohio Landsat data from their multiyear archive. Similarly, WisconsinView provides fresh MODIS imagery of Wisconsin every day, and the other AmericaView member states make many data

FIGURE 15.8

The current membership composition of AmericaView. *[Source: AmericaView]*

Hands-On Application 15.3

AmericaView and the StateView Programs

To see if your state is part of the AmericaView consortium (and to see the current version of the program's membership), open your Web browser to the interactive map located at **https://americaview .org/about-us/membership-map**. If it is a full member, click on the state to bring up the contact info and check out the specific StateView's Website (that is, MichiganView's Website or VirginiaView's Website) to see who that state's members are and what type of data and services they provide. If your state's not yet a full member—or is currently inactive—take a look at

some of your neighboring states and see what they offer as a StateView program.

Expansion Questions

➤ What is the status of your home state as a StateView program?

➤ What types of data and geospatial services does your StateView provide? (If your state is not yet a full member of AmericaView, look at the StateView Websites of neighboring states that are.)

products (such as high-resolution aerial photography and Landsat imagery) freely available. In addition, TexasView offers free access to remotely sensed images and provides links to GIS data of Texas. See *Hands-On Application 15.3: AmericaView and the StateView Programs* to investigate further the resources available through the various StateView programs.

How Is Geospatial Technology Used in K–12 Educational Efforts?

Groups such as NCGE, OhioView, and AmericaView are dedicated to promoting education in the ever-growing field of geospatial technology, especially in K–12 classrooms. With geospatial technologies becoming increasingly utilized in today's society, introducing concepts early in the grammar-school classroom and using them for illustrating topics such as geography, spatial thinking, or earth science is becoming increasingly common. These efforts are aided by the increasing availability of free software programs (like those used in this book), and plenty of freely available geospatial data. Remotely sensed images from the Earth Observing System (EOS) can help to provide clear, graphic explanations of environmental or climatic conditions, such as changing global temperatures, while easily comprehensible Landsat imagery can be used to evaluate the health of vegetation or the loss of forested lands and calculate the subsequent effects. NASA and Google have released teacher kits and supplemental lesson plans to accompany their products, and many other examples of geospatial technology being integrated into K–12 curricula can be found on the Internet and in our schools (see *Hands-On Application 15.4: Educational Resources and Lesson Plans* for several examples).

Hands-On Application 15.4

Educational Resources and Lesson Plans

The widespread nature of geospatial technologies and the availability and access to geospatial software have allowed educators to integrate GIS, GPS, and remote sensing into lesson plans in multiple ways. Examine some of the following examples to see how geospatial technology is being incorporated into K–12 curricula.

1. AmericaView Educational Resource Portal: **https://americaview.org/program-areas/education/resources**

2. Educaching (GPS lesson plans): **www.educaching.com**

3. GIS 2 GPS (GIS and GPS lesson plans): **http://gis2gps.com**

4. Google Maps Education: **https://www.google.com/help/maps/education**

5. Integrated Geospatial Education and Technology Training (IGETT) (labs and other educational materials for GIS and remote sensing): **http://www.igettremotesensing.org/about.html**

6. Landsat Science Education (for use with Landsat imagery): **http://landsat.gsfc.nasa.gov/?page_id=11/teacherkit**

7. Science Spot (GPS lesson plans): **http://sciencespot.net/Pages/classgpslsn.html**

8. VirginiaView Educational Resources (lesson plans incorporating remote sensing concepts and trackable geocoins): **http://virginiaview.cnre.vt.edu/education.html**

Expansion Questions

➤ What kinds of activities are teachers using in the classroom that involve geospatial technology?

➤ When you've examined these examples, locate some other examples of how geospatial technologies are being applied in the classroom. What other types of activities and lesson plans are being used in classroom settings?

Geographic Alliance Network a National Geographic–sponsored organization set up in each state to promote geographic education

SATELLITES Students and Teachers Exploring Local Landscapes to Interpret the Earth from Space; an OhioView initiative involving K–12 teachers and students with geospatial technologies

Beyond using Google Earth (and other software) for interactive classroom investigations, many other programs are aimed at promoting geospatial technology at the K–12 level. The National Council for Geographic Education (NCGE) is at the forefront of geographic education, supporting teachers and furthering the understanding of geographic concepts at the K–12 level. Another organization dedicated to promoting geographic education is the **Geographic Alliance Network,** a national organization of geography educators at the K–12 and higher education levels. Geographic Alliances are set up in each state as a way of hosting workshops, professional development courses, meetings, events, involvement with geography bees, lesson plans, and more at the statewide level.

Among OhioView's initiatives is **SATELLITES** (Students and Teachers Exploring Local Landscapes to Interpret the Earth from Space), a program designed to instruct K–12 teachers about geospatial technologies and to help them to introduce them into their classrooms. SATELLITES hosts weeklong "teachers' institutes" during the summer in locations throughout Ohio. These institutes are free for teachers to attend, and teachers can earn free graduate credits, when available, through successful participation. During the week, teachers receive introductory instruction in several of this book's topics (including GIS, GPS, remote sensing, landscapes, and how to use remotely sensed environment and climate data) and participate in hands-on work with the technology itself. Teachers also receive free equipment to take back to their classrooms, including a GPS receiver and an infrared thermometer. The aim is for the teachers to develop a research project that will incorporate geospatial technologies and an environmental theme (previous SATELLITES institutes have engaged with global research initiatives such as the International Polar Year).

The inquiry-based projects the K–12 students become involved with are part of **GLOBE** (Global Learning and Observations to Benefit the Environment), an international initiative sponsored by NASA and NOAA that incorporates over 20,000 schools from 114 countries. GLOBE is dedicated to the study of Earth's environments and ecosystems. Using specific data-collection protocols, students involved with GLOBE collect data related to land cover, temperature, water, soils, and the atmosphere. This data is shared with other students around the world via the GLOBE Website, creating rich and far-ranging datasets for use by everyone. Teachers attending the SATELLITES institutes learn GLOBE protocols and engage their students through field data collection coupled with geospatial technology applications.

GLOBE Global Learning and Observations to Benefit the Environment; an education program aimed at incorporating user-generated environmental data observations from around the world

An example of a SATELLITES success story is the "Satellite Girls" (a group of four middle school students from Akron, Ohio). Their teacher attended the SATELLITES program in the summer of 2007 and involved a group of his students with data collection and analysis through GLOBE. The girls competed with their project at state and national levels, and were selected to be one of five teams of students to represent the United States at the international GLOBE Learning Expedition in Cape Town, South Africa, in 2008. In 2010 and 2012, GLOBE projects developed by students as a result of their teachers' participation in the SATELLITES program were presented by the students at the White House Science Fair in Washington, D.C.

What Types of Educational Opportunities Are Available with Geospatial Technology?

Beyond K–12 involvement, geospatial technology is integrated into the curricula of higher education programs internationally. The U.S. Department of Labor's Geospatial Technology Competency Model (**GTCM**) provides a framework for geospatial skills that are key for today's workforce and workforces of the future—a framework that has been used as the basis for educational curriculum development by the GeoTech Center. Geospatial technology course offerings are increasingly to be found in

GTCM the Geospatial Technology Competency Model set up by the U.S. Department of Labor college and university geography departments, but also within civil engineering, natural resources, environmental sciences, and geology programs. Bachelor's degrees programs may offer some sort of concentration in geospatial technology, or at least provide the option to take several courses as part of the major. A minor in GIS or geospatial technologies is also a common option in geography departments. Graduate programs (both master's and doctoral degrees) in geography often have an option to focus on aspects of geospatial technology.

However, with the growth of the geospatial field, it's now possible to get a higher education degree in geospatial technology (or GIS or geographic information science, depending on the name of the program). Bachelor's, master's, and even doctoral programs have become available at several universities. Some of these are interdisciplinary programs, involving elements of various social sciences or computer science, and they often integrate elements of information technology, such as programming, creation of databases, or Web design. For instance, Youngstown State University offers a bachelor's degree in Spatial Information Systems, which combines elements from the geospatial side of the geography discipline with computer-related coursework in programming and databases, classes in CAD and professional writing, and an option for the student and advisors to design a set of courses for specific applications (such as environmental science, biology, or archeology).

Certificate programs are also a common option for geospatial education. These programs vary from school to school, but typically involve taking a structured set of geospatial classes (perhaps four to seven classes total). These programs tend to be focused squarely on completing the requirements for a set of courses, and are often available at both the undergraduate and graduate levels. For example, the Geospatial Certificate at Youngstown State University involves students taking four required courses—an introductory mapping or geospatial class, an introductory GIS class, an introductory remote sensing course, and either an advanced GIS or advanced remote sensing class. Students take two additional courses, chosen from a pool of options, including field methods, GPS and GIS, a geospatial-related internship, object-oriented programming, and database design. Certificate and degree programs are also becoming increasingly available at the community and technical college level (see **Figure 15.9** and *Hands-On Application 15.5: Degree Programs and Certificates for Geospatial Technology* for some further information).

Professional certification in geospatial technologies is also available. GIS Professional (**GISP**) certification is offered by the GIS Certification Institute to applicants who pass an exam as well as a portfolio review that includes contributions to the GIS field, evidence of candidates' professional experience, and their level of achievement in education. Esri offers technical certification based on an exam. ASPRS offers certification status to candidates in several geospatial fields, including Certified Photogrammetrist,

GISP the title given for professional GIS certification

Certified Mapping Scientist in GIS/LIS, and Certified Mapping Scientist in remote sensing.

With geospatial technology being used in so many different fields, students from outside the discipline of geography may seek out a couple of courses or aim at certification to give them the kind of formal background in geospatial technology that will assist them when they apply for positions in their own field (be it business, real estate, archeology, geology, or human ecology). Online courses are offered by a variety of schools to allow students to complete a geospatial certificate or degree remotely, and "virtual universities" offer students at one school the option to take a geospatial course with a professor at another school.

Geospatial technology is playing an increasingly significant part in our daily lives, whether in our work, travel, or recreation. We have even

Hands-On Application 15.5

Degree Programs and Certificates for Geospatial Technology

An online utility exists for examining the available geospatial programs at community colleges and technical colleges throughout the United States—open your Web browser and go to **http://www.geotechcenter.org/geotech-center-national-program-locator.html**. This map (part of the Geo-Tech Center Website at **www.geotechcenter.org**) shows those community and technical colleges that are involved with geospatial education. Schools are mapped by categories, such as whether they offer a degree in an aspect of geospatial technology, a certificate program, or have classes available.

Check your local area to see which schools have geospatial courses, and find out what kinds of degree programs they offer (select the option for CC Info on GIS Offerings in the cube menu icon to get the names of the schools).

Expansion Questions

➤ What community or technical colleges in your local area offer some form of geospatial education?

➤ What kinds of programs are available at these schools—certificates, degrees, or classes?

FIGURE 15.9

The GeoTech Center's Web map of geospatial institutions showing two-year colleges (in green) and four-year colleges and universities (in blue).

GIS Day
the Wednesday of Geography Awareness Week (observed in November), dedicated to GIS

Earth Observation Day
a day dedicated to remote sensing research and education held in conjunction with Earth Sciences week (observed in October)

established observance days dedicated to showcasing and celebrating these vital new technologies. The National Geographic Society kicks off Geography Awareness Week each November, and since 1998 the Wednesday of this week has been **GIS Day**. GIS Day is sponsored by a number of prominent agencies and groups (including the National Geographic Society, Esri, the USGS, and the Library of Congress). It is aimed at promoting GIS for students, educators, and professionals in the field. Each year on GIS Day, there are GIS-related events at schools, universities, and workplaces across the world. Similarly, 2006 saw the launch of **Earth Observation Day** (currently observed in conjunction with Earth Sciences week in October of each year) through AmericaView to showcase remote sensing research and education across the globe.

Thinking Critically with Geospatial Technology 15.2

What's Next for Geospatial Technologies?

Right now, you can get satellite imagery from anywhere on a smartphone with a free Google Earth app, download government satellite imagery free of charge, create and view interactive Web maps in no time flat, interact with landscapes and cities in immersive 3D using only a Web browser, earn a doctorate in the geospatial field, and buy a commercial-level drone off the shelf for less than $500 then fly it to create a highly detailed 3D model of the landscape in under a day. If that's what you can do today, what's going to happen tomorrow? Where do you see the geospatial field going? As access to information becomes faster and easier, as the power and importance of computers advances, and as our lives become more and more tied together in a global web, what's going to be the next stage for geospatial technology? What do you see the geospatial world looking like five years from now?

Chapter Wrap-Up

So this is it—the end of the last chapter. Over these 15 chapters, we've looked at many aspects of geospatial technologies, both from the theoretical side and also from working hands-on with various aspects of the technologies themselves. The geospatial world has changed a lot in the last few years (remember, Google Earth only debuted back in 2005), and it's going to keep changing. That next "killer app" for geospatial technology is probably being coded by someone out there as you read this. GIS, GPS, remote sensing, and all of their applications are only going to become more important in our lives, and even though many people still may not be familiar with the term "geospatial technology," by now you've probably got a pretty good idea of what it is and what it can do.

See *Geospatial Cloud and Organizational Apps* for a mobile app for ArcGIS Online, and also check out *Geospatial Organizations and the Geospatial Cloud in Social Media* for several blogs, Facebook pages, Twitter accounts, and YouTube videos related to using the cloud as well as this chapter's look at some of the people and groups out on the frontiers of geospatial technologies. Also, there's still one last Geospatial Lab Application. In *Geospatial Lab Application 15.1* you'll be using Esri's ArcGIS Online to create and publish your own Web maps in the cloud and also look at some of the organizational information we've seen in this chapter.

Important note: The references for this chapter are part of the online companion for this book and can be found at **www.macmillanlearning.com/catalog/shellitoigt4e**.

Geospatial Cloud and Organizational Apps

Here's a sampling of available representative apps related to this chapter's contents for your smartphone or tablet. Note that some apps are for Android, some are for Apple iOS, and some may be available for both.

➤ **ASPRS:** An app that gives you access to ASPRS events, publications, and conference information

➤ **Explorer for ArcGIS:** An app that allows you to access ArcGIS Online Web maps in the cloud (made by others or within your organization)

Geospatial Organizations and the Geospatial Cloud in Social Media

Here's a sampling of some representative Facebook and Twitter accounts, along with some YouTube videos and blogs related to this chapter's concepts.

 On Facebook, check out the following:

➤ AmericaView:
   www.facebook.com/AmericaView

➤ Earth Observation Day:
   www.facebook.com/EarthObsDay

➤ Esri Geo Developers:
   www.facebook.com/EsriGeoDev

➤ GeoTech Center:
   www.facebook.com/geotechcenter20

> **GIS Day:**
> www.facebook.com/gisday
> **GLOBE Program:**
> www.facebook.com/TheGLOBEProgram

 Become a Twitter follower of:

> **AAG:** @theAAG
> **AmericaView:** @AmericaView
> **ArcGIS Online:** @ArcGISOnline
> **ASPRS:** @ASPRSorg
> **CARTO:** @CARTO
> **Earth Observation Day:** @EarthObsDayAV
> **Esri GIS Education:** @GISEd
> **Esri Story Maps:** @EsriStoryMaps
> **Extension** (America's Research-Based Learning Network)—Geospatial Technology: @exGeospatial
> **GeoTech Center:** @GeoTechCenter
> **GIS Day:** @gisday
> **Globe Program:** @GLOBEProgram
> **Joseph Kerski** (geospatial educational news from Esri's Education program): @josephkerski
> **Leaflet:** @LeafletJS
> **Mapbox:** @Mapbox
> **NCGE:** @NCGE1915
> **OhioView:** @OHViewGeo
> **Open Geospatial Consortium:** @opengeospatial
> **URISA:** @URISA

On YouTube, watch the following videos:

> **Geographyuberalles** (a YouTube channel featuring all types of GIS and geographic videos): **www.youtube.com/user/geographyuberalles**
> **GIS Education** (an Esri YouTube channel devoted to educational concepts using GIS): **www.youtube.com/user/ESRIEdTeam**
> **GLOBE Program** (The GLOBE Program's YouTube channel featuring videos of students' and teachers' research): **www.youtube.com/user/globeprogram**

For further up-to-date info, read up on these blogs:

➤ **AmericaView Blog** (the latest news and geospatial applications from the members of AmericaView): **http://blog.americaview.org**

➤ **ArcGIS Blog—ArcGIS Online Category** (blog posts concerning topics related to working with ArcGIS Online resources): **http://blogs.esri .com/esri/arcgis/category/arcgis-online**

➤ **ArcGIS Blog—Esri Story Maps Category** (blog posts about using ArcGIS Online and Story Maps): **http://blogs.esri.com/esri/arcgis/ category/story-maps**

➤ **Esri GIS Education Community** (a blog about education applications of GIS): **http://blogs.esri.com/esri/gisedcom**

➤ **Google Maps APIs Blog** (a blog dealing with new developments and applications of Google's geospatial tools): **https://maps-apis .googleblog.com**

➤ **Letters from the SAL** (a blog about geospatial research from the Spatial Analysis Laboratory at the University of Vermont): **http:// letters-sal.blogspot.com**

➤ **Spatial Reserves** (a blog about publicly available geospatial data): **http://spatialreserves.wordpress.com**

Key Terms

Creating Web Maps with ArcGIS Online

This chapter's lab will introduce you to the use of Esri's cloud-based GIS mapping and analysis package, ArcGIS Online along with some data related to the organizational topics discussed in the chapter. As noted in this chapter, there are two different versions of ArcGIS Online: one that requires you to be part of an organization and purchase credits for use, and a free version that doesn't use the credits system. This lab will use the free version, so all data, methods, and techniques you'll use in this lab are free of charge. During this lab, you'll use the cloud as your platform for working with GIS data, remotely sensed imagery, and geospatial applications. You'll create two Web maps—one from data available online and one from data within a spreadsheet—and submit the URLs for these Web maps to your instructor for grading.

Objectives

The goals of this lab are:

➤ To learn the basics of the ArcGIS Online environment

➤ To create a Web map from data available through ArcGIS Online

➤ To adjust the symbology of features on a Web map

➤ To save and share a Web map

➤ To create a Web application from a Web map

➤ To perform geocoding using ArcGIS Online

➤ To use different basemaps in creating a Web map

➤ To create a Web map and a Web application from your own data

Using Geospatial Technologies

The concepts you'll be working with in this lab are used in a variety of real-world applications, including:

➤ Engineering, where Web maps are utilized in the planning of new roads or highways by examining terrain conditions such as elevation, steepness, and land cover

➤ County auditing, where assessors examine Web maps and streaming high-resolution imagery of residential and commercial properties to assist them in determining property taxes

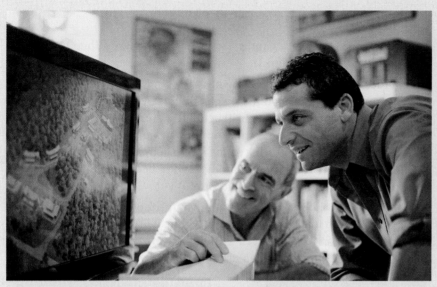

[Mint Images-Tim Pannell/Getty Images]

Obtaining Software

No software or data is necessary to download for this lab—everything you will be working with can be done through a Web browser.

Important note: Software and online resources can change fast. This lab was designed with the most recently available version of the software at the time of writing. However, if the software or Websites have changed significantly between then and now, an updated version of this lab (using the newest versions) will be available online at **www.macmillanlearning.com/catalog/shellitoigt4e**.

Lab Data

Copy the folder **Chapter15**—it contains an Excel file (converted to CSV format) called OhioViewAddresses.csv, which contains the names and addresses of the colleges and universities that comprise the membership of OhioView.

Localizing This Lab

This lab starts at the broad scale of the entire United States and then narrows its focus to areas within Ohio, specifically the members of the OhioView consortium. However, the same techniques used in this lab can be used for your own local area. When you're creating the Web map from your own data in Step 15.4, you can substitute a .csv file consisting of the names and addresses of the schools from your own StateView (such as the schools within the VirginiaView consortium, for instance). Alternatively, you could use a .csv file containing the names and addresses of the state universities for your state.

15.1 Getting Started with ArcGIS Online

1. Open your Web browser and go to **www.arcgis.com**. When you arrive at this page, click on **Sign In**.

2. In order to save and share your work with ArcGIS Online, you'll need to obtain a free Esri account. You can do this by clicking on **Create a Public Account**. Follow the instructions to create your account, then sign in to ArcGIS Online using your new account. After you have signed in, you'll see your first name in the upper right-hand corner of the screen where the words "Sign In" used to be—this indicates you're properly signed in to your Esri account.

3. After signing in, click on the **Map** link at the top of the Web page. The basic Web map interface for ArcGIS Online will appear. You'll see the controls to zoom in and out of the map in the upper left-hand corner of the map view (you can also use the mouse for this).

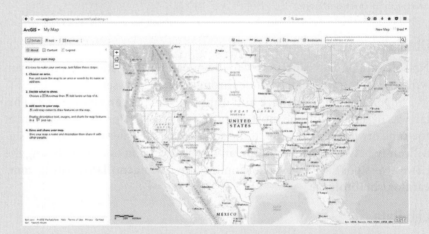

15.2 Creating a Web Map from Online Data

1. We'll begin by making a Web map from data that's already available online. A dataset showing the 2017 AmericaView membership has already been created and shared through ArcGIS Online. To retrieve this dataset and begin developing your Web map, click the **Add** button, and then select **Search for Layers**.

2. In the **Find:** box, type **AmericaView**.

3. In the **In:** box, select **ArcGIS Online**.

4. Remove the checkmark from the **Within map area** box.

5. Click **GO** to search for the AmericaView layer in ArcGIS Online.

6. ArcGIS Online will respond that it has found the result—the layer called AmericaView (by bradbook). Click the text that says **Add** next to this result. A new layer showing the current membership of states in AmericaView will be added to the Web map. Click on **DONE ADDING LAYERS** to continue making your map.

7. You'll see the AmericaView layer listed under Contents along with the current basemap being used (by default it should be the Topographic basemap). These are the two layers that make up your map. However, you'll see that the layer showing the AmericaView states is covering up the basemap. What you'll do next is change the color of the AmericaView states layer and also make it semi-transparent so that you can see both the states layer and the basemap underneath.

8. To begin, click on the small ellipsis symbol ... next to AmericaView in the Contents box. A new group of options will appear.

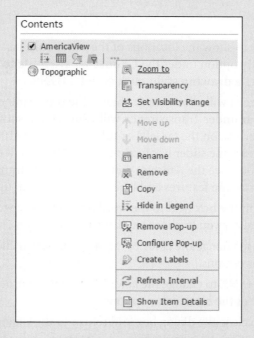

9. These options will allow you to affect the layer, rename it, remove it, or change some of its properties and appearance. To begin changing the appearance of the layer, select the **Change Style** icon.

10. As you want to show the locations of the states only, just use the **Show location only** option under the **Choose an attribute to show** menu. Under **Select a drawing style**, click on **OPTIONS**.

11. The next screen will allow you to change the transparency of the layer. The slider bar under Transparency will allow you to show the layer fully opaque (0%, on the far left) or fully transparent (100%, on the far right). Move the slider bar to about the **40%** percent transparency range and you'll see the AmericaView layer become transparent enough to view the features and labels of the basemap underneath.

12. Next, click on the **box** next to the word Symbols. A new menu will appear and allow you to choose a different color from the default.

13. Select **FILL** to change the color of the states and then click on one of the red color options.

14. Select **OUTLINE** to change the color of the states' borders and then click on one of the black color options.

15. Click **OK** to close the dialog box and return to the Change Style menu.

16. You'll see your chosen colors shown on the Web map. If they're not what you want, return to the color choice options and choose some new colors. When the map is looking the way you want it

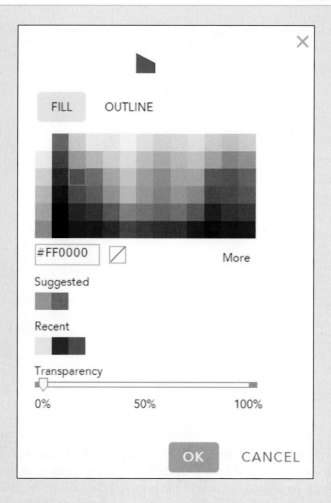

(semi-transparent red states with black borders), click **OK** in the Change Style menu. Back in the initial Change Style menu, click **DONE** to return to the Contents panel.

17. Next, click on one the AmericaView states and you'll see a pop-up box appear filled with attributes. These attributes are all the items that were part of the initial U.S. Census Bureau TIGER file that serves as the source for this AmericaView layer. However, for the purposes of the Web map, all you want to display is the name of each state, not such information as the area of the state or its FIPS code. To do this, you'll have to tell ArcGIS Online which attributes you want to display.

18. To do so, again select the ellipsis icon **. . .** next to the AmericaView layer under Contents and select **Configure pop-up**.

19. Under Pop-up Title, change the default text to the following: **AmericaView StateView: {NAME}**. What this means is that when you click on a state, the title of the pop-up box will say "AmericaView StateView:" and then display whatever text is assigned to the NAME field of the layer's attribute table. For instance, clicking on Ohio will cause the title of the pop-up box to be "AmericaView StateView: Ohio."

20. Because name of the state is displayed in the title of the box, under Display select **No attribute information**. If you wanted to show other attributes from the layer's attribute table they could be selected from here.

21. As you have no graphics or charts to display along with the name of the state, just use the **No images or charts option** under **Pop-up media**. If you did have a graphic or picture to accompany the states, here is where you would add it.

22. Next, click on **OK** to save your changes.

23. Click on one of the AmericaView states in your Web map to see its pop-up. Also, adjust your view so that you can see the lower 48 states of the United States.

24. Finally, to save your map, click on the **Save** button on the toolbar.

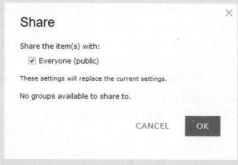

25. In the Save Map dialog box, under Title, give your Web map an appropriate name (such as "AmericaView Membership" or something similar).

26. Under Tags, use *IGT4e* as your tag (for **Introduction to Geospatial Technologies**, 4th edition). This will be one of the search terms that you can use to find other maps like this.

27. Under Summary, give the map an appropriate description of what it contains.

28. The **Save in folder** option will show your own ArcGIS Online account, using the user name that you created back in Section 15.1.

29. When you're ready, click **SAVE MAP**.

15.3 Creating and Sharing a Web Application

Your Web map is now saved. You can reopen it at any time by selecting the **ArcGIS** pull-down menu in the upper-left corner of the screen, choosing **My Content**. This will show you all of the Web maps you've created and any other services, Web applications, or anything else you've created using ArcGIS Online. By selecting your Web map, you'll have another option to open the saved map. However, you're the only one who can access it until you share it with others.

1. To get started sharing your map, first click on the name of your Web map itself.

2. A new screen will appear with a number of different options. From the choices on the right-hand side, click on the **Share** button.

3. In the Share dialog box, put a checkmark in the box next to **Everyone (public)**. This will share your Web map with everyone out there. Then click **OK**.

4. One thing to keep in mind is that any Web map created with ArcGIS Online will have the basic appearance of the ArcGIS Online interface. To distinguish it, change the overall appearance of the interface, and customize it further, you'd want to create a Web application. To get started with this, click on **Create A New Web App**, then select **Using a Template**.

5. There are numerous types of Web applications, each with its own distinct appearance and features. For the basic map you're creating, some of the Web application templates are not appropriate (such as the Elevation Profile or the Story Map options). Try out some of the Web application templates by clicking on one of them and then choosing **Preview** from the new options that appear. A new tab or pop-up window will open showing what the final Web application will look like. Many of the templates would be appropriate for the AmericaView Membership map you've created, but in particular check out and evaluate the following templates: Basic Viewer, Information Lookup, Map Tools, Public Information, and Simple Map Viewer.

6. When you have settled on a template for what your final Web application will look like, double-click on it, then choose **CREATE WEB APP**.

7. On the next screen, you'll see the Title, Tags, and Summary you added for the Web map. Adjust these as you wish, and when you're ready, click **DONE**.

8. Your Web application will open in your Web browser. A new pane will open on the right-hand side, allowing you to make other changes or adjustments. Take a look at the many options available to you, including which widgets can be used on the Web application. However, for this basic Web application, just accept all of the defaults, and then click **LAUNCH**.

9. Your Web application will open in a tab in your browser. If you want to share it with someone, you could copy the URL from your browser, then post it on Facebook or e-mail it to someone. Examine your final Web application, then answer Question 15.1.

> **Question 15.1** Submit the URL for your final Web application to your instructor, who will check over your work for completeness and technical merit (do you have the pop-ups configured properly, do you have the map zoomed to the proper extent, etc.) as well as its appearance (did you choose an appropriate color scheme, did you select an appropriate Web application template, etc.) to make sure you get credit for this exercise.

15.4 Creating a Web Map from Your Own Data

1. Next, we'll make a second Web map, but using your own data (i.e., data not available through ArcGIS Online). Back in the main Web browser tab, return to **www.arcgis.com**, select the option for **MAP**. When it returns to your basic Web map, select **NEW MAP**, then select **Create New Map** to effectively clear out your previous map and provide a fresh mapping canvas with which to work.

2. Bring up Windows Explorer and navigate to the location to which you copied the **Chapter15** folder. Inside of it is a spreadsheet called **OhioViewAddresses.csv**. Open this file and you'll see the names of the 14 colleges and universities that make up the membership of OhioView. What we're going to do next is geocode these addresses (see Chapter 8) and turn them into a point layer on the map. In the end, you'll have a Web map of Ohio featuring these 14 different locations of the OhioView member schools.

3. Geocoding with ArcGIS Online is a snap. Resize your Windows Explorer window and your Web browser window so that you can see both of them side by side. Then simply drag and drop the **OhioViewAddresses.csv** file from Windows Explorer into the map shown in ArcGIS Online.

4. The **Add CSV Layer** dialog box will appear. This lets you specify how you want to add this data to ArcGIS Online. Under **Locate features using:**, select the radio button for **Address** and make sure that in the Field Name column, the entry for Address is using the Location Fields of **Street**. This lets ArcGIS Online know that you will be geocoding addresses (rather than plotting points based on coordinates) and that the attribute field called "Address" contains the necessary street information to be used in geocoding. Click **ADD LAYER** to proceed with geocoding your points.

5. The next screen will allow you to specify how you want the points to appear on the map. Make sure that the **OhioView school** field

is selected for. Choose an attribute to show (as you want to show the name of each school on the map), and under Select a drawing style, click on **Options**. You'll see that of the 14 schools, ArcGIS Online gives a unique color to 10 of them, then puts the last four in a category called "Others." To give each of the 14 schools its own unique color, scroll through the list of schools and click on the **blue arrows** symbol next to "Others." This will expand the option and show the other four schools as unique points (see the graphic below for a guide).

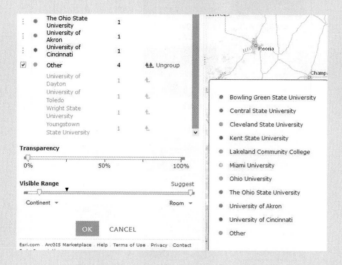

6. Move the left-hand arrow of the **Visible Range** slider bar to **Countries** so that you'll be able to see the points when you're zoomed in to the level of the United States or closer.

7. Click on **OK** to close the dialog box and then click on **DONE** to complete the mapping of the schools as points on the map.

8. In the Web map, zoom out so that you can see all of Ohio and your 14 geocoded schools.

9. Click on the **Show Legend** icon in the table of contents and you'll see the names and colors that correspond with each point displayed. Click on a school on the map and you'll see a pop-up with its name and address.

15.5 Using Basemaps in Web Mapping

1. The next thing we'll do is select a different basemap for this Web map. To see your basemap options, click on the **Basemap** button.

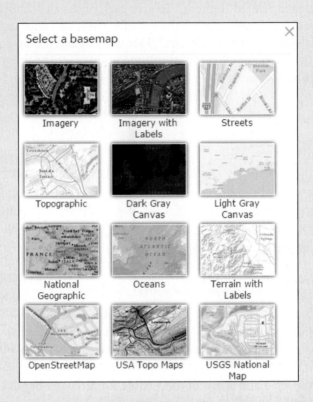

2. In the **Select a basemap** dialog box, you'll see several different options for the basemap (which is the graphics layer underneath your data). By default, ArcGIS Online will start you off with the Topographic basemap.

3. To switch to the Imagery basemap, click on its icon in the Select a basemap dialog box. The basemap will switch to high-resolution satellite imagery of Ohio (and the surrounding areas).

4. While this looks nice, it doesn't put your geocoded points in much context. Click on **Basemap** again, and this time choose **Imagery with Labels**. The map will change to now have borders for the state as well as some labeled cities. As you zoom in on the map, you'll see other cities appear as well as county borders.

5. Your final map will have a single basemap underlying it, so try out all of the other choices for basemaps. You'll find that some of them are less appropriate for the kind of map you're making (such as Imagery, Dark Grey Canvas, Light Grey Canvas, and Oceans), but other choices will be more appropriate. Keep in mind that you'll want to select a

basemap that looks good while examining Ohio at the state scale as well as when zooming in on particular school points. Move on to the next step when you've chosen which basemap you want to use in your final map (just make sure it's not Topographic).

15.6 Creating a Web Application from Your Own Data

1. When you have your Web map looking the way you want it, the last step will be to turn it into a Web application (as you did to the AmericaView map back in Section 15.3—you may want to refer back to that section for how to do each of the following things).

2. **Save** your Web map (giving it an appropriate title and summary, and using IGT4e for a tag).

3. **Share** your Web map with **Everyone**.

4. Make a Web application from your Web map, choosing an appropriate Web application template.

5. When you have created your final Web application and have a URL assigned to it, open that URL in a new tab, examine your final Web application, and then answer Question 15.2.

Question 15.2 Submit the URL for your final OhioView Web application to your instructor, who will check over your work for completeness and technical merit (do you have all 14 schools labeled properly, do you have the map zoomed to the proper extent, etc.) as well as its appearance (did you choose an appropriate basemap, did you select an appropriate Web application template, etc.) to make sure you get credit for this exercise.

Closing Time

ArcGIS Online is a great tool for creating Web maps from your own data or preexisting data online. There are a number of other types of data that can be used in ArcGIS Online, including GPX files (see Chapter 4) and a shapefile within a zipped file (see Chapter 5), as well as numerous types of adjustments and visualization options for your Web maps.

In addition, there are numerous types of Web applications that can be created from the templates, including Story Maps. While the organizational version of ArcGIS Online adds many other features, you can create professional-looking Web maps quickly and easily by using only the free version. When you're finished, simply exit your Web browser—your Web maps and Web apps will remain safely stored in your space in the cloud that came with your free Esri account. Then you can say goodbye to lab applications for the rest of the book.

Glossary

3D modeling (p. 503) designing and visualizing data that contains a third dimension (a z-value) in addition to x- and y-values

3D trilateration (p. 103) finding a location on Earth's surface in relation to the positions of three satellites

3DEP (p. 481) the 3D Elevation Program, a U.S. government program that provides digital elevation data for the entire United States

8-bit imagery (p. 374) a digital image that carries a range of brightness values from 0 to 255

AAG (p. 551) the American Association of Geographers; the national organization for the field of geography

absorption (p. 368) when light is trapped and held by a target

across-track scanning (p. 397) a scanning method using a rotating mirror to collect data by moving the device back and forth across the width of the satellite's swath

address matching (p. 301) another term for geocoding

address standardization (p. 302) setting up the components of an address in a regular format

aerial photography (p. 2, 330) taking photographs of objects on the ground from an airborne platform

affine transformation (p. 76) a linear mathematical process by which data can be altered to align with another data source

Afternoon Constellation (p. 436) a set of satellites that pass the Equator in the afternoon during their orbits

AIRS (p. 448) the Advanced Infrared Sounder instrument onboard Aqua

algorithm (p. 307) a set of steps used in a process designed to solve a particular type of problem (for example, the steps used in computing a shortest path)

almanac (p. 101) data concerning the status of a GPS satellite, which is included in the information being transmitted by the satellite

along-track scanning (p. 397) a scanning method using a linear array to collect data directly on a satellite's path

AmericaView (p. 552) a national organization dedicated to remote sensing research and education

AMSR2 (p. 446) the Advanced Microwave Scanning Radiometer 2 instrument onboard GCOM-W1

anaglyph (p. 517) a new image created by slightly offsetting the features in an existing image from each other and highlighting them in two different colors

AND (p. 194) the Boolean operation that corresponds with an intersection operation

API (p. 549) Application Programming Interface; the functions and commands used when making requests of software

Aqua (p. 436) an Earth observing satellite whose mission is to monitor Earth's water cycle

ArcGIS (p. 144) Esri's GIS platform

ArcGIS Desktop (p. 145) the component of ArcGIS that runs on a desktop computer

ArcGIS Online (p. 145, 544) Esri's cloud-based GIS platform where data and Web mapping software can be accessed via the Internet

ArcGIS Pro (p. 145) a new 64-bit stand-alone GIS program that comes as part of ArcGIS Desktop

ArcGIS Server (p. 548) an Esri utility for distribution of geospatial data in an interactive mapping format on the Internet

ArcGlobe (p. 512) a 3D visualization tool used to display data in a global environment, which is part of Esri's ArcGIS Desktop program

ArcMap (p. 145) the component of ArcGIS Desktop used for viewing and analyzing data

ArcScene (p. 512) the 3D visualization and design component of Esri's ArcGIS Desktop program

ASPRS (p. 551) the American Society for Photogrammetry and Remote Sensing; a professional organization in the geospatial and remote sensing field

atmospheric windows (p. 367) those wavelengths of electromagnetic energy in which most of the energy passes through Earth's atmosphere

A-Train (p. 436) another term for the Afternoon Constellation

attribute table (p. 141) a spreadsheet-style form where the rows consist of individual objects and the columns are the attributes associated with those objects

attributes (p. 140) the non-spatial data that can be associated with a spatial location

Aura (p. 436) a satellite dedicated to monitoring Earth's atmospheric chemistry

AVHRR (p. 446) Advanced Very High Resolution Radiometer; the six-band sensor onboard NOAA satellites

band (p. 365) a narrow range of wavelengths that may be measured by a remote sensing device

base heights (p. 486, 505) the elevation values assigned to the terrain upon which the objects will be placed

basemap (p. 547) an image layer that serves as a backdrop for the other layers that are available in ArcGIS Online

batch geocoding (p. 304) matching a group of addresses together at once

Beidou-2 (p. 110) China's GNSS, currently in development

block (p. 503) a flat polygon that has been extruded to transform it into an object with a *z*-value

blue band (p. 365) the range of wavelengths between 0.4 and 0.5 micrometers

Boolean operator (p. 194) one of the four connectors (AND, OR, NOT, XOR) used in building a compound query

brightness values (p. 373) the energy measured at a single pixel according to a pre-determined scale; also referred to as digital numbers (DNs)

buffer (p. 196) a polygon of spatial proximity built around a feature

C/A code (p. 101) the digital code broadcast on the L1 frequency, which is accessible by all GPS receivers

cartographic generalization (p. 246) the simplification of representing items on a map

cartography (p. 243) the art and science of creating and designing maps

Catalog (p. 145) the component of ArcMap used for managing data

CERES (p. 446) the Clouds and Earth's Radiant Energy System instruments onboard Terra, Aqua, Suomi NPP, and JPSS-1

CGIS (p. 131) the Canadian Geographic Information System—a large land inventory system developed in Canada, and the first system of this type to use the term "GIS"

channels (p. 100) the number of satellite signals a GPS unit can receive

choropleth map (p. 252) a type of thematic map in which data is displayed according to one of several different classifications

CIR photo (p. 338) color infrared photo—a photo in which infrared reflection is shown in shades of red, red reflection is shown in shades of green, and green reflection is shown in shades of blue

citizen science (p. 17) the activities of untrained volunteers conducting science

CityEngine (p. 513) a powerful rule-based 3D visualization program designed for quickly modeling 3D cities

cloud (p. 542) a computer structure wherein a user's data, resources, or applications are kept at another location and made available to the user over the Internet

CMYK (p. 255) a color scheme based on using the colors cyan, magenta, yellow, and black

collar (p. 475) the white, information-filled border around a US Topo

color composite (p. 375) an image formed by placing a band of imagery into each of the three color guns (red, green, and blue) to view an image in color rather than grayscale

color gun (p. 374) equipment used to display a color pixel on a screen through the use of the colors red, green, and blue

color ramp (p. 256) a particular range of colors applied to the thematic data on a map

components (p. 509) pre-designed (or user-designed) objects in SketchUp

compound query (p. 194) a query that contains more than one operator

connectivity (p. 297) the linkages between edges and junctions of a network

constellation (p. 99) the full complement of satellites comprising a GNSS

continuous field view (p. 136) a conceptualization of the world in which all items vary across Earth's surface as constant fields, and values are available at all locations along the field

contour interval (p. 474) the vertical difference between two adjacent contour lines drawn on a map

contour line (p. 473) an imaginary line drawn on a map to connect points of common elevation

control points (p. 72) point locations where the coordinates are known; used to align the unreferenced image to the source

control segment (p. 100) one of the three segments of GPS, consisting of the control stations that monitor the signals from the GPS satellites

Corona (p. 394) a U.S. government remote sensing program, which utilized film-based camera equipment mounted in a satellite in Earth's orbit; Corona was in operation from 1960 to 1972

CORS (p. 108) Continuously Operating Reference Stations; a system operated by the National Geodetic Survey to provide a ground-based method of obtaining more accurate GPS positioning

credits (p. 544) A system used by Esri to control the amount of content that can be served by an organization in ArcGIS Online

crowdsourcing (p. 17) the activities of untrained volunteers to create content and resources that can be utilized by others

CubeSat (p. 414) a type of satellite with a weight of less than 3 pounds and a length of about 4 inches

data classification (p. 252) various methods used for grouping together (and displaying) values on a choropleth map

datum (p. 42) a reference surface of Earth

datum transformation (p. 43) changing measurements from one datum to measurements in another datum

decimal degrees (DD) (p. 45) the fractional decimal equivalent to coordinates found using degrees, minutes, and seconds

degrees, minutes, and seconds (DMS) (p. 44) the measurement system used in GCS

DEM (p. 478) digital elevation model; a representation of the terrain surface, created by measuring a set of equally spaced elevation values

DGPS (p. 107) differential GPS; a method of using a ground-based correction in addition to the satellite signals to determine position

digitizing (p. 134) the creation of vector objects through sketching or tracing representations from a map or image source

Dijkstra's Algorithm (p. 307) an algorithm used in calculating the shortest path between an origin node and other destination nodes in a network

discrete object view (p. 132) a conceptualization of the world in which all reality can be represented by a series of separate objects

dissolve (p. 196) the ability of the GIS to combine polygons with the same features

DLG (p. 473) Digital Line Graph; the features (such as roads, rivers, or boundaries) digitized from USGS maps

DOQ (p. 342) a Digital Orthophoto Quad—orthophoto that covers an area of 3.75 minutes of latitude by 3.75 minutes of longitude, or one-fourth of a 7.5 minute USGS quad

Dove (p. 414) the name for the CubeSat series launched by Planet Labs

DPI (p. 257) dots per inch—a measure of how coarse (lower values) or sharp (higher values) an image or map resolution will be when exported to a graphical format

draping (p. 485) a process in which an image is given z-values to match the heights in a digital terrain model

DRG (p. 472) Digital Raster Graphic; a scanned version of a USGS topographic map

DSM (p. 480) digital surface model; a measurement of the heights of ground elevations as well as the objects on top of the ground as captured by lidar

DTM (p. 478) a representation of a terrain surface calculated by measuring elevation values at a series of locations

dual frequency (p. 100) a GPS receiver that can pick up both the L1 and L2 frequencies

Earth Observation Day (p. 558) a day dedicated to remote sensing research and education held in conjunction with Earth Sciences week (observed in October)

easting (p. 52) a measurement of so many units east (or west) of some principal meridian

edge (p. 297) a term used for the links of a network

EGNOS (p. 110) an SBAS that covers Europe

electromagnetic spectrum (p. 364) the light energy wavelengths and the properties associated with them

ellipsoid (p. 42) a model of the rounded shape of Earth

ephemeris (p. 101) data referring to the GPS satellite's position in orbit

equal interval (p. 253) a data classification method that selects class break levels by taking the total span of values (from highest to lowest) and dividing by the number of desired classes

Equator (p. 44) the line of latitude that runs around the center of Earth and serves as the 0 degree line from which to make latitude measurements

EROS (p. 398) the Earth Resources Observation Science Center, located outside Sioux Falls, South Dakota, that serves (among many other things) as a downlink station for satellite imagery

ESA (p. 435) the European Space Agency, an organization made up of multiple European countries, dedicated to space exploration

Esri (p. 144) a key developer and industry leader of GIS products

ETM+ (p. 402) the Enhanced Thematic Mapper sensor onboard Landsat 7

exclusive or (p. 195) the operation wherein the chosen features are all of those that meet the first criterion as well as all of those that meet the second criterion, except for the features that have both criteria in common in the query

extrusion (p. 503) the extending of a flat object to have a z-value

face (p. 505) one side of a block

false color composite (p. 378) an image arranged when the distribution of bands differs from placing the red band in the red color gun, the green band in the green color gun, and the blue band in the blue color gun

false easting (p. 54) a measurement made east (or west) of an imaginary meridian set up for a particular zone

false northing (p. 53) a measurement made north (or south) of an imaginary line such as is used in measuring UTM northings in the southern hemisphere

feature class (p. 147) a single data layer in a geodatabase

feature dataset (p. 147) a grouping of multiple feature classes together within a geodatabase

fields (p. 141) the columns of an attribute table

file geodatabase (p. 148) a single folder that can hold multiple files, with nearly unlimited storage space

Flock-1 (p. 414) a constellation of high-resolution remote sensing CubeSats launched by Planet Labs in 2014

fonts (p. 249) various styles of lettering used on maps

Galileo (p. 110) the European Union's GNSS, currently in development

GCOM-W1 (p. 437) the Global Change Observation Mission—Water, a JAXA satellite mission dedicated to studying Earth's water cycle

geocaching (p. 112) using GPS and a set of coordinates to find locations of hidden objects

geocoding (p. 301) the process of using the text of an address to plot a point at that location on a map

geocoin (p. 113) a small coin used in geocaching with a unique ID number that allows its changing location to be tracked and mapped

geodatabase (p. 147) a single item that can contain multiple layers, each as its own feature class

geodesy (p. 42) the science of measuring Earth's shape

GeoEye-1 (p. 412) a satellite launched in 2008 that features a panchromatic sensor with a spatial resolution of 0.46 meters

Geographic Alliance Network (p. 554) a National Geographic–sponsored organization set up in each state to promote geographic education

geographic coordinate system (GCS) (p. 44) a set of global latitude and longitude measurements used as a reference system for finding locations

geographic information system (GIS) (p. 2, 129) a computer-based set of hardware and software used to capture, analyze, manipulate, and visualize geospatial information

geographic scale (p. 244) the real-world size or extent of an area

geoid (p. 42) a model of Earth using mean sea level as a base

GEOINT (p. 8) Geospatial Intelligence

geolocation (p. 17) the technique of determining where something is in the real world

GeoPDF (p. 258) a format that allows maps to be exported to a PDF format, yet contain multiple layers of geographic information

geoprocessing (p. 198) the term that describes when an action is taken to a dataset that results in a new dataset being created

georeferencing (p. 70) a process of aligning an unreferenced dataset with one that has spatial reference information

geospatial data (p. 8) items that are tied to specific real-world locations

geospatial technology (p. 2) a number of different high-tech systems that acquire, analyze, manage, store, or visualize various types of location-based data

geostationary orbit (p. 395) an orbit in which an object precisely follows the direction and speed of Earth's rotation and is therefore always directly above the same point on Earth's surface

geotag (p. 18) connecting real-world location information to an item

GeoTIFF (p. 472) a graphical file format that can also carry spatial referencing

GeoWall (p. 518) a powerful computer tool for displaying data and imagery in stereoscopic 3D

GIS Day (p. 558) the Wednesday of Geography Awareness Week (observed in November), dedicated to GIS

GIS model (p. 205) a representation of the factors used for explaining the processes that underlie an event or for predicting results

GISP (p. 556) the title given for professional GIS certification

Global Positioning System (GPS) (p. 2) acquisition of real-time location information from a series of satellites in Earth's orbit

GLOBE (p. 555) Global Learning and Observations to Benefit the Environment; an education program aimed at incorporating user-generated environmental data observations from around the world

GLONASS (p. 110) the former USSR's (now Russia's) GNSS

GNSS (p. 98) the global navigation satellite system, an overall term for the technologies that use signals from satellites to find locations on Earth's surface

Google Earth (p. 19) a freely available virtual globe program first released in 2005 by Google

GPS (p. 97) the Global Positioning System, a technology using signals broadcast from satellites for navigation and position determination on Earth

GPX (p. 111) a standard file format for working with GPS data collected by GPS receivers

graduated symbols (p. 250) the use of different-sized symbology to convey thematic information on a map

great circle distance (p. 45) the shortest distance between two points on a spherical surface

green band (p. 365) the range of wavelengths between 0.5 and 0.6 micrometers

grid cell (p. 137) a square unit, representing some real-world size, which contains a single value

GTCM (p. 555) the Geospatial Technology Competency Model set up by the U.S. Department of Labor

hillshade (p. 484) a shaded relief map of the terrain created by modeling the position of the Sun in the sky relative to the landscape

hyperspectral imagery (p. 374) remotely sensed imagery created from the bands collected by a sensor capable of sensing hundreds of bands of energy at once

hyperspectral sensor (p. 399) a sensor that can measure hundreds of different wavelength bands simultaneously

identity (p. 200) a type of GIS overlay that retains all features from the first layer along with the features it has in common with a second layer

IKONOS (p. 412) a satellite launched in 1999 that features sensors with a multispectral spatial resolution of 3.2 meters and panchromatic resolution of 0.8 meters

incident energy (p. 369) the total amount of energy (per wavelength) that interacts with an object

International Charter (p. 417) an agreement among multiple international space agencies and companies to provide satellite imagery and resources to help in times of disaster

International Date Line (p. 49) a line of longitude that uses the 180th meridian as a basis (but changes away from a straight line to accommodate geography)

intersect (p. 199) a type of GIS overlay that retains the features that are common to two layers

intersection (p. 194) the operation wherein the chosen features are those that meet both criteria in the query

interval data (p. 140) a type of numerical data in which the difference between numbers is significant, but there is no fixed non-arbitrary zero point associated with the data

IR (p. 365) infrared; the portion of the electromagnetic spectrum with wavelengths between 0.7 and 100 micrometers

Jason-3 (p. 452) a satellite mission dedicated to measuring ocean topography

JAXA (p. 435) the Japanese Aerospace Exploration Agency, Japan's national space exploration organization

join (p. 142) a method of linking two (or more) tables together

JPEG (p. 257) the Joint Photographic Experts Group image or graphic file format

JPSS-1 (p. 437) a joint satellite mission of NASA and NOAA, intended as the first in a series of Earth observing missions

junction (p. 297) a term used for the nodes (or places where edges come together) in a network

key (p. 142) the field that two tables have to have in common with each other in order for the tables to be joined

KML (p. 510) Keyhole Markup Language—the file format used for Google Earth data

KMZ (p. 510) the compressed version of a KML file

label (p. 249) text placed on a map to identify features

Landsat (p. 401) a long-running and ongoing U.S. remote sensing project that launched its first satellite in 1972

Landsat 7 (p. 402) the seventh Landsat mission, launched in 1999, which carries the ETM+ sensor

Landsat 8 (p. 402) the latest Landsat mission, launched in 2013

Landsat scene (p. 405) a single image obtained by a Landsat satellite sensor

large-scale map (p. 245) a map with a higher value for its representative fraction; such maps will usually show a smaller geographic area

LAS (p. 480) the industry standard data format used for lidar data

latitude (p. 44) imaginary lines on a globe north and south of the Equator that serve as a basis of measurement in GCS

layout (p. 249) the assemblage and placement of various map elements used in constructing a map

legend (p. 248) a graphical device used to explain what the various map symbols and colors represent

lidar (p. 479) a process in which a series of laser beams fired at the ground from an aircraft is used both to create highly accurate elevation models and also to measure the elevation of objects from the ground

line segment (p. 298) a single edge of a network that corresponds to one portion of a street (for instance, the edge between two junctions)

linear interpolation (p. 303) a method used in geocoding to place an address location among a range of addresses along a segment

lines (p. 132) one-dimensional vector objects

longitude (p. 44) imaginary lines on a globe east and west of the Prime Meridian that serve as a basis of measurement in GCS

LST (p. 445) Land surface temperature; a measurement of the temperature of Earth's land cover surface

map (p. 243) a representation of geographic data

map algebra (p. 200) combining datasets together using simple mathematical operators

map projection (p. 49) the translation of locations on the three-dimensional (3D) Earth to a two-dimensional (2D) surface

map scale (p. 245) a metric used to determine the relationship between measurements made on a map and their real-world equivalents

map templates (p. 249) pre-made arrangements of items in a map layout

mashup (p. 550) the combination of two or more map layers thereby creating one new application

MCE (p. 207) multi-criteria evaluation—the use of several factors (weighted and combined) to determine the suitability of a site for a particular purpose

metadata (p. 143) descriptive information about geospatial data

micrometer (p. 364) a unit of measurement equal to one-millionth of a meter, abbreviated μm

Mie scattering (p. 368) scattering of light caused by atmospheric particles the same size as the wavelength being scattered

MISR (p. 451) the Multi-angle Imaging SpectroRadiometer instrument aboard Terra

MODIS (p. 440) the Moderate Resolution Imaging Spectroradiometer instrument onboard Terra and Aqua

MOPITT (p. 450) the Measurements of Pollution in the Troposphere instrument onboard Terra

Morning Constellation (p. 436) a set of satellites (including Terra, Landsat 7 and two others), each one passing the Equator in the morning during its own orbit

MSAS (p. 111) an SBAS that covers Japan and nearby regions

MSI (p. 410) Multispectral Instrument, the multispectral sensor onboard the Sentinel-2 satellites

MSS (p. 402) the Multispectral Scanner aboard Landsat 1 through 5

multipath (p. 106) an error caused by a delay in the signal due to reflecting from surfaces before reaching the receiver

multispectral imagery (p. 374) remotely sensed imagery created from the bands collected by a sensor capable of sensing several bands of energy at once

multispectral sensor (p. 399) a sensor that can measure multiple wavelength bands simultaneously

NAD27 (p. 43) the North American Datum of 1927

NAD83 (p. 43) the North American Datum of 1983

nadir (p. 338) the location on the ground that lies directly below the camera in aerial photography

NAIP (p. 342) the National Agriculture Imagery Program, which produces orthophotos of the United States

nanometer (p. 364) a unit of measurement equal to one-billionth of a meter, abbreviated nm

nanosatellite (p. 414) a type of satellite that weighs between 2.2 and 22 pounds

NASA (p. 394) the National Aeronautics and Space Administration, established in 1958; it is the U.S.

government's space exploration and aerospace development branch

natural breaks (p. 253) a data classification method that selects class break levels by searching for spaces in the data values

NAVD88 (p. 471) the North American Vertical Datum of 1988, the vertical datum used for much geospatial data in North America

NAVSTAR GPS (p. 98) the U.S. Global Positioning System

NCGE (p. 551) the National Council for Geographic Education; a professional organization dedicated to geographic education

NDGPS (p. 108) Nationwide Differential GPS, which consists of ground-based DGPS locations around the United States

NDVI (p. 371) Normalized Difference Vegetation Index; a method of measuring the health of vegetation using near-infrared and red energy measurements

near-polar orbit (p. 395) an orbital path close to the North and South Poles that carries an object around Earth at an unvarying elevation

negation (p. 195) the operation wherein the chosen features are those that meet all of the first criteria and none of the second criteria (including where the two criteria overlap) in the query

network (p. 297) a series of junctions and edges connected together for modeling concepts such as streets

NHD (p. 135) The National Hydrography Dataset—a collection of GIS data of water resources for the United States

NIR (p. 365) near infrared; the portion of the electromagnetic spectrum with wavelengths between 0.7 and 1.3 micrometers

NLCD (p. 137) the National Land Cover Database is a raster-based GIS dataset that maps the land-cover types for the entire United States at 30-meter resolution

NOAA (p. 434) the National Oceanic and Atmospheric Administration, a U.S. federal agency focused on weather, oceans, and the atmosphere

nominal data (p. 140) a type of data that is a unique identifier of some kind—if numerical, the differences between numbers are not significant

non-selective scattering (p. 368) scattering of light caused by atmospheric particles larger than the wavelength being scattered

non-spatial data (p. 9) data that is not directly linked to a geospatial location (such as tabular data)

normalized (p. 254) altering count data values so that they are at the same level of representation (such as using them as percentages instead of regular count values)

north arrow (p. 248) a graphical device used to show the orientation of the map

northing (p. 52) a measurement of so many units north (or south) of a baseline

NOT (p. 194) the Boolean operator that corresponds with a negation operation

oblique photo (p. 342) an aerial photo taken at an angle

OCO-2 (p. 437) the Orbital Carbon Observatory, a satellite dedicated to measuring global carbon dioxide

off-nadir viewing (p. 401) the capability of a satellite to observe areas other than the ground directly below it

offset (p. 504) a value applied to an object that is not attached to the ground, but is entirely above ground level

OGC (p. 551) the Open Geospatial Consortium; a group involved with developing new standards of geospatial data interoperability

OhioView (p. 552) a research and education consortium consisting of the state universities in Ohio, along with related partners

OLI (p. 403) the Operational Land Imager, the multispectral sensor onboard Landsat 8

OMI (p. 450) the Ozone Monitoring Instrument onboard Aura

OMPS (p. 450) the Ozone Monitoring Profile Suite onboard Suomi NPP and JPSS-1

OR (p. 194) the Boolean operator that corresponds with a union operation

ordinal data (p. 140) a type of data that refers solely to a ranking of some kind

orthophoto (p. 341) an aerial photo with uniform scale

orthorectification (p. 341) a process used on aerial photos to remove the effects of relief displacement and give the image uniform scale

overlay (p. 199) the combining of two or more layers in the GIS

P code (p. 101) the digital code broadcast on the L1 and L2 frequencies, which is accessible by the military

panchromatic (p. 338) black and white aerial imagery

panchromatic imagery (p. 374) black and white imagery formed by viewing the entire visible portion of the electromagnetic spectrum

panchromatic sensor (p. 399) a sensor that can measure one range of wavelengths

pan-sharpening (p. 400) the technique of fusing a higher-resolution panchromatic band with lower-resolution multispectral bands to improve the clarity and detail seen in the image

parsing (p. 302) breaking an address up into its component parts

pattern (p. 345) the arrangement of objects in an image, used as an element of image interpretation

PDOP (p. 106) the position dilution of precision; describes the amount of error due to the geometric position of the GPS satellites

personal geodatabase (p. 148) a single file that can contain a maximum of 2 GB of data

perspective view (p. 483) the oblique angle view of a digital terrain model from which the model takes on a "three-dimensional" appearance

photo scale (p. 348) the representation used to determine how many units of measurement in the real world are equivalent to one unit of measurement on an aerial photo

photogrammetry (p. 348) the process of making measurements using aerial photos

Pléiades (p. 415) a series of satellites operated by Airbus Defence and Space, which feature sensors with a multispectral spatial resolution of 2 meters and panchromatic resolution of 0.50 meters

point cloud (p. 480) the name given to the massive number of elevation data measurements collected by lidar

points (p. 132) zero-dimensional vector objects

polygons (p. 132) two-dimensional vector objects

Poseidon-3B (p. 452) a radar altimeter instrument onboard Jason-3

Prime Meridian (p. 44) the line of longitude that runs through Greenwich, England, and serves as the 0 degree line of longitude from which to base measurements

principal point (p. 340) the center point of an aerial photo

prism map (p. 507) a 3D thematic map from whose surface shapes (such as the polygons that define states or nations) are extruded to different heights, which reflect different values for the same type of data

prisms (p. 507) the extruded shapes on a prism map

pseudo-3D (p. 483) a term often used to describe the perspective view of a terrain model because it is often a 2.5D model, not a full 3D model

pseudorange (p. 101) the calculated distance between a GPS satellite and a GPS receiver

Python (p. 146) a free computer programming language used for writing scripts in ArcGIS

QGIS (p. 148) an open source GIS program

quantile (p. 253) a data classification method that attempts to place an equal number of data values in each class

query (p. 192) the conditions used to retrieve data from a database

QuickBird (p. 412) a satellite launched in 2001 by DigitalGlobe with sensors that have 2.4-meter multispectral resolution and 0.61-meter panchromatic resolution

radiometric resolution (p. 400) the degree of a sensor's ability to determine fine differences in a band of energy measurements

RapidEye (p. 416) a constellation of five satellites operated by Planet Labs that feature sensors with 5-meter multispectral sensors and a daily global revisit

raster data model (p. 137) a way of representing spatial data that utilizes a series of equally spaced and sized grid cells

ratio data (p. 140) a type of numerical data in which the difference between numbers is significant, but there is a fixed non-arbitrary zero point associated with the data

Rayleigh scattering (p. 367) scattering of light caused by atmospheric particles smaller than the wavelength being scattered

records (p. 141) the rows of an attribute table

red band (p. 365) the range of wavelengths between 0.6 and 0.7 micrometers

reference database (p. 302) the base network data used as a source for geocoding

reference map (p. 250) a map that serves to show the location of features, rather than thematic information

relational operator (p. 194) one of the six connectors ($=, <>, <, >, >=,$ or $<=$) used to build a query

relief displacement (p. 340) the effect seen in aerial imagery where tall items appear to "bend" outward from the photo's center toward the edges

remote sensing (p. 2, 329, 361) the process of collecting information related to the electromagnetic energy reflected or emitted by a target on the ground, using a device a considerable distance away from the target on board an aircraft or spacecraft

reproject (p. 68) changing a dataset from one map projection (or measurement system) to another

RF (p. 245) representative fraction—a value indicating how many units of measurement in the real world are equivalent to how many of the same units of measurement on a map

RGB (p. 255) a color scheme based on using three primary colors: red, green, and blue

root mean square error (RMSE) (p. 77) an error measure used in determining the accuracy of the overall transformation of the unreferenced data

rotated (p. 76) when the unreferenced image is turned during the transformation

SaaS (p. 544) Software as a Service; a method by which the software and data for an application are hosted in the cloud and accessed through a Web browser

satellite imagery (p. 2) digital images of Earth acquired by sensors onboard orbiting spaceborne platforms

satellite navigation system (p. 295) a device used to plot the user's position on a map using GPS technology

SATELLITES (p. 554) Students and Teachers Exploring Local Landscapes to Interpret the Earth from Space; an OhioView initiative involving K–12 teachers and students with geospatial technologies

SBAS (p. 109) satellite-based augmentation system; a method of using correction information sent from an additional satellite to improve GPS position determination

scale bar (p. 247) a graphical device used on a map to represent map scale

scaled (p. 76) when the unreferenced image is altered or resized during the transformation

Script (p. 146) a short piece of computer code

selective availability (SA) (p. 105) the intentional degradation of the timing and position information transmitted by a GPS satellite

Sentinel (p. 410) a satellite program operated by the European Space Agency

Sentinel-2 (p. 410) a Sentinel mission, consisting of two satellites with characteristics similar to Landsat

service (p. 544) a format for GIS data and maps to be distributed (or "served") to others via the Internet

shadow (p. 346) the dark shapes in an image caused by a light source shining on an object, used as an element of image interpretation

shape (p. 346) the distinctive form of an object, used as an element of image interpretation

shapefile (p. 147) a series of files (with extensions such as .shp, .shx, and .dbf) that make up one vector data layer

shortest path (p. 307) the route that corresponds to the lowest cumulative transit cost between stops in a network

single frequency (p. 100) a GPS receiver that can pick up only the L1 frequency

site and association (p. 345) the information referring the location of objects and their related attributes in an image, used as elements of image interpretation

site suitability (p. 203) the determination of the "useful" or "non-useful" locations based on a set of criteria

size (p. 346) the physical dimensions (length, width, and area on the ground) of objects, used as an element of image interpretation

SketchUp (p. 508) a 3D modeling and design software program distributed online by Trimble

skewed (p. 76) when the unreferenced image is distorted or slanted during the transformation

SKP file (p. 510) the file format and extension used by an object created in SketchUp

SkySat (p. 413) a high-resolution, small satellite program operated by Terra Bella

SLC (p. 407) the Scan Line Corrector in the ETM+ sensor; its failure in 2003 caused Landsat 7 ETM+ imagery to not contain all data from a scene

slope (p. 482) a measurement of the rate of elevation change at a location, found by dividing the vertical height (the rise) by the horizontal length (the run)

slope aspect (p. 482) the direction that a slope is facing

small satellite (p. 414) a type of satellite that weighs between 220 and 1100 pounds

small-scale map (p. 245) a map with a lower value for its representative fraction; such maps will usually show a larger geographic area

SMAP (p. 444) Soil Moisture Active Passive; a satellite mission dedicated to monitoring soil moisture levels

sounder (p. 448) the term given to a remote sensing instrument that measures atmospheric phenomena as a function of vertical distribution

space segment (p. 98) one of the three segments of GPS, consisting of the satellites and the signals they broadcast from space

spatial analysis (p. 191) examining the characteristics or features of spatial data, or how features spatially relate to one another

spatial query (p. 198) selecting records or objects from one layer based upon their spatial relationships with other layers (rather than using attributes)

spatial reference (p. 70) the use of a real-world coordinate system for identifying locations

spatial resolution (p. 373, 399) the ground area represented by one pixel of satellite imagery

SPCS zone (p. 56) one of the divisions of the United States set up by the SPCS

spectral reflectance (p. 369) the percentage of the total incident energy that was reflected from that surface

spectral resolution (p. 398) the bands and wavelengths measured by a sensor

spectral signature (p. 370) a unique identifier for a particular item, generated by charting the percentage of reflected energy per wavelength against a value for that wavelength

SPOT (p. 414) a high-resolution imagery satellite program operated by Airbus Defence and Space

SQL (p. 193) Structured Query Language—a formal setup for building queries

SRTM (p. 479) the Shuttle Radar Topography Mission, flown in February 2000, which mapped Earth's surface from orbit for the purpose of constructing digital elevation models of the planet

SST (p. 446) Sea surface temperature; a measurement of the temperature of the surface of Earth's oceans

standard deviation (p. 253) a data classification method that computes class break values by using the mean of the data values and the average distance a value is away from the mean

standard false color composite (p. 378) an image arranged by placing the near-infrared band in the red color gun, the red band in the green color gun, and the green band in the blue color gun

State Plane Coordinate System (SPCS) (p. 55) a grid-based system for determining coordinates of locations within the United States

StateView (p. 552) the term used to describe each of the programs affiliated with AmericaView in each state

stereoscopic 3D (p. 518) an effect that creates the illusion of depth or immersion in an image by utilizing stereo imaging techniques

stops (p. 308) destinations to visit on a network

Story Map (p. 545) an interactive Web map created using ArcGIS Online resources that can add descriptive text, photos, or videos to locations on a map to inform the user about the map's topic

street centerline (p. 299) a file containing line segments representing roads

Street View (p. 312) a component of Google Maps and Google Earth that allows the viewer to see 360-degree imagery around an area on a road

sUAS (p. 334) small unmanned aircraft systems that weigh less than 55 pounds

suitability index (p. 205) a system whereby locations are ranked according to how well they fit a set of criteria

Sun altitude (p. 484) the value between 0 and 90 used in constructing a hillshade to model the Sun's elevation above the terrain

Sun azimuth (p. 484) the value between 0 and 360 used in constructing a hillshade to model the Sun's position in the sky to show the direction of the Sun's rays striking the surface

sun-synchronous orbit (p. 397) an orbital path set up so that the satellite always crosses the same areas at the same local time

Suomi NPP (p. 437) a joint satellite mission of NASA, NOAA, and the U.S. Department of Defense

swath width (p. 395) the width of the ground area the satellite is imaging as it passes over Earth's surface

SWIR (p. 366) shortwave infrared; the portion of the electromagnetic spectrum with wavelengths between 1.3 and 3.0 micrometers

symmetrical difference (p. 200) a type of GIS overlay that retains all features from both layers except for the features that they have in common

tasking (p. 412) the ability to direct the sensors of a satellite on demand to collect imagery of a specific area

temporal resolution (p. 401) the length of time a sensor takes to come back and image the same location on the ground

Terra (p. 435) the flagship satellite of the NASA Earth Science program

texture (p. 346) repeated shadings or colors in an image, used as an element of image interpretation

textures (p. 506) graphics applied to 3D objects to give them a more realistic appearance

The National Map (p. 10) an online basemap and data viewer with downloadable geospatial data maintained and operated by the USGS and part of the National Geospatial Program

thematic map (p. 250) a map that displays a particular theme or feature

three-dimensional (3D) model (p. 478) a model of the terrain that allows for multiple z-values to be assigned to each x/y coordinate location

TIFF (p. 257) the Tagged Image File Format used for graphics or images

TIGER/Line (p. 299) a file produced by the U.S. Census Bureau that contains (among other items) the line segments that correspond with roads all over the United States

time zones (p. 45) a method of measuring time around the world, created by dividing the world into subdivisions of longitude and relating the time in that subdivision to the time in Greenwich, England

TIN (p. 478) Triangulated Irregular Network; a terrain model that allows for non-equally spaced elevation points to be used in the creation of the surface

TIR (p. 366) thermal infrared; the portion of the electromagnetic spectrum with wavelengths between 3.0 and 14.0 micrometers

TIRS (p. 403) the Thermal Infrared Sensor, the instrument that acquires thermal imagery onboard Landsat 8

TM (p. 402) the Thematic Mapper sensor onboard Landsat 4 and 5

Tomnod (p. 418) an online crowdsourcing initiative that applies satellite imagery for a variety of real-world issues

tone (p. 347) the grayscale levels (from black to white), or range of intensity of a particular color discerned as a characteristic of particular features present in an image, used as an element of image interpretation

topographic map (p. 472) a map created by the USGS to show landscape and terrain as well as the location of features on the land

topology (p. 135) how vector objects relate to each other (in terms of their adjacency, connectivity, and containment) independently of the objects' coordinates

transit cost (p. 307) a value that represents how many units (of time or distance, for example) are used in moving along a network edge

translated (p. 76) when the unreferenced image is shifted during the transformation

transmission (p. 368) when light passes through a target

trilateration (p. 102) finding a location in relation to three points of reference

true color composite (p. 378) an image arranged by placing the red band in the red color gun, the green band in the green color gun, and the blue band in the blue color gun

true orthophoto (p. 341) an orthophoto in which all objects look as if they're being seen from directly above

two-and-a-half–dimensional (2.5D) model (p. 478) a model of the terrain that allows for a single z-value to be assigned to each x/y coordinate location

type (p. 248) the lettering used on a map

UAS (p. 334) unmanned aircraft systems—a reconnaissance aircraft that is piloted from the ground via remote control

union (p. 194) the operation wherein the chosen features are all that meet the first criterion as well as all that meet the second criterion in the query

union (overlay) (p. 200) a type of GIS overlay that combines all features from both layers

United States National Grid (USNG) (p. 54) a grid system of identifying locations in the United States

Universal Transverse Mercator (UTM) (p. 52) the grid system of locating coordinates across the globe

UN-SPIDER (p. 418) the United Nations Space-Based Information for Disaster Emergency and Response program

URISA (p. 551) the Urban and Regional Information Systems Association; an association for GIS professionals

US Topo (p. 474) a digital topographic map series created by the USGS to allow multiple layers of data to be used on a map in GeoPDF file format

user segment (p. 100) one of the three segments of GPS, consisting of the GPS receivers on the ground that pick up the signals from the satellites

UTM zone (p. 52) one of the 60 divisions of the world set up by the UTM system, each zone being 6 degrees of longitude wide

UV (p. 365) ultraviolet; the portion of the electromagnetic spectrum with wavelengths between 0.01 and 0.4 micrometers

vector data model (p. 133) a conceptualization of the world that represents spatial data as a series of vector objects (points, lines, and polygons)

vector objects (p. 133) points, lines, and polygons that are used to model real-world phenomena using the vector data model

vertical datum (p. 471) a baseline used as a starting point in measuring elevation values (which are either above or below this value)

vertical exaggeration (p. 487) a process whereby the z-values are artificially enhanced for purposes of terrain visualization

vertical photo (p. 338) an aerial photo in which the camera is looking directly down at a landscape

VGI (p. 15) volunteered geographic information, a term used to describe user-generated geospatial content and data

viewshed (p. 482) a data layer that indicates what an observer can and cannot see from a particular location due to terrain

VIIRS (p. 442) the Visible Infrared Imaging Radiometer Suite instrument onboard Suomi NPP

virtual globe (p. 19) a software program that provides an interactive three-dimensional map of Earth

visible light spectrum (p. 365) the portion of the electromagnetic spectrum with wavelengths between 0.4 and 0.7 micrometers

visual hierarchy (p. 244) how features are displayed on a map to emphasize their level of prominence

visual image interpretation (p. 345) the process of examining information to identify objects in an aerial (or other remotely sensed) image

W3C Geolocation API (p. 17) the commands and techniques used in programming to obtain the geolocation of a computer or mobile device

WAAS (p. 109) Wide Area Augmentation System; a satellite-based augmentation system that covers the United States and other portions of North America

wavelength (p. 363) the distance between the crests of two waves

Web map (p. 544) an interactive online representation of geospatial data, which can be accessed via a Web browser

WGS84 (p. 43) the World Geodetic System of 1984 datum (used by the Global Positioning System)

wiki (p. 15) a database available for everyone to utilize and edit

WorldView-1 (p. 412) a satellite launched in 2007 by DigitalGlobe whose panchromatic sensor has 0.5-meter spatial resolution

WorldView-2 (p. 412) a satellite launched in 2009 that features an eight-band multispectral sensor with 1.84-meter resolution and a panchromatic sensor of 0.46-meter resolution

WorldView-3 (p. 413) a satellite launched in 2014 by DigitalGlobe that features shortwave infrared and CAVIS bands, plus a panchromatic sensor of 0.31-meter resolution and eight multispectral bands with 1.24-meter resolution

WorldView-4 (p. 413) a satellite launched in 2016 that features a panchromatic sensor of 0.31-meter resolution and 4 multispectral bands with 1.24-meter resolution

Worldwide Reference System (p. 405) the global system of Paths and Rows that is used to identify what area on Earth's surface is present in which Landsat scene

XOR (p. 195) the Boolean operator that corresponds with an exclusive or operation

Y code (p. 101) an encrypted version of the P code

z-value (p. 471, 503) the elevation assigned to an x/y coordinate

Index

Note: Page numbers with f indicate figures; those with t indicate tables.